Linear and Complex Analysis

Dedicated to V. P. Havin on the Occasion of His 75th Birthday

Victor Havin near the main building
of St. Petersburg State University.

American Mathematical Society

TRANSLATIONS

Series 2 • Volume 226

Advances in the Mathematical Sciences—63

(*Formerly Advances in Soviet Mathematics*)

Linear and Complex Analysis

Dedicated to V. P. Havin on the Occasion of His 75th Birthday

Alexei Alexandrov
Anton Baranov
Sergey Kislyakov
Editors

American Mathematical Society
Providence, Rhode Island

2000 *Mathematics Subject Classification.* Primary 30–06.

Library of Congress Card Number 91-640741
ISBN-13: 978-0-8218-4801-2
ISBN-10: 0-8218-4801-1
ISSN 0065-9290

♾ The paper used in this book is acid-free and falls within the guidelines
established to ensure permanence and durability.
Visit the AMS home page at `http://www.ams.org/`

10 9 8 7 6 5 4 3 2 1 14 13 12 11 10 09

Contents

Preface

Victor Petrovich Havin was born in Leningrad on 7 March, 1933. His academic career is linked entirely with the Department of Mathematics and Mechanics of St. Petersburg State University, which he entered as a student in 1950. (On many occasions he talked about the dramatic choice between linguistics and mathematics he faced at that time, and about the role of his father who strongly dissuaded him from linguistics — because to be a mathematician was much safer in the Soviet era, especially under the Stalin regime.) He got his Ph. D. degree in 1958 under supervision of L. V. Kantorovich and (informally) V. I. Smirnov; he obtained the Doctor of Science degree in 1969 and full professorship in 1970. He headed the division of mathematical analysis at St. Petersburg State University from 1997 until 2004. V. P. Havin supervised 27 Ph. D. theses (in particular, two of us were his students at different times; see Mathematics Genealogy Project site `http://genealogy.math.ndsu.nodak.edu` for more details) and some uncontrollable number of master theses. Three of his former students have won the Salem Prize. He received several honorary awards and degrees from his home university and from some foreign universities. His publication list totals more than 100 items, among which there are at least 70 research papers, a seminal book on the uncertainty principle in harmonic analysis (written jointly with Burglind Jöricke, also his former student), an introductory university course of Analysis (*not* to be named "calculus"), etc.

However, this bare listing of facts hides something. It only reflects an ordinary career of a very successful university professor. In reality, it is to a great extent due to various kinds of activity of Victor Petrovich and to his personality that Mathematical Analysis in St. Petersburg has developed to its present state. A brilliant scientist and absolutely brilliant teacher, he has succeeded in making difficult issues of modern analysis attractive to many young researchers. Since 1963, he has been running the Linear and Complex Analysis Seminar (joint for the University and the V. A. Steklov Mathematical Institute in St. Petersburg; before 1991, this responsibility was shared with N. K. Nikolski). For many years, this seminar has been the center of activity of an informal (and very large) research group; it plays this role even now, when many members of this group live and work elsewhere.

This volume is a gift, made by the contributors and editors for Victor Petrovich Havin on the occasion of his 75th birthday.

The list of Havin's publications follows this Preface. We refrain from a detailed analysis of his scientific achievements, which are well known. We only indicate some fields of his interest usually emphasized by Havin himself (we do not name coauthors and collaborators):

- separation of singularities of (bounded) analytic functions;
- divisibility by the inner factor in certain classes of analytic functions, and the drop of the smoothness of a function relative to that of its modulus;
- integral analogs of Vitushkin's theorem on uniform approximation by analytic functions;
- nonlinear potential theory;
- free interpolation;
- approximation by harmonic vector fields;
- uncertainty principle in harmonic analysis;
- questions related to the Beurling–Malliavin theorem.

Alexei Alexandrov
Anton Baranov
Sergey Kislyakov

Victor Havin with his fellow students near the
Department of Mathematics and Mechanics. Leningrad, 1953.

Giving a lecture at a conference. Ufa, 1980.

Doctor Honoris Causa. Linköping, 1993.

With Jean-Pierre Kahane and Nikolai Nikolski.

Publications Authored by V. P. Havin

Books

[1] *Tha basic of mathematical analysis, Part* 1. *Differential and integral calculus of functions of one real variable.* Leningrad. Univ., Leningrad, 1989. MR1007979 (90j:00001)

[2] *The uncertainty principle in harmonic analysis* (with B. Jöricke), Ergebnisse der Mathematik und ihrer Grenzgebiete (3), **28**, Springer-Verlag, Berlin, 1994. MR1303780 (96c:42001)

Research papers

[1] *On analytic functions representable by an integral of Cauchy–Stieltjes type*, Vestnik Leningrad. Univ. Ser. Mat. Meh. Astr. **13** (1958), no. 1, 66–79. MR0095256 (20 #1762)

[2] *The separation of the singularities of analytic functions*, Dokl. Akad. Nauk SSSR **121** (1958), 239–242. MR0098168 (20 #4630)

[3] *Analytic continuation of power series and Faber polynomials*, Dokl. Akad. Nauk SSSR (N.S.) **118** (1958), 879–881. MR0098175 (20 #4637)

[4] *On a problem of V. V. Golubev*, Dokl. Akad. Nauk SSSR **126** (1959), 511–513. MR0106292 (21 #5026)

[5] *The rate of growth of functions of the H_p class and the multiplication of integrals of the Cauchy–Stieltjes type*, Dokl. Akad. Nauk SSSR **127** (1959), 757–759. MR0107010 (21 #5738)

[6] *On the norms of some operations in the space of polynomials*, Vestnik Leningrad. Univ. **14** (1959), no. 19, 47–59. MR0111826 (22 #2686)

[7] *On the space of bounded regular functions*, Dokl. Akad. Nauk SSSR **131** (1960), 40–43; translated in Soviet Math. Dokl. **1** (1960), 202–204. MR0120525 (22 #11277)

[8] *Some estimates of analytic capacity* (with S. Ja. Havinson), Dokl. Akad. Nauk SSSR **138** (1961), 789–792. MR0131529 (24 #A1379)

[9] *On the space of bounded regular functions*, Sibirsk. Mat. Z. **2** (1961), 622–638. MR0138985 (25 #2425)

[10] *Relations between certain classes of functions regular in the unit circle*, Vestnik Leningrad. Univ. **17** (1962), no. 1, 102–110. MR0152660 (27 #2635)

[11] *Analytic representation of linear functionals in spaces of harmonic and analytic functions which are continuous in a closed region*, Dokl. Akad. Nauk SSSR **151** (1963), 505–508. MR0152661 (27 #2636)

[12] *Boundary properties of integrals of Cauchy type and of conjugate harmonic functions in regions with rectifiable boundary*, Mat. Sb. (N.S.) **68 (110)** (1965), 499–517. MR0192059 (33 #286)

[13] *Removability of singularities of analytic functions and transfer of masses* (with R. E. Val′skii), Sibirsk. Mat. Z. **7** (1966), 55–60. MR0203441 (34 #3293)

[14] *Duality of spaces of analytic functions and certain of its applications*, Contemporary Problems in Theory Anal. Functions (Proc. Internat. Conf., Erevan, 1965), "Nauka", Moscow, 1966, pp. 311–314. MR0208356 (34 #8166)

[15] *Continuity in L_p of an integral operator with the Cauchy kernel*, Vestnik Leningrad. Univ. **22** (1967), no. 7, 103–108. MR0218537 (36 #1623)

[16] *Approximation in the mean by harmonic functions* (with V. G. Maz′ya), Zap. Naučn. Sem. Leningrad. Otdel. Mat. Inst. Steklov. (LOMI) **5** (1967), 196–200. MR0226046 (37 #1636)

[17] *Approximation by analytic functions in the mean*, Dokl. Akad. Nauk SSSR **178** (1968), 1025–1028. MR0224830 (37 #429)

[18] *Approximation in the mean by analytic functions* (with V. G. Maz′ya), Vestnik Leningrad. Univ. **23** (1968), no. 13, 62–74. MR0235131 (38 #3443)

[19] *The Cauchy problem for Laplace's equation* (with V. G. Maz′ya), Vestnik Leningrad. Univ. **23** (1968), no. 7, 146–147. MR0236534 (38 #4829)

[20] *Polynomial approximation in the mean in certain non-Carathéodory regions.* I, Izv. Vyssh. Uchebn. Zaved. Matematika **9 (76)** (1968), 86–93. MR0237796 (38 #6077)

[21] *Polynomial approximation in the mean in certain non-Carathéodory regions.* II, Izv. Vyssh. Uchebn. Zaved. Matematika **10 (77)** (1968), 87–94. MR0247104 (40 #373)

[22] *On the uniqueness theorem of L. Carleson for analytic functions with finite Dirichlet integral* (with V. G. Maz′ya), Problems of Math. Anal. **2**: Linear Operators and Operator Equations, 153–156, Izdat. Leningrad. Univ., Leningrad, 1969. MR0269843 (42 #4738)

[23] *A nonlinear analogue of the Newtonian potential, and metric properties of $(p,\, l)$-capacity* (with V. G. Maz′ya), Dokl. Akad. Nauk SSSR **194** (1970), 770–773. MR0273045 (42 #7926)

[24] *Analytic functions with a Lipschitzian modulus of the boundary values* (with F. A. Shamoyan), Zap. Naučn. Sem. Leningrad. Otdel. Mat. Inst. Steklov. (LOMI) **19** (1970), 237–239. MR0289784 (44 #6971)

[25] *Multipliers and divisors of Cauchy–Stieltjes type integrals* (with S. A. Vinogradov and M. G. Goluzina), Zap. Naučn. Sem. Leningrad. Otdel. Mat. Inst. Steklov. (LOMI) **19** (1970), 55–78. MR0291471 (45 #562)

[26] *A generalization of the Privalov–Zygmund theorem on the modulus of continuity of the conjugate function*, Izv. Akad. Nauk Armjan. SSR Ser. Mat. **6** (1971), no. 2–3, 252–258; ibid. **6** (1971), no. 4, 265–287. MR0302923 (46 #2066)

[27] *Impossibility of the Carleman approximation of functions continuous on the unit circle by the boundary values of functions analytic and uniformly continuous in the unit disk*, Zap. Naučn. Sem. Leningrad. Otdel. Mat. Inst. Steklov. (LOMI) **22** (1971), 161–170. MR0291813 (45 #903)

[28] *The factorization of analytic functions that are smooth up to the boundary*, Zap. Naučn. Sem. Leningrad. Otdel. Mat. Inst. Steklov. (LOMI) **22** (1971), 202–205. MR0289783 (44 #6970)

[29] *Approximation in the mean by harmonic functions* (with V. G. Maz′ya), Investigations on linear operators and the theory of functions, III. Zap. Naučn. Sem. Leningrad. Otdel. Mat. Inst. Steklov. (LOMI) **30** (1972), 91–105. MR0328107 (48 #6449)

[30] *On the theory of nonlinear potentials and (p, l)-capacity* (with V. G. Maz′ya), Vestnik Leningrad. Univ. Mat. Meh. Astronom. **13** (1972), no. 3, 46–51. MR0342720 (49 #7465)

[31] *A nonlinear potential theory* (with V. G. Maz′ya), Uspehi Mat. Nauk **27** (1972), no. 6, 67–138. MR0409858 (53 #13610)

[32] *Weak completeness of the space L^1/H_0^1*, Vestnik Leningrad. Univ. Mat. Meh. Astronom. **13** (1973), no. 3, 77–81. MR0344863 (49 #9602)

[33] *Application of the $(p,\, l)$-capacity to certain problems of the theory of exceptional sets* (with V. G. Maz′ya), Mat. Sb. (N.S.) **90 (132)** (1973), 558–591. MR0367207 (51 #3449)

[34] *The spaces H^∞ and L^1/H_0^1*, Investigations on linear operators and the theory of functions, IV. Zap. Naučn. Sem. Leningrad. Otdel. Mat. Inst. Steklov. (LOMI) **39** (1974), 120–148. MR0348467 (50 #965)

[35] *The solutions of the Cauchy problem for the Laplace equation (uniqueness, normality, approximation)* (with V. G. Maz′ya), Trudy Moskov. Mat. Obshch. **30** (1974), 61–114. MR0385135 (52 #6000)

[36] *Free interpolation in H^∞ and in certain other classes of functions.* I (with S. A. Vinogradov), Investigations on linear operators and the theory of functions, V. Zap. Naučn. Sem. Leningrad. Otdel. Mat. Inst. Steklov. (LOMI) **47** (1974), 15–54. MR0393497 (52 #14307)

[37] *Approximation in L^p by the solutions of certain systems of linear differential equations*, Collection of articles dedicated to the memory of Academician V. I. Smirnov. Vestnik Leningrad. Univ. Mat. Meh. Astronom. **1** (1975), no. 1, 150–158. MR0390752 (52 #11575)

[38] *Free interpolation in H^∞ and in certain other classes of functions.* II (with S. A. Vinogradov), Investigations on linear operators and theory of functions, VI. Zap. Naučn. Sem. Leningrad. Otdel. Mat. Inst. Steklov. (LOMI) **56** (1976), 12–58. MR0492283 (58 #11425)

[39] *On the Carleson formula for the Dirichlet integral of an analytic function* (with A. B. Aleksandrov and A. E. Dzrbasjan), Vestnik Leningrad. Univ. Mat. Mekh. Astronom. **1979**, no. 4, 8–14. MR0564567 (81d:30051)

[40] *The uncertainty principle for operators that commute with a shift*, I (with B. Jöricke), Investigations on linear operators and the theory of functions, IX. Zap. Nauchn. Sem. Leningrad. Otdel. Mat. Inst. Steklov. (LOMI) **92** (1979), 134–170. MR0566746 (81e:42021)

[41] *On a class of uniqueness theorems for convolutions* (with B. Jöricke), Complex analysis and spectral theory (Leningrad, 1979/1980), Lecture Notes in Math., **864**, Springer, Berlin-New York, 1981, 143–170, MR0643382 (83c:44008)

[42] *The uncertainty principle for operators that commute with a shift.* II (with B. Jöricke), Investigations on linear operators and the theory of functions, XI. Zap. Nauchn. Sem. Leningrad. Otdel. Mat. Inst. Steklov. (LOMI) **113** (1981), 97–134. MR0629836 (84j:47050)

[43] *The uncertainty principle for one-dimensional Riesz potentials*, Dokl. Akad. Nauk SSSR **264** (1982), no. 3, 559–563. MR0659763 (84j:31002)

[44] *Free interpolation and the Dirichlet problem* (with A. O. Derviz), Vestnik Leningrad. Univ. Mat. Mekh. Astronom. **1983**, 84–86. MR0702634 (84j:31001)

[45] *Analogues of the Carleman–Goluzin–Krylov interpolation formula* (with I. V. Videnskii and E. M. Gavurina), Operator theory and function theory, no. 1, pp. 21–32, Leningrad. Univ., Leningrad, 1983. MR0768773 (86a:30057)

[46] *Free interpolation and the Dirichlet problem* (with A. O. Derviz), Investigations on linear operators and the theory of functions, XIII. Zap. Nauchn. Sem. Leningrad. Otdel. Mat. Inst. Steklov. (LOMI) **135** (1984), 51–65. MR0741694 (85j:31002)

[47] *Equivalent norms in L^p-spaces of analytic functions and free interpolation of harmonic functions* (with B. Jöricke), Vestnik Leningrad. Univ. Mat. Mekh. Astronom. **1984**, no. 4, 103–105. MR0776964 (86d:30080)

[48] *Traces of harmonic functions and comparison of L^p-norms of analytic functions* (with B. Jöricke), Math. Nachr. **123** (1985), 225–254. MR0809347 (87e:31003)

[49] *The uncertainty principle for operators that commute with a shift.* III (with B. Jöricke), Geometric questions in the theory of functions and sets, Kalinin. Gos. Univ., Kalinin, 1985, pp. 62–80. MR0829929 (88c:47053)

[50] *A remark on Taylor series of harmonic functions*, Multidimensional complex analysis, Akad. Nauk SSSR Sibirsk. Otdel., Inst. Fiz., Krasnoyarsk, 1985, pp. 192–197. MR0899349 (88g:31007)

[51] *The uncertainty principle in harmonic analysis* (with B. Jöricke), Proc. GDR mathematicians' congress (Rostock, 1986), pp. 72–74, Wilhelm-Pieck-Univ., Rostock, 1986. MR0845290

[52] *The length of a harmonic vector* (with M. Giesecke), Vestnik Leningrad. Univ. Mat. Mekh. Astronom. **1987**, no. 3, 33–38. MR0928158 (88m:58175)

[53] *The Poisson kernel is the only approximate identity that is asymptotically multiplicative on H^∞* (with H. Wolff), Zap. Nauchn. Sem. Leningrad. Otdel. Mat. Inst. Steklov. (LOMI) **170** (1989), 82–89; translation in J. Soviet Math. **63** (1993), no. 2, 159–163. MR1039574 (91i:46058)

[54] *Approximation properties of harmonic vector fields and differential forms* (with A. Presa Sague), Methods of approximation theory in complex analysis and mathematical physics (Leningrad, 1991), pp. 149–156, Lecture Notes in Math., 1550, Springer, Berlin. MR1322297

[55] *Weighted approximation by trigonometric sums and the Carleman–Goluzin–Krylov formula* (with V. A. Bart), Zap. Nauchn. Sem. St. Peterburg. Otdel. Mat. Inst. Steklov. (POMI) **206** (1993), Issled. po Linein. Oper. i Teor. Funktsii. **21**, pp. 5–14; translation in J. Math. Sci. **80** (1996), no. 4, 1873–1879. MR1255311 (95f:42004)

[56] *Szegő–Kolmogorov–Krein theorems on weighted trigonometric approximation, and Carleman-type formulas* (with V. A. Bart), Ukrain. Mat. Zh. **46** (1994), no. 1-2, 100–127; translation in Ukrainian Math. J. **46** (1994), no. 1–2, 101–132. MR1294817 (95k:42004)

[57] *Uniqueness and free interpolation for logarithmic potentials and the Cauchy problem for the Laplace equation in R^2* (with A. Aleksandrov, J. Bourgain, M. Giesecke, and Yu. Vymenets), Geom. Funct. Anal. **5** (1995), no. 3, 529–571. MR1339816 (96f:31007)

[58] *Uniform approximation by harmonic differential forms in Euclidean space* (with A. Presa Sage), Algebra i Analiz **7** (1995), no. 6, 104–152; translation in St. Petersburg Math. J. **7** (1996), no. 6, 943–977. MR1381980 (97b:58003)

[59] *Approximation properties of harmonic vector fields and differential forms*, Potential theory—ICPT **94** (Kouty, 1994), pp. 91–102, de Gruyter, Berlin, 1996. MR1404702 (97f:58005)

[60] *Uniform approximation by harmonic differential forms. A constructive approach* (with E. V. Malinnikova), Algebra i Analiz **9** (1997), no. 6, 156–196; translation in St. Petersburg Math. J. **9** (1998), no. 6, 1149–1180. MR1610176 (99g:58005)

[61] *Approximation and extension problems for some classes of vector fields* (with S. K. Smirnov), Algebra i Analiz **10** (1998), no. 3, 133–162; translation in St. Petersburg Math. J. **10** (1999), no. 3, 507–528. MR1628034 (99f:41046)

[62] *Bounded separation of singularities of analytic functions* (with A. H. Nersessian), Entire functions in modern analysis (Tel-Aviv, 1997), pp. 149–171, Israel Math. Conf. Proc., 15, Bar-Ilan Univ., Ramat Gan, 2001. MR1890536 (2002m:31018)

[63] *On the uncertainty principle for M. Riesz potentials* (with D. B. Beliaev), Ark. Mat. **39** (2001), no. 2, 223–243. MR1861059 (2003a:31002)

[64] *Admissible majorants for model subspaces of H^2, I. Slow winding of the generating inner function* (with J. Mashreghi), Canad. J. Math. **55** (2003), no. 6, 1231–1263. MR2016246 (2004i:30029a)

[65] *Admissible majorants for model subspaces of H^2*, II. *Fast winding of the generating inner function* (with J. Mashreghi), Canad. J. Math. **55** (2003), no. 6, 1264–1301. MR2016247 (2004i:30029b)

[66] *Separation of singularities of analytic functions with preservation of boundedness*, Algebra i Analiz **16** (2004), no. 1, 293–319; translation in St. Petersburg Math. J. **16** (2005), no. 1, 259–283. MR2068355 (2005d:30081)

[67] *On a theorem of I. I. Privalov on the Hilbert transform of Lipschitz functions* (with Yu. S. Belov), Mat. Fiz. Anal. Geom. **11** (2004), no. 4, 380–407. MR2114001 (2005k:26006)

[68] *The Beurling–Malliavin multiplier theorem: the seventh proof* (with J. Mashreghi and F. L. Nazarov), Algebra i Analiz **17** (2005), no. 5, 3–68; translation in St. Petersburg Math. J. **17** (2006), no. 5, 699–744. MR2241422 (2007g:42028)

[69] *Admissible majorants for model subspaces and arguments of inner functions* (with A. D. Baranov), Funktsional. Anal. i Prilozhen. **40** (2006), no. 4, 3–21; translation in Funct. Anal. Appl. **40** (2006), no. 4, 249–263. MR2307699 (2008c:30058)

[70] *Majorants of meromorphic functions with fixed poles* (with A. D. Baranov and A. A. Borichev), Indiana Univ. Math. J. **56** (2007), no. 4, 1595–1628. MR2354693 (2008i:30034)

[71] *Uniform estimates in the Poincaré–Aronszajn theorem on the separation of singularities of analytic functions* (with A. H. Nersessian and J. Ortega-Cerdà), J. Anal. Math. **101** (2007), 65–93. MR2346540 (2008k:30061)

Surveys or similar

[1] *Spaces of analytic functions*, Math. Analysis, Akad. Nauk SSSR Inst. Naučn. Inform., Moscow, 1966, pp. 76–164. MR0206694 (34 #6512)

[2] P. Koosis, *Introduction to H^p spaces. With an appendix on Wolff's proof of the corona theorem.* Translated from the English by V. V. Peller and A. G. Tumarkin. Translation edited and with a preface and appendixes by V. P. Havin. "Mir", Moscow, 1984. MR0777505 (86b:30053)

[3] *Methods and structure of commutative harmonic analysis*, Commutative harmonic analysis, I. Current problems in mathematics. Fundamental directions, **15**, pp. 6–133, Vsesoyuz. Inst. Nauchn. i Tekhn. Inform., Moscow, 1987. Translation in Encyclopaedia Math. Sci., **15**, pp. 1–111, Springer-Verlag, Berlin, 1991. MR1134136

[4] *The uncertainty principle in harmonic analysis* (with B. Jöricke). Commutative harmonic analysis, III, Current problems in mathematics. Fundamental directions, pp. 181–260, Vsesoyuz. Inst. Nauchn. i Tekhn. Inform., Moscow, 1991. Translation in Encyclopaedia Math. Sci., **72**, pp. 177–259, Springer-Verlag, Berlin, 1995. MR1129019 (93e:42001)

[5] *On some potential theoretic themes in function theory*, The Mazya anniversary collection, Vol. 1 (Rostock, 1998), pp. 99–110, Oper. Theory Adv. Appl., **109**, Birkhäuser, Basel, 1999. MR1747866 (2001f:31007)

[6] P. Koosis, *Introduction to H_p spaces.* Second edition. With two appendixes by V. P. Havin. Cambridge Tracts in Mathematics, **115**. Cambridge University Press, Cambridge, 1998. MR1669574 (2000b:30052)

[7] *On the uncertainty principle in harmonic analysis*, Twentieth century harmonic analysis—a celebration (Il Ciocco, 2000), pp. 3–29, NATO Sci. Ser. II Math. Phys. Chem., **33**, Kluwer Acad. Publ., Dordrecht, 2001. MR1858777 (2002k:42015)

Translations and editing

[1] W. K. Hayman, *Multivalent functions.* Translated from the English by V. P. Havin. Edited by I. E. Bazilevic. Izdat. Inostr. Lit., Moscow, 1960. MR0121492 (22 #12230)

[2] J. A. Jenkins, *Univalent functions and conformal mapping.* Translated from the English by V. P. Havin. Izdat. Inostr. Lit., Moscow, 1962.

[3] L. Carleson, *Selected problems in the theory of exceptional sets.* Translated from the English by V. P. Havin. Edited by V. G. Maz′ya. "Mir", Moscow, 1971. MR0352445 (50 #4932)

[4] W. Rudin, *Principles of mathematical analysis.* Translated from the English by V. P. Havin. "Mir", Moscow, 1976.

[5] *Investigations on linear operators and the theory of functions. 99 unsolved problems in linear and complex analysis.* Compiled and edited by N. K. Nikolskii, V. P. Havin, and S. V. Hrushchev. Zap. Nauchn. Sem. Leningrad. Otdel. Mat. Inst. Steklov. (LOMI) **81** (1978). "Nauka", Leningrad, 1978. MR0524720 (82e:46002)

[6] *Complex analysis and spectral theory.* Papers from the Seminar held in Leningrad, 1979/1980. Edited by V. P. Havin and N. K. Nikolskii. Lecture Notes in Mathematics, **864**. Springer, Berlin, 1981. MR0643379 (82m:46001)

[7] J. Garnett, *Bounded analytic functions.* Translated from the English by E. M. Dyn′kin. Translation edited and with a preface by V. P. Havin. "Mir", Moscow, 1984. MR0767451 (85j:30072)

[8] *Linear and complex analysis problem book.* 199 *research problems.* Edited by V. P. Havin, S. V. Hrushchev, and N. K. Nikolskii. Lecture Notes in Mathematics, **1043**, Springer-Verlag, Berlin, 1984. MR0734178 (85k:46001)

[9] V. I. Smirnov, *Selected works. Complex analysis. Mathematical theory of diffraction.* With a biography of Smirnov by O. A. Ladyzhenskaya. Compiled and with supplementary material by V. M. Babich, N. K. Nikolskii and V. P. Havin. Leningrad. Univ., Leningrad, 1988. MR1064111 (91g:01040)

[10] *Commutative harmonic analysis,* I. *General survey. Classical aspects.* Current problems in mathematics. Fundamental directions, **15**. Vsesoyuz. Inst. Nauchn. i Tekhn. Inform., Moscow, 1987. Translation by D. Khavinson and S. V. Kislyakov. Translation edited by V. P. Havin and N. K. Nikolskii. Encyclopaedia of Mathematical Sciences, **15**, Springer-Verlag, Berlin, 1991. MR1134135 (93b:42001)

[11] *Commutative harmonic analysis,* II. *Group methods in commutative harmonic analysis.* Current problems in mathematics. Fundamental directions, **25**, Vsesoyuz. Inst. Nauchn. i Tekhn. Inform., Moscow, 1988. Translated by D. Dynin and S. Dynin. Translation edited by V. P. Havin N. K. Nikolskii. Encyclopaedia of Mathematical Sciences, **25**. Springer-Verlag, Berlin, 1998. MR1622489 (99h:43002)

[12] *Commutative harmonic analysis.,* IV. *Harmonic analysis in* R^n. Current problems in mathematics. Fundamental directions. **42**. Vsesoyuz. Inst. Nauchn. i Tekhn. Inform., Moscow, 1989. Translation by J. Peetre. Translation edited by V. P. Havin and N. K. Nikolskii. Encyclopaedia of Mathematical Sciences, **42**, Springer-Verlag, Berlin, 1992. MR1299534 (95d:42001)

[13] *Commutative harmonic analysis,* III. *Generalized functions. Applications.* Current problems in mathematics. Fundamental directions. Vsesoyuz. Inst. Nauchn. i Tekhn. Inform., Moscow, 1991. Translation by R. Cooke. Translation edited by V. P. Havin and N. K. Nikolskii. Encyclopaedia of Mathematical Sciences, **72**, Springer-Verlag, Berlin, 1995. MR1447630 (98b:42001)

[14] *Complex analysis,* I. *Entire and meromorphic functions. Polyanalytic functions and their generalizations.* Akad. Nauk SSSR, Vsesoyuz. Inst. Nauchn. i Tekhn. Inform., Moscow, 1991. Translation by V. I. Rublinetskii and V. A. Tkachenko. Translation edited by A. A. Gonchar, V. P. Havin, and N. K. Nikolskii. Encyclopaedia of Mathematical Sciences, **85**, Springer-Verlag, Berlin, 1997. MR1464197 (98d:30002)

[15] *Linear and complex analysis. Problem book* 3. *Part* I. Edited by V. P. Havin and N. K. Nikolskii. Lecture Notes in Mathematics, **1573**, Springer-Verlag, Berlin, 1994. MR1334345 (96c:00001a)

[16] *Linear and complex analysis. Problem book* 3. *Part* II. Edited by V. P. Havin and N. K. Nikolskii. Lecture Notes in Mathematics, **1574**, Springer-Verlag, Berlin, 1994. MR1334346 (96c:00001b)

[17] *Complex analysis, operators, and related topics. The S. A. Vinogradov memorial volume.* Edited by V. P. Havin and N. K. Nikolskii. Operator Theory: Advances and Applications, **113**, Birkhäuser Verlag, Basel, 2000. MR1771746 (2001b:00031)

[18] S. A. Vinogradov, *Interpolation problems for analytic functions continuous in the closed disk and for functions whose sequence of coefficients is in* ℓ^p. Translated from the 1968 Russian original by V. P. Havin. Oper. Theory Adv. Appl., **113**, Complex analysis, operators, and related topics, pp. 23–29, Birkhäuser, Basel, 2000. MR1771749

[19] S. A. Vinogradov, *Free interpolation in spaces of analytic functions.* Translated from the 1983 Russian original by V. P. Havin. Oper. Theory Adv. Appl., **113**, Complex analysis, operators, and related topics, pp. 31–42, Birkhäuser, Basel, 2000. MR1771750

Biographical notes

[1] *Boris Zaharovič Vulih* (*on the occasion of his sixtieth birthday*) (with D. A. Vladimirov, M. K. Gavurin, G. I. Natanson, I. V. Romanovskii), Vestnik Leningrad. Univ. Mat. Meh. Astronom. **7** (1973), no. 2, 158–159. MR0325338 (48 #3685)

[2] *Grigorii Mikhailovich Fikhtengolts* (*on the* 100*th anniversary of his birth*) (with S. A. Vinogradov, D. A. Vladimirov, M. K. Gavurin, S. M. Ermakov, B. M. Makarov, G. I. Natanson, M. Z. Solomyak, and D. K. Faddeev), Vestnik Leningrad. Univ. Mat. Mekh. Astronom. **1988**, no. 3, 3–6. MR0974784 (89m:01112)

[3] *Gleb Pavlovich Akilov* (with A. D. Aleksandrov, A. M. Vershik, V. V. Ivanov, A. G. Kusraev, S. S. Kutateladze, B. M. Makarov, and Yu. G. Reshetnyak), Uspekhi Mat. Nauk **43** (1988), no. 1, 181–182; translation in Russian Math. Surveys **43** (1988), no. 1, 221–223. MR0937023 (89b:01031)

[4] *Recollections of G. P. Akilov* (*Leningrad, the* 1950*s and* 1960*s*) (with A. I. Veksler, D. A. Vladimirov, and B. M. Makarov), Optimizatsiya **48 (65)** (1990), 35–40. MR1177106 (93g:01080b)

[5] *Stanislav Aleksandrovich Vinogradov, his life and mathematics* (with N. K. Nikolskii), Complex analysis, operators, and related topics, pp. 1–18, Oper. Theory Adv. Appl., **113**, Birkhäuser, Basel, 2000. MR1771747 (2001b:01025a)

[6] *Viktor Solomonovich Videnskii* (*on the occasion of his eightieth birthday*) (with V. V. Zhuk, V. N. Malozemov and G. I. Natanson), Uspekhi Mat. Nauk **57** (2002), no. 5 (347), 182–186; translation in Russian Math. Surveys **57** (2002), no. 5, 1033–1038. MR1992100 (2004d:01042)

[7] *Garald Isidorovich Natanson* (with O. L. Vinogradov, V. V. Zhuk, and V. L. Fainshmidt), Proceedings of the St. Petersburg Mathematical Society, Vol. X, pp. 237–252, Amer. Math. Soc. Transl. Ser. 2, **214**, Amer. Math. Soc., Providence, RI, 2005. MR2181517

[8] *Semyon Yakovlevich Khavinson* (*May* 17, 1927*–January* 30, 2003), Selected topics in complex analysis, pp. 1–22, Oper. Theory Adv. Appl., **158**, Birkhäuser, Basel, 2005. MR2147585

Amer. Math. Soc. Transl.
(2) Vol. **226**, 2009

Zeros of Functions in Weighted Bergman Spaces

Alexandru Aleman and Carl Sundberg

1. Introduction

Let $\mathbb{D} = \{z \in \mathbb{C} : |z| < 1\}$ be the unit disk in the complex plane and denote by $H(\mathbb{D})$ the set of functions analytic in $\mathbb{D}$. Given an $f \in H(\mathbb{D})$ we let Z_f denote the zero sequence of f, with multiple zeros repeated according to their multiplicity. For two such functions f, g we say that Z_g is a subset of Z_f, $Z_g \subset Z_f$, if every point in Z_g appears in Z_f at least the same number of times, i.e. if $\frac{f}{g} \in H(\mathbb{D})$.

A well-known result of Weierstrass tells us that any sequence in $\mathbb{D}$ with no limit point in $\mathbb{D}$ is Z_f for some $f \in H(\mathbb{D})$, but things get more complicated if we require f to lie in some restricted subspace of $H(\mathbb{D})$. The best known and best understood examples are the *Hardy classes*

$$H^p = \left\{ f \in H(\mathbb{D}) : \sup_{0<r<1} \int_0^{2\pi} |f(re^{i\theta})|^p \frac{d\theta}{2\pi} < \infty \right\},$$

where $0 < p < \infty$. As is well known, a sequence $Z = \{z_1, z_2, \ldots\}$ is Z_f for some $f \in H^p$ if and only if $\sum(1 - |z_n|) < \infty$. This characterization makes it immediately clear that if a sequence Z is included in Z_f, where $f \in H^p$, then there is a $g \in H^p$ such that $Z = Z_g$. In other words, "subsets of zero sets are zero sets".

In this paper we will be concerned with the zero sequences of functions in *weighted Bergman spaces*: let ω be a positive measure in $\mathbb{D}$; then the weighted Bergman space $L^p_a(\omega)$ is just $H(\mathbb{D}) \cap L^p(\omega)$. In order to assure the completeness of $L^p_a(\omega)$ we require that ω is such that the evaluation functionals $f \to f(z)$ are $L^p(\omega)$-bounded for all $z \in \mathbb{D}$. This will be the case e.g. if $d\omega = w\,dA$, where $w > 0$ is continuous and dA is normalized Lebesgue measure on $\mathbb{D}$.

The *standard weighted Bergman spaces* are those for which $d\omega(z) = d\omega_\alpha(z) = \frac{1}{\alpha+1}(1 - |z|^2)^\alpha\, dA(z)$ for $\alpha > -1$. Note that as $\alpha \to -1$ we recover the Hardy spaces, and for $\alpha = 0$ we have "the" Bergman space. In spite of much effort (see, e.g. [**2, 3, 4**]) there is as yet no characterization of the zero sequences of functions in the standard weighted Bergman spaces, but nevertheless C. Horowitz [**1**] showed in 1974 that subsets of zero sets are zero sets for these spaces.

2000 *Mathematics Subject Classification.* Primary 30H05, 47B35.

Key words and phrases. Zero sequences, weighted Bergman spaces.

The second author was supported by a grant from the National Science Foundation.

In this paper we will construct the first known example of a weighted Bergman space, with in fact a rotationally symmetric weight, for which subsets of zero sets are not necessarily zero sets. We have some degree of freedom in the construction, e.g. in how quickly our weights go to zero near the boundary of $\mathbb{D}$.

Our main result, Theorem 3.2, is proven in Section 3 after some preliminary work in Section 2. In Section 4 we discuss some natural questions raised by our work.

2. Preliminary estimates

Throughout in what follows we consider increasing sequences of positive integers (N_j), (M_j), (m_j) tending to $+\infty$ such that (N_j) and (M_j) are strictly increasing and $m_j > 4$ for all $j \geq 1$. If we also assume that

$$\lim_{j\to\infty} \frac{\log(m_j N_j)}{\sqrt{M_j}} = 0, \tag{2.1}$$

then the product

$$Q(z) = \prod_{j=1}^{\infty} (1 + m_j z^{M_j})^{N_j} \tag{2.2}$$

converges uniformly on compact subsets of the unit disk. We begin with the following simple but fundamental estimate for this product.

PROPOSITION 2.1. *Assume that* (2.1) *holds and that*

$$\lim_{j\to\infty} \frac{M_j \log m_{j+1}}{M_{j+1}} = 0. \tag{2.3}$$

For $(\frac{1}{2})^{1/M_1} \leq r < 1$ *let* $l(r)$ *be the least integer* l *such that* $r \leq (\frac{1}{2})^{1/M_l}$. *Then there exists* $r_0 \in (0,1)$ *such that for* $|z| = r > r_0$ *we have*

$$\begin{aligned} \big|\log|Q(z)| - &N_{l(r)} \log|1 + m_{l(r)} z^{M_{l(r)}}|\big| \\ &\leq \sum_{k<l(r)} N_k \log(1 + m_k) + C \sum_{k>l(r)} N_k m_k 2^{-M_k/M_{l(r)}}, \end{aligned}$$

where

$$C = \sup_{|w|\leq 1/2} \frac{\big|\log|1+w|\big|}{|w|}.$$

PROOF. Note first that if $k < l(r)$, then $r^{M_k} > \frac{1}{2}$; hence, for $|z| = r$ we have

$$1 + m_k \geq |1 + m_k z^{M_k}| > \frac{m_k}{2} - 1 > 1,$$

which implies that

$$\left|\sum_{k<l(r)} N_k \log|1 + m_k z^{M_k}|\right| \leq \sum_{k<l(r)} N_k \log(1 + m_k).$$

By (2.3) there exists $0 < r_0 < 1$ such that for $r > r_0$, and $k > l(r)$ we have

$$m_k r^{M_k} \leq m_k 2^{-M_k/M_{l(r)}} \leq \frac{1}{2}.$$

If C is the constant given in the statement, then for $|z| = r > r_0$ we have

$$\sup_{k>l(r)} \frac{|\log|1 + m_k z^{M_k}||}{m_k r^{M_k}} \le C,$$

which gives

$$\sum_{k>l(r)} N_k |\log|1 + m_k z^{M_k}|| \le C \sum_{k>l(r)} N_k m_k r^{M_k} \le C \sum_{k>l(r)} N_k m_k 2^{-M_k/M_{l(r)}}$$

and the result follows. □

Motivated by this estimate, we consider the following conditions on our sequences:

1) $M_{j+1} \ge M_j^2$ for all $j \ge 1$.
2) $m_j \log M_j = o(N_j)$ as $j \to \infty$.
3) $\log N_j = o(\sqrt{M_j}\,)$ as $j \to \infty$.
4) $\sum_{k<j} N_k \log m_k = o(m_{j+1}^{-2} N_{j+1})$ as $j \to \infty$.

It is very easy to construct such sequences. A concrete example is

$$M_j = 2^{2^j}, \quad N_j = 2^{j^2}, \quad m_j = 5j, \quad j \ge 1.$$

Note that these conditions imply (2.1) and (2.3).

We now introduce the radii r_j, $s_j \in (0, 1)$ which are defined by

$$r_j^{M_j} = m_j^{-2} \quad \text{and} \quad s_j^{M_j} = \frac{1}{2}. \tag{2.4}$$

Given sequences of positive numbers (a_j), (b_j) we shall write

$$a_j \approx b_j$$

if

$$\frac{a_j}{b_j} \to 1 \quad \text{as} \quad j \to \infty.$$

For sequences dependent on some other parameter s, we say that $a_j \approx b_j$ uniformly in s if the above convergence is uniform with respect to this parameter.

For example, we have

$$1 - r_j \approx -\log r_j = 2\frac{\log m_j}{M_j}$$

and

$$1 - s_j \approx -\log s_j = \frac{\log 2}{M_j}.$$

Let us also consider the companion product

$$P(z) = \prod_{j=1}^{\infty} (1 + m_j^2 z^{2M_j})^{N_j}, \tag{2.5}$$

which converges uniformly on compact subsets of the unit disk.

COROLLARY 2.1. (i) *If* $\lambda_j = \exp(\frac{2\pi i}{M_j})$ *then*

$$\log|Q(r_j\lambda_j^k)| \approx N_j m_j^{-1}$$

uniformly in $k \in \mathbb{Z}$.

(ii) *We have*

$$\min_{|z|=s_j} \log|Q(z)| \approx \max_{|z|=s_j} \log|Q(z)| \approx N_j \log m_j.$$

(iii) *If* $\frac{1}{2} < \alpha < 1$ *then*

$$\log P(r_j^\alpha) \approx N_j m_j^{2-4\alpha}$$

and

$$\log P(s_j^\alpha) \approx 2N_j \log m_j.$$

(iv) *If we denote by* $l_P(r)$ *the number defined in Proposition* 2.1 *corresponding to the product* P, *then*

$$\log P(r) = O(N_{l_P(r)} \log m_{l_P(r)}),$$

as $r \to 1^-$.

PROOF. (i) For sufficiently large j, by (2.3) we have $l(r_j) = j$, and Proposition 2.1 combined with 1) gives

$$|\log|Q(r_j\lambda_j^k)| - N_j\log(1+m_j^{-1})| < \sum_{k<j} N_k \log(1+m_k) + C\sum_{k>j} N_k m_k 2^{-\sqrt{M_k}}.$$

Now apply 3) and 4) to conclude that

$$\sum_{k<j} N_k \log m_k + C\sum_{k>j} N_k m_k 2^{-\sqrt{M_k}} = o(N_j m_j^{-2})$$

as $j \to \infty$, which implies the estimate in the statement.

(ii) As for (i), we note first that $l(s_j) = j$, $j \geq 2$, and apply again Proposition 2.1 together with the above considerations to conclude that

$$\big|\log|Q(s_je^{it})| - N_j\log|1 + m_j(s_je^{it})^{M_j}|\big| = o(N_j m_j^{-2})$$

as $j \to \infty$. Then the result follows from the simple fact that

$$\lim_{j\to\infty} \frac{\log|1 + \frac{m_j}{2}e^{itM_j}|}{\log m_j} = 1$$

uniformly in $t \in [0, 2\pi]$.

(iii) It is easy to verify that Proposition 2.1 applies to the product P as well. Note that

$$r_j^{2\alpha M_j} = m_j^{-4\alpha}, \quad s_j^{2\alpha M_j} = 2^{-2\alpha},$$

and again, for sufficiently large j we deduce with these parameters that $l_P(r_j^\alpha) = l_P(s_j^\alpha) = j$. If we apply Proposition 2.1 we obtain as above the estimates

$$|\log P(r_j^\alpha) - N_j\log(1 + m_j^{2-4\alpha})| = o(N_j m_j^{-2})$$

and

$$|\log P(s_j^\alpha) - N_j\log(1 + m_j^2 2^{-2\alpha})| = o(N_j m_j^{-2}),$$

and the result follows.

(iv) Applying Proposition 2.1 to P again, we get the estimate

$$\log P(r) \leq N_{l_P(r)} \log(1+m_{l_P(r)}^2) + \sum_{k<l_P(r)} N_k \log(1+m_k^2) + C \sum_{k>l_P(r)} N_k m_k^2 2^{-2M_k/M_{l_P(r)}}$$

for the fixed positive constant C and all r sufficiently close to 1. Then the result follows with the argument used in the previous cases. □

3. Estimates of weighted Bergman norms

We shall consider positive radial measures carried by the unit disk $\mathbb{D}$, that is, measures of the form $\mu = \nu \times m$, where ν is a positive regular Borel measure carried by $[0,1]$ and m is the normalized Lebesgue measure on $[0, 2\pi]$. The corresponding Bergman spaces are denoted by $L_a^p(\mu)$ and consist of analytic functions in $L^p(\mu)$. All the measures considered here satisfy

$$\mu(\{\rho < |z| < 1\}) > 0$$

for all $\rho \in [0,1)$, so that $L_a^p(\mu)$ is a closed subspace of $L^p(\mu)$. This is a consequence of the following basic estimate, which will be frequently used in what follows. For the sake of completeness we have included a sketch of the proof.

LEMMA 3.1. *Let μ be a radial measure as above, and let $0 < \eta < \delta < 1$, $r \in (0,1)$. Then for a nonnegative subharmonic function u in $\mathbb{D}$ and $z \in \mathbb{D}$ with $|z| \leq r$ we have*

$$u(z)\mu(\{r^\delta \leq |\zeta| \leq r^\eta\}) \leq \frac{2}{r^\delta - |z|} \int_{\{r^\delta \leq |\zeta| \leq r^\eta\}} u(\zeta) d\mu(\zeta).$$

PROOF. If $\mu = \nu \times m$, $r^\delta \leq \rho \leq r^\eta$, we can use the subharmonicity of u to obtain

$$u(z) \leq \frac{\rho + |z|}{\rho - |z|} \int_0^{2\pi} u(\rho e^{it})\, dm \leq \frac{2}{r^\delta - |z|} \int_0^{2\pi} u(\rho e^{it})\, dm.$$

If we integrate this inequality on $[r^\delta, r^\eta]$ w. r. t. ν, then the result follows. □

The next estimate for harmonic functions will play an important role in the proof of the main results.

LEMMA 3.2. *Let $0 < r < s < 1$, $A > 0$, and let M be a positive integer. If u is a real-valued harmonic function in $\mathbb{D}$ with $u(0) = 0$ and*

$$\min_{|z|=r} u(z^M) < -A,$$

then

$$\max_{|z|=s} u(z^M) \geq \frac{A}{2} \left[\left(\frac{s}{r}\right)^M - 1 \right].$$

PROOF. Assume the contrary, that is,

$$\max_{|z|=s} u(z^M) \leq (1-\varepsilon) \frac{A}{2} \left[\left(\frac{s}{r}\right)^M - 1 \right]$$

for some $\varepsilon > 0$. Consider the harmonic function v defined on $\mathbb{D}$ by

$$v(z) = \frac{2u(s^M z)}{(1-\varepsilon)A[(\frac{s}{r})^M - 1]}.$$

We have $v(0) = 0$ and by the maximum principle, $v(z) \le 1$ for all $z \in \mathbb{D}$. Then $1-v$ is a positive harmonic function in $\mathbb{D}$ with the value 1 at the origin, so it satisfies Harnack's inequality

$$1 - v(\zeta) \le \frac{1+|\zeta|}{1-|\zeta|}$$

or equivalently,

$$v(\zeta) \ge -\frac{2|\zeta|}{1-|\zeta|}, \quad \zeta \in \mathbb{D}.$$

Given $w = re^{it}$ set $\zeta = (\frac{r}{s})^M e^{iMt}$ and use the last inequality to obtain

$$\frac{2u(w^M)}{(1-\varepsilon)A[(\frac{s}{r})^M - 1]} = v(\zeta) \ge -\frac{2(\frac{r}{s})^M}{1-(\frac{r}{s})^M} = -\frac{2}{(\frac{s}{r})^M - 1}.$$

This implies that $u(w^M) \ge -(1-\varepsilon)A$, which contradicts our assumption and completes the proof. □

Our next result is rather technical but it provides a useful necessary condition for a measure μ such that if Q is the product considered in the previous section, Z_Q is a zero sequence for the weighted Bergman space $L^p_a(\mu)$.

THEOREM 3.1. *Assume that the sequences (m_j), (M_j), (N_j) satisfy the conditions 1)–4), let Q be the product defined by (2.2), and let (r_j), (s_j) be defined by (2.4). If μ is a radial measure carried by $\mathbb{D}$ such that Z_Q is a zero sequence for $L^p_a(\mu)$, $p > 0$, then for every α, β, η, $\delta \in (0,1)$ with $\alpha < \beta$ and $\eta < \delta$ we have*

$$-\limsup_{j\to\infty} \frac{\log \mu(\{s_j^\delta \le |z| \le s_j^\eta\})}{m_j N_j} \ge \frac{p}{4} + \frac{1}{4} \liminf_{j\to\infty} \frac{m_j \log \mu(\{r_j^\beta \le |\zeta| \le r_j^\alpha\})}{N_j}.$$

PROOF. Observe first that if the right-hand side is less than or equal to zero, there is nothing to prove. Thus if $0 < \alpha < \beta < 1$ are given, we can assume that

$$L = -\liminf_{j\to\infty} \frac{m_j \log \mu(\{r_j^\beta \le |\zeta| \le r_j^\alpha\})}{N_j} < p.$$

The fact that Z_Q is a zero sequence for $L^p_a(\mu)$ means that there exists a zero-free analytic function ψ in $\mathbb{D}$ such that $\psi(0) = 1$ and

$$K = \int_{\mathbb{D}} |\psi Q|^p d\mu < \infty.$$

Given $j \ge 1$, let $\lambda_j = \exp(\frac{2\pi i}{M_j})$ and denote by ϕ_j the analytic function in $\mathbb{D}$ defined by the equality

$$\phi_j(z^{M_j}) = \left(\prod_{k=1}^{M_j} \psi(\lambda_j^k z)\right)^{1/M_j}, \quad z \in \mathbb{D}.$$

Consider the subharmonic functions

$$z \mapsto |\phi_j(z^{M_j})|^p \left(\prod_{k=1}^{M_j} |Q(\lambda_j^k z)|\right)^{p/M_j}$$

and apply Lemma 3.1 together with Hölder's inequality to obtain for $|z| = r_j$,

$$\begin{aligned}
\mu(\{r_j^\beta \le |\zeta| \le r_j^\alpha\})|\phi_j(z^{M_j})|^p \Big(\prod_{k=1}^{M_j} |Q(\lambda_j^k z)|\Big)^{p/M_j} \\
&\le \frac{2}{r_j^\beta - r_j} \int_{r_j^\beta \le |\zeta| \le r_j^\alpha} |\phi_j(\zeta^{M_j})|^p \Big(\prod_{k=1}^{M_j} |Q(\lambda_j^k \zeta)|\Big)^{p/M_j} d\mu(\zeta) \\
&\le \frac{2}{r_j^\beta - r_j} \Big(\prod_{k=1}^{M_j} \int_{r_j^\beta \le |\zeta| \le r_j^\alpha} |\psi Q|^p d\mu\Big)^{1/M_j} \\
&\le \frac{2K}{r_j^\beta - r_j}.
\end{aligned}$$

This implies that

$$p \log |\phi_j(r_j^{M_j})| \le -\log \mu(\{r_j^\beta \le |\zeta| \le r_j^\alpha\}) - \frac{p}{M_j} \sum_{k=1}^{M_j} \log |Q(\lambda_j^k r_j)| + \log \frac{2K}{r_j^\beta - r_j}.$$

By Corollary 2.1 (i) we have

$$\frac{1}{M_j} \sum_{k=1}^{M_j} \log |Q(\lambda_j^k r_j)| \approx N_j m_j^{-1}.$$

Recall also that $1 - r_j \approx 2 \frac{\log m_j}{M_j}$ and apply the conditions 2) and (2.1) to conclude that

$$\limsup_{j\to\infty} \frac{m_j}{N_j} \log |\phi_j(r_j^{M_j})| \le -1 + \frac{L}{p}.$$

Equivalently, for every $\varepsilon > 0$ there exists $j_\varepsilon \ge 1$ such that

$$\log |\phi_j(r_j^{M_j})| \le -\Big(1 - \frac{L}{p} - \varepsilon\Big) N_j m_j^{-1} \tag{3.1}$$

whenever $j > j_\varepsilon$. Now by assumption we have $1 - \frac{L}{p} > 0$; hence, for $0 < \varepsilon < 1 - \frac{L}{p}$ we can apply Lemma 3.2 to the harmonic functions $u_j(z) = \log |\phi_j(z)|$ with $M = M_j$, in order to deduce that there exists a point z_j with $|z_j| = s_j$ such that

$$\begin{aligned}
\log |\phi_j(z_j^{M_j})| &\ge \frac{1}{2}\Big(1 - \frac{L}{p} - \varepsilon\Big) N_j m_j^{-1} \Big[\Big(\frac{s_j}{r_j}\Big)^{M_j} - 1\Big] \\
&= \frac{1}{2}\Big(1 - \frac{L}{p} - \varepsilon\Big) N_j m_j^{-1} \Big(\frac{m_j^2}{2} - 1\Big).
\end{aligned}$$

From the definition of ϕ_j we see immediately that there exists w_j with $|w_j| = s_j$ such that

$$\log |\psi(w_j)| \ge \frac{1}{2}\Big(1 - \frac{L}{p} - \varepsilon\Big) N_j m_j^{-1} \Big(\frac{m_j^2}{2} - 1\Big).$$

Using again Lemma 3.1 for the function $|\psi Q|^p$ we obtain the estimate

$$\begin{aligned}\log \mu(\{s_j^\delta \le |\zeta| \le s_j^\eta\}) &+ \frac{p}{2}\Big(1 - \frac{L}{p} - \varepsilon\Big) N_j m_j^{-1}\Big(\frac{m_j^2}{2} - 1\Big) \\ &\le \log[\mu(\{s_j^\delta \le |\zeta| \le s_j^\eta\})|\psi Q(w_j)|^p] - p \log \min_{|z|=s_j} |Q(z)| \\ &\le \log \frac{2K}{s_j^\delta - s_j} - p \log \min_{|z|=s_j} |Q(z)|.\end{aligned}$$

We have $1 - s_j \approx \frac{\log 2}{M_j}$ and by Corollary 2.1 (ii)

$$\log \min_{|z|=s_j} |Q(z)| \approx N_j \log m_j;$$

hence, if we let $j \to \infty$, the result follows from the condition 2). □

Given a radial measure ω that is absolutely continuous w.r.t. the Lebesgue measure dA on $\mathbb{D}$, we denote $w = \frac{d\omega}{dA}$. The function w depends only on $|z|$ and is called a radial weight.

THEOREM 3.2. *Let $p > 0$ and let φ be a positive increasing function on $[0, 1)$ with $\lim_{r\to 1^-} \varphi(r) = \infty$. Then there exists a radial weight w such that*

(i) *w is real-analytic on $[0, 1)$;*

(ii) *w is strictly decreasing and $r \mapsto \log w$ is a concave function of $\log r$, $r \in [0, 1)$;*

(iii) *$-\log w(r) = o(-\varphi(r) \log(1 - r))$ as $r \to 1$;*

(iv) *there are zero sequences for $L^p_a(w\, dA)$ which contain subsequences that are not a zero sequence for any $L^q_a(w\, dA)$, $q > 0$.*

PROOF. Let (m_j), (M_j), (N_j) satisfy 1)–4), let P be the product defined by (2.5), and set

$$w(r) = P(r)^{-p}, \quad r \in [0, 1).$$

Since $P(r) = \max_{|z|=r} |P(z)|$, the properties (i) and (ii) in the statement are immediately verified. The second part of (ii) follows from the fact that

$$P(r) = \max_{|z|\le r} |P(z)|.$$

Let us now verify the condition (iv). To this end, we consider the infinite product

$$\tilde{P}(z) = \prod_{j=1}^{\infty} (1 - m_j^2 z^{2M_j})^{N_j} \tag{3.2}$$

and note that

$$|\tilde{P}(z)| \le P(|z|),$$

which implies that $\tilde{P} \in L^p_a(w\, dA)$. Clearly, Z_Q is contained in $Z_{\tilde{P}}$, and so it suffices to show that Z_Q is not a zero sequence for any weighted Bergman space $L^q_a(w\, dA)$, $q > 0$. Set $d\mu = w\, dA$ and note that if $\frac{3}{4} < \alpha < \beta < 1$, then by part (iii) of Corollary 2.1 and the remarks preceding this corollary we have an estimate of the form

$$\mu(\{r_j^\beta \le |z| \le r_j^\alpha\}) \ge \pi(r_j^{2\alpha} - r_j^{2\beta}) P(r_j^\alpha)^{-p} \ge A \frac{\log m_j}{M_j} \exp(-B N_j m_j^{2-4\alpha})$$

for some positive constants A, B depending only on p, α, β, and all j sufficiently large. Using also the condition 2) and the fact that $\alpha > \frac{3}{4}$ we see that

$$\lim_{j\to\infty} \frac{m_j \log \mu(\{r_j^\beta \le |\zeta| \le r_j^\alpha\})}{N_j} = 0.$$

Similarly, for $\frac{1}{2} < \eta < \delta < 1$ we have the estimate

$$\mu(\{s_j^\delta \le |z| \le s_j^\eta\}) \ge \pi(s_j^{2\eta} - s_j^{2\delta})P(s_j^\eta)^{-p} \ge \frac{A'}{M_j} \exp(-B' N_j \log m_j)$$

and, as above, we see that

$$\lim_{j\to\infty} \frac{\log \mu(\{s_j^\delta \le |z| \le s_j^\eta\})}{m_j N_j} = 0.$$

Then by Theorem 3.1 it follows that Z_Q cannot be a zero sequence for $L_a^q(w\,dA)$ for any $q > 0$.

It remains to prove that the sequences (m_j), (M_j), (N_j) can be chosen such that w satisfies (iii). Assume without loss of generality that the function φ from the statement satisfies $\varphi(r) > e^{12}$, $r \in (0,1)$ and define the sequence (M_j) inductively by $M_1 = 2$ and

$$\log M_{j+1} = [\varphi(2^{-1/M_j})^{1/2}] \log M_j,$$

where $[x]$ denotes the integer part of $x \in \mathbb{R}$. To simplify the notation, write

$$\varphi_j = [\varphi(2^{-1/M_{j-1}})^{1/3}]$$

and set

$$N_j = \varphi_j [\log M_j], \quad m_j = \log \varphi_j.$$

From these definitions we see that the conditions 1)–3) are satisfied. To verify 4), we first observe that

$$N_j \log m_j = o(N_{j+1} \log m_{j+1}),$$

which implies that

$$\sum_{k<j} N_k \log m_k = O(N_{j-1} \log m_{j-1}).$$

Again from the definition of these sequences we deduce that

$$\frac{m_j^2 N_{j-1} \log m_{j-1}}{N_j} = O\left(\frac{\log\log \varphi_{j-1} (\log \varphi_j)^2 \varphi_{j-1}}{\varphi_j^{5/2}}\right) = o(1),$$

which gives 4). Finally, for $r \in (0,1)$, let $l_P(r)$ be the integer defined in Corollary 2.1 (see also Proposition 2.1) and recall that $2^{-1/M_{l_P(r)}} \ge r > 2^{-1/M_{l_P(r)-1}}$; hence,

$$\varphi_{l_P(r)} \le \varphi(r^2)^{1/3} \le \varphi(r)^{1/3}, \quad \log M_{l_P(r)-1} < \log \frac{\log 2}{\log \frac{1}{r}}.$$

Then by Corollary 2.1 (iv) we have

$$\begin{aligned} -\log w(r) &= p \log P(r) = O(N_{l_P(r)} \log m_{l_P(r)}) \\ &= O(\varphi_{l_P(r)}^{5/2} \log M_{l_P(r)-1} \log\log \varphi_{l_P(r)}), \end{aligned}$$

and from above

$$\varphi_{l_P(r)}^{5/2} \log M_{l_P(r)-1} \log\log \varphi_{l_P(r)}) = o\left(\varphi(r) \log \frac{1}{1-r}\right)$$

which completes the proof. □

The weight in the theorem can be chosen to be logarithmically-concave on $[0, 1)$. Indeed, a direct calculation shows that if $w = P^{-p}$ as above, then $\frac{d^2}{dr^2} \log w(r) \leq 0$ on $[0, 1)$, provided that $2M_j - 1 \geq m_j^2$ for all j. Now by construction, this inequality does hold for sufficiently large j, say $j \geq k$. Then the weight

$$w_k(r) = \prod_{j=1}^{k} (1 + m_j^2 r^{2M_j})^{pN_j} w(r)$$

satisfies the conditions (i)–(iii) in Theorem 3.2, $L_a^p(w_k \, dA) = L_a^p(w \, dA)$, and $\log w_k$ is concave. We are grateful to the referee for his remarks related to this issue.

4. Concluding remarks

The asymptotics in Horowitz's proof [**1**] as $\alpha \to \infty$ suggests that if $w(r) \to 0$ as $r \to 1^-$ fast enough, namely faster than $(1 - r)^\alpha$ for any $\alpha > 0$, then subsets of zero sets will not necessarily be zero sets. Our result is an example of this phenomenon. On the other hand, note that for our method to work it is necessary that $\frac{\log w(r)}{\log(1-r)} \to \infty$, i.e. that $w(r) \to 0$ faster than $(1 - r)^\alpha$ — this is easily seen from Corollary 2.1 (iii).

These considerations suggest the following questions. Suppose $w(r) < 1$ is a positive, continuous, decreasing function of r and $d\omega(z) = w(|z|) \, dA(z)$.

Question 1. If

$$\limsup_{r \to 1^-} \frac{\log w(r)}{\log(1 - r)} < \infty,$$

are subsets of $L_a^p(\omega)$-zero sets necessarily $L_a^p(\omega)$-zero sets?

Question 2. If

$$\limsup_{r \to 1^-} \frac{\log w(r)}{\log(1 - r)} = \infty,$$

will there be subsets of $L_a^p(\omega)$-zero sets that are not $L_a^p(\omega)$-zero sets?

A particularly natural example of the situation of Question 2 is the exponentially decreasing weight $d\omega(z) = \exp(-\frac{1}{1-|z|^2}) \, dA(z)$.

References

[1] C. Horowitz, *Zeros of functions in the Bergman spaces*, Duke Math. J. **41** (1974), 693–710.

[2] B. Korenblum, *An extension of the Nevanlinna theory*, Acta Math. **135** (1975), no. 3–4, 187–219.

[3] K. Seip, *On a theorem of Korenblum*, Ark. Mat. **32** (1994), no. 1, 237–243.

[4] ———, *On Korenblum's density condition for the zero sequences of $A^{-\alpha}$*, J. Anal. Math. **67** (1995), 307–322.

Department of Mathematics, Lund University, P.O. Box 118, S-221 00 Lund, Sweden

E-mail address: aleman@maths.lth.se

Department of Mathematics, University of Tennessee, Knoxville, TN 37996-1300, USA

E-mail address: sundberg@math.utk.edu

Amer. Math. Soc. Transl.
(2) Vol. **226**, 2009

A Characterization of the Fourier Transform and Related Topics

Semyon Alesker, Shiri Artstein-Avidan, and Vitali Milman

To my friend, Victor Havin, with warmest memories and enormous respect. — Vitali Milman

1. Introduction

In this note we show that the Fourier transform is essentially, up to a simple adjustment, the only transform on the corresponding space which interchanges convolutions and products.

To state various results we first need to recall some well-known definitions and simple observations. For a reference on the standard definitions and results stated below, see e.g. [**GS1**], the elementary introduction to the subject [**SS**], or the more advanced book [**S**].

DEFINITION 1.1. One says that an infinitely smooth function $f\colon \mathbb{R}^n \to \mathbb{C}$ is *rapidly decreasing* (also called *Schwartz function*) if for any $l \in \mathbb{Z}_+$ and any multi-index $\alpha = (\alpha_1, \dots, \alpha_n)$ of non-negative integers one has

$$\sup_{x\in\mathbb{R}^n}\left|\frac{\partial^\alpha f(x)}{\partial x^\alpha}(1+|x|^l)\right| < \infty,$$

where as usual $\frac{\partial^\alpha f(x)}{\partial x^\alpha} := \frac{\partial^{|\alpha|} f}{\partial x_1^{\alpha_1}\dots\partial x_n^{\alpha_n}}$, $|\alpha| := \sum_{i=1}^n \alpha_i$.

The space of all rapidly decreasing real-valued functions on $\mathbb{R}^n$ is denoted by $\mathcal{S}(n)$, and is called the Schwartz space. Below we denote $\mathcal{S}(n) = \mathcal{S}$ when the dependence on n is obvious.

The space $\mathcal{S} = \mathcal{S}(n)$ becomes a Fréchet space when equipped with the system of norms:

$$\|f\|_N := \sup\left\{\left|\frac{\partial^\alpha f(x)}{\partial x^\alpha}(1+|x|^N)\right| : x \in \mathbb{R}^n,\ |\alpha| \le N\right\}.$$

2000 *Mathematics Subject Classification.* Primary 42A38.

Key words and phrases. Fourier transform, product, convolution.

The authors were partially supported by ISF grants (nos. 1369/04, 865/07, and 491/04). The second and third authors were also supported in part by BSF grant no. 2006079.

One of the main properties of the Schwartz space $\mathcal{S}$ is that the Fourier transform $\mathbb{F}\colon \mathcal{S} \to \mathcal{S}$, defined by

$$(\mathbb{F}f)(\xi) = \int_{\mathbb{R}^n} f(x) e^{-2\pi i \langle x, \xi \rangle}\, dx,$$

is a linear topological isomorphism.

The space $\mathcal{S}$ has two structures of an algebra given by the pointwise product and the convolution,

$$\cdot\colon \mathcal{S} \times \mathcal{S} \to \mathcal{S},$$
$$*\colon \mathcal{S} \times \mathcal{S} \to \mathcal{S}.$$

Both operations are continuous with respect to both arguments simultaneously.

Let $\mathcal{S}'(n)$ be the topological dual of $\mathcal{S}(n)$. (Again, we will write $\mathcal{S}'(n) = \mathcal{S}'$ whenever there is no possibility for confusion.) It will be equipped with the weak topology. Elements of $\mathcal{S}'$ are called *distributions of tempered growth.* We have the canonical continuous map $\mathcal{S} \to \mathcal{S}'$ given by

$$\langle \phi, f \rangle = \int_{\mathbb{R}^n} f(x)\phi(x)\, dx.$$

This map is injective and has a dense image in the weak topology. We will identify $\mathcal{S}$ with its image in $\mathcal{S}'$: $\mathcal{S} \subset \mathcal{S}'$.

The proof of the following claim is straightforward.

CLAIM 1. (i) *The pointwise product on $\mathcal{S}$ extends (uniquely) to a separately continuous map $\mathcal{S} \times \mathcal{S}' \to \mathcal{S}'$ which is given explicitly by*

$$\langle \phi, \psi \cdot f \rangle = \langle \phi \cdot \psi, f \rangle$$

for any $f \in \mathcal{S}'$, $\phi, \psi \in \mathcal{S}$. Then $\mathcal{S}'$ becomes a module over $\mathcal{S}$.

(ii) *The convolution on $\mathcal{S}$ extends (uniquely) to a separately continuous map $\mathcal{S} \times \mathcal{S}' \to \mathcal{S}'$ which is given explicitly by*

$$\langle \phi, \psi * f \rangle = \langle \phi * (-Id)^* \psi, f \rangle$$

for any $f \in \mathcal{S}'$, $\phi, \psi \in \mathcal{S}$. Then $\mathcal{S}'$ becomes a module over $\mathcal{S}$. (Here $(-Id)^*$ denotes the operator defined by $((-Id)^*\psi)(x) = \psi(-x)$, a special case of the definition of u^* below.)

(iii) *The Fourier transform extends (uniquely) to an isomorphism of linear topological spaces $\mathbb{F}\colon \mathcal{S}' \tilde{\to} \mathcal{S}'$. Moreover, $\mathbb{F}$ is an isomorphism of $\mathcal{S}$-modules: for any $\phi \in \mathcal{S}$, $f \in \mathcal{S}'$,*

$$\mathbb{F}(\phi \cdot f) = \mathbb{F}(\phi) * \mathbb{F}(f),$$
$$\mathbb{F}(\phi * f) = \mathbb{F}(\phi) \cdot \mathbb{F}(f).$$

For a proof see [**GS1**], Chapter 2.

To give a sample of our main results, we start with one which is not the strongest, but which we find to be a true 'characterization of the Fourier transform':

THEOREM 2. *Assume we are given a bijective transform $\mathcal{F} : \mathcal{S} \to \mathcal{S}$ that admits an extension $\mathcal{F}' : \mathcal{S}' \to \mathcal{S}'$ which is also bijective, and such that for every $f \in \mathcal{S}$ and $g \in \mathcal{S}'$ we have $\mathcal{F}'(f \cdot g) = (\mathcal{F}f) * (\mathcal{F}'g)$. Assume also that $\mathcal{F}' \circ \mathcal{F}' = (-Id)^*$ (that is, for every $f \in \mathcal{S}'$, $\mathcal{F}'(\mathcal{F}'(f))(x) = f(-x)$). Then, there exists $B \in GL_n(\mathbb{R})$*

with $B = B^*$ *and* $|\det(B)| = 1$ *such that either for every* $f \in \mathcal{S}$, $\mathcal{F}f = \mathbb{F}(f \circ B)$, *or, for every* $f \in \mathcal{S}$, $\mathcal{F}f = \overline{\mathbb{F}(f \circ B)}$.

In particular we see that the theorem implies that a transform satisfying the condition *must* be $\mathbb{R}$-linear and continuous, which we did not assume a priori. Clearly the linear map B introduces a scalar product (not necessarily positive definite) in which our map $\mathcal{F}$ becomes the standard Fourier transform.

REMARK 3. Another useful and standard notation for $\mathcal{T}f = f \circ u$ is $\mathcal{T}f = u^* f$, the same way as we used $(-Id)^* f$ where $((-Id)^* f)(x) = f(-x)$ before. We will use both types of notation.

Note that since $\mathcal{F}' \circ \mathcal{F}' = (-Id)^*$, not only are products mapped to convolutions, but also vice versa. Thus, the theorem is a consequence of the following.

THEOREM 4. *Assume we are given a bijective transform* $\mathcal{F} : \mathcal{S} \to \mathcal{S}$ *that admits an extension* $\mathcal{F}' : \mathcal{S}' \to \mathcal{S}'$ *which is also bijective, and such that:*

1. *For every* $f \in \mathcal{S}$ *and* $g \in \mathcal{S}'$ *we have* $\mathcal{F}'(f * g) = (\mathcal{F}f) \cdot (\mathcal{F}'g)$.
2. *For every* $f \in \mathcal{S}$ *and* $g \in \mathcal{S}'$ *we have* $\mathcal{F}'(f \cdot g) = (\mathcal{F}f) * (\mathcal{F}'g)$.

Then, $\mathcal{F}$ *is essentially the Fourier transform* $\mathbb{F}$, *that is, for some* $B \in GL_n(\mathbb{R})$ *with* $|\det(B)| = 1$, *we have either* $\mathcal{F}(f) = \mathbb{F}(f \circ B)$ *for all* f *or* $\mathcal{F}(f) = \mathbb{F}(\overline{f \circ B})$ *for all* f.

This result, in turn, follows from the next theorem.

THEOREM 5. *Assume we are given a bijective transform* $\mathcal{T} : \mathcal{S} \to \mathcal{S}$ *that admits a bijective extension* $\mathcal{T}' : \mathcal{S}' \to \mathcal{S}'$ *such that*

1. *For every* $f \in \mathcal{S}$ *and* $g \in \mathcal{S}'$ *we have* $\mathcal{T}'(f * g) = (\mathcal{T}f) * (\mathcal{T}'g)$.
2. *For every* $f \in \mathcal{S}$ *and* $g \in \mathcal{S}'$ *we have* $\mathcal{T}'(f \cdot g) = (\mathcal{T}f) \cdot (\mathcal{T}'g)$.

Then, for all $f \in \mathcal{S}$ *we have* $\mathcal{T}f = f \circ B$, *or for all* f *we have* $\mathcal{T}f = \overline{f \circ B}$, *for some* $B \in GL_n(\mathbb{R})$ *with* $|\det(B)| = 1$.

Note that in each of the results above, the linearity of the transform $\mathcal{T}$ is not assumed in advance, but is a consequence of the multiplicativity assumption.

Theorem 5 may be stated also for real-valued functions. In this case we prove a much stronger statement. Denote by $\mathcal{S}_{\mathbb{R}}$ the subspace of $\mathcal{S}$ consisting precisely of real-valued functions. Let $\mathcal{S}'_{\mathbb{R}}$ denote its real dual. We have the following theorem.

THEOREM 6. *Assume we are given a bijective transform* $\mathcal{T} : \mathcal{S}_{\mathbb{R}} \to \mathcal{S}_{\mathbb{R}}$ *which admits an extension* $\mathcal{T}' : \mathcal{S}'_{\mathbb{R}} \to \mathcal{S}'_{\mathbb{R}}$ *which is also bijective, and such that for every* $f \in \mathcal{S}_{\mathbb{R}}$ *and* $g \in \mathcal{S}'_{\mathbb{R}}$ *we have* $\mathcal{T}'(f \cdot g) = (\mathcal{T}f) \cdot (\mathcal{T}'g)$.

Then, there exists a diffeomorphism $u : \mathbb{R}^n \to \mathbb{R}^n$ *such that for all* $f \in \mathcal{S}_{\mathbb{R}}$, $\mathcal{T}f = f \circ u$.

Note that under non-extremely restrictive conditions on u, for example that it and its inverse are C^∞ and that it acts non-trivially only on a set with compact closure, a transform $\mathcal{T}$ defined in one of these ways is product-preserving.

REMARK 7. The conclusion of the above theorem remains valid not only for $f \in \mathcal{S}_{\mathbb{R}}$, but also for delta functions, $\mathcal{T}\delta_x = \delta_{u(x)}$, and for constant functions, $\mathcal{T}C = C$. This follows from the proof, and is used in some of the corollaries below. In fact, these are the only functions in $\mathcal{S}'_{\mathbb{R}} \setminus \mathcal{S}_{\mathbb{R}}$ which are important to us in the proof.

Before we pass to the proofs of our results, let us briefly comment on where these theorems originated from. For reasons connected with the topic of convex analysis, we were interested in the characterization of a very basic concept in convexity: duality and the Legendre transform. In the paper [**AM1**] it was shown that the Legendre transform can be characterized as follows: up to linear terms, it is the only involution on the class of convex lower semicontinuous functions on $\mathbb{R}^n$ which reverses the (partial) order of functions. Since the Legendre transform has another special property, namely, it exchanges summation of functions with their inf-convolution (for definitions and details see [**AM2**]), this in fact implied that an involution on lower semicontinuous convex functions which reverses order *must* have this special property. It turns out that also the opposite is true, namely, any involutive transform (on this class) which exchanges summation with inf-convolution, must reverse order, and, in fact, be up to linear terms the Legendre transform (see [**AM2**] for proofs and a discussion). Thus, already at this stage we observed that very minimal basic properties essentially uniquely define some classical transform, which traditionally is defined in a concrete and quite involved form.

It looks very intriguing to determine how far this point of view can be extended. It turns out that also the classical Fourier transform may be defined essentially uniquely by very minimal and basic conditions, namely, by the condition of exchanging convolution with product (together with, for example, the form of the square of the transform). This is what we show in this paper. The methods of proof are different (for the Legendre transform convexity is used very strongly), but the ideology is similar, although in the case of Fourier transform the proofs seem to be, for the moment, more involved.

2. Some remarks on the proofs, and a more general theorem

It turns out that one can prove the following modified version of Theorem 6 along exactly the same lines: everywhere the space of rapidly decreasing real-valued functions $\mathcal{S}_\mathbb{R}$ can be replaced by the space of compactly supported smooth real-valued functions $\mathcal{D}_\mathbb{R}$, and the space $\mathcal{S}'_\mathbb{R}$ can be replaced by the dual space $\mathcal{D}'_\mathbb{R}$. With this modification, Theorem 6 can be generalized to arbitrary smooth manifolds instead of $\mathbb{R}^n$ as follows.

Let X be a smooth manifold. Let $\mathcal{D}_\mathbb{R}(X)$ denote the space of compactly supported infinitely smooth real-valued functions on X. Let $\mathcal{M}_\mathbb{R}(X)$ denote the space of compactly supported infinitely smooth real-valued measures on X. The space $\mathcal{M}_\mathbb{R}(X)$ is equipped with the standard linear locally convex topology (which is the strict inductive limit of Fréchet spaces provided X is countable at infinity). Let $\mathcal{D}'_\mathbb{R}(X)$ denote the topological dual of $\mathcal{M}_\mathbb{R}(X)$. Then we have a natural imbedding $\mathcal{D}_\mathbb{R}(X) \hookrightarrow \mathcal{D}'_\mathbb{R}(X)$ with dense image in the weak topology given by

$$f \mapsto \Big[\phi \mapsto \int_X f\cdot\phi\Big].$$

Naturally $\mathcal{D}'_\mathbb{R}(X)$ is a $\mathcal{D}_\mathbb{R}(X)$-module.

THEOREM 8. *Let $\mathcal{T}\colon \mathcal{D}_\mathbb{R}(X) \to \mathcal{D}_\mathbb{R}(X)$ be a bijective (not necessarily linear or/and continuous) map which admits a bijective extension $\mathcal{T}\colon \mathcal{D}'_\mathbb{R}(X) \to \mathcal{D}'_\mathbb{R}(X)$ such that for every $f \in \mathcal{D}_\mathbb{R}(X)$ and $g \in \mathcal{D}'_\mathbb{R}(X)$ one has*

$$\mathcal{T}'(f\cdot g) = (\mathcal{T}f)\cdot(\mathcal{T}'g).$$

Then there exists a C^∞-diffeomorphism $u\colon X \to X$ such that $\mathcal{T}f = u^*f$ for all $f \in \mathcal{D}_{\mathbb{R}}(X)$.

In particular, a map $\mathcal{T}$ satisfying the above conditions must be linear and with strong continuity properties, although a priori the only property assumed was the preservation of products.

We omit the proof of this theorem, as it is almost identical to the proof of Theorem 6.

In a more general case, when instead of smooth functions one considers a different class of functions, we can usually—if we include linear functionals such as delta functions—prove the localization part, and if we include a continuity-type assumption, prove a power-like form. Clearly if we do not assume some differentiability of the class of functions involved, then we must tolerate transformations of the type $(\mathcal{T}f)(x) = \operatorname{sgn}(f(u(x)))|f(u(x))|^{a(x)}$ which are multiplicative (here $a(x)$ is, say, a continuous function).

3. Proof of Theorem 6

We consider Theorem 6 to be the central ideological fact of this paper, and so we start with the proof of this theorem. In this section we assume that all function spaces consist of real-valued functions. To simplify the notation, in this specific section we denote the space $\mathcal{S}_{\mathbb{R}}$ of real-valued Schwartz functions by $\mathcal{S}$, and similarly its real dual space $\mathcal{S}'_{\mathbb{R}}$ is denoted by $\mathcal{S}'$.

We may of course assume without loss of generality that the transformation $\mathcal{T}$ is defined over the whole space $\mathcal{S}'$.

Before starting the proof, a short discussion is in order of what it means to evaluate generalized functions from $\mathcal{S}'$ at a point when needed. Clearly, generalized functions in $\mathcal{S}'$ are defined via their action on functions in $\mathcal{S}$. However, there are subclasses $A \subset \mathcal{S}'$ which are themselves, say, continuous functions, and so evaluating them at a point is meaningful. Consider the following interesting subclass $\mathcal{S}_1 \subset \mathcal{S}'$ consisting of all generalized functions $\phi \in \mathcal{S}'$ such that for every $f \in \mathcal{S}$ we have that $\phi \cdot f \in \mathcal{S}$. For example, all constant functions belong to $\mathcal{S}_1$. It is also easy to see that any function in $\mathcal{S}_1$ is infinitely smooth. One can easily check that $\mathcal{S}_1$ consists precisely of all infinitely smooth functions f with the following growth condition: any partial derivative of f is bounded by $C(1+|x|)^k$ for some constant C and integer k.

Note that for $\phi \in \mathcal{S}_1$ and $\psi \in \mathcal{S}'$, the product $\psi \cdot \phi$ is well defined and belongs to $\mathcal{S}'$. Indeed, for $f \in \mathcal{S}$ we see that $\langle \psi \cdot \phi, f\rangle = \langle \psi, (\phi \cdot f)\rangle$ is the action of a function in $\mathcal{S}'$ on a function in $\mathcal{S}$, which is well defined (because for $\phi \in \mathcal{S}_1$ the operator $\mathcal{S} \to \mathcal{S}$ given by $[f \mapsto \phi \cdot f]$ is continuous).

It is not difficult to see that if $\phi \in \mathcal{S}_1$ then also $\mathcal{T}\phi \in \mathcal{S}_1$, since for every $f \in \mathcal{S}$ there is some $g \in \mathcal{S}$ such that $\mathcal{T}g = f$ so that $(\mathcal{T}\phi)\cdot f = \mathcal{T}(\phi \cdot g) \in \mathcal{S}$.

Moreover, this implies that for $\phi \in \mathcal{S}_1$ and $\psi \in \mathcal{S}'$ we have $\mathcal{T}(\psi \cdot \phi) = (\mathcal{T}\psi)\cdot(\mathcal{T}\phi)$, both sides being well defined: indeed, note that for any $g \in \mathcal{S}$, letting $f = \mathcal{T}^{-1}g$ we have

$$\mathcal{T}(\psi \cdot \phi)\cdot g = \mathcal{T}(\psi \cdot \phi \cdot f) = \mathcal{T}(\psi)\cdot \mathcal{T}(\phi \cdot f)$$

and likewise

$$(\mathcal{T}\psi \cdot \mathcal{T}\phi)\cdot g = \mathcal{T}(\psi)\cdot \mathcal{T}(\phi \cdot f).$$

Thus the product of the two with any function $g \in \mathcal{S}$ is the same (and well defined), which easily implies that they are the same element in $\mathcal{S}'$. We will use this observation mainly for the case where ϕ is a constant function.

Finally, note that for functions ϕ in $\mathcal{S}_1$, evaluation at a point is meaningful since as noted before these may be identified with a subclass of C^∞ functions and, in particular, are continuous.

PROOF OF THEOREM 6. The proof is divided into eleven steps.

Step 1. The following identities hold, despite the fact that a priori both functions in each identity belong to $\mathcal{S}' \setminus \mathcal{S}$:

$$\mathcal{T}(C\delta_x) = \mathcal{T}(\delta_x) \cdot \mathcal{T}(C),$$
$$\mathcal{T}(C_1 \cdot C_2) = \mathcal{T}(C_1) \cdot \mathcal{T}(C_2).$$

This can be shown directly, but follows from the above discussion and the fact that the constant functions belong to $\mathcal{S}_1$.

Step 2. Next we determine the image of the constant functions 0 and 1. We show that $\mathcal{T}(1) = 1$ and $\mathcal{T}(0) = 0$.

Proof of Step 2. Indeed, for all $f \in \mathcal{S}$, and in fact all $f \in \mathcal{S}'$, we have $1 \cdot f = f$, hence $\mathcal{T}1 \cdot g = g$ for all $g \in \mathcal{S}'$, and hence (taking e.g. $g = 1$) we see that $\mathcal{T}1 = 1$. Similarly, $0 \cdot f = 0$ for all $f \in \mathcal{S}'$, so $\mathcal{T}(0) \cdot g = \mathcal{T}(0)$ for all $g \in \mathcal{S}'$ (e.g., for $g = 0$) and so $\mathcal{T}(0) = 0$.

DEFINITION 3.1. For $f \in \mathcal{S}$ the support of f, denoted $\mathrm{supp}(f)$, is defined, as usual, to be the closure of the set of points where $f \neq 0$. For $\phi \in \mathcal{S}'$, the support of ϕ, denoted $\mathrm{supp}(\phi)$, is defined as follows: $x \notin \mathrm{supp}(\phi)$ if there exists an open neighborhood U with $x \in U$ and such that for any $f \in \mathcal{S}$ with $\mathrm{supp}(f) \subset U$ one has $\langle \phi, f \rangle = 0$. (When $\phi \in \mathcal{S}$, the second definition coincides with the first.)

Step 3. At this step we show that any constant multiple of the delta function at a point x is mapped to a function supported at one point.

Proof of Step 3. Indeed, assume that there are two different points $z, y \in \mathrm{supp}(\mathcal{T}(C\delta_x))$. Take disjoint neighborhoods of y and z, U_y and U_z.

By the definition of the support of a function in $\mathcal{S}'$, we may find two functions, $g_y, g_z \in \mathcal{S}$, supported on U_y and U_z respectively, which satisfy $\langle \mathcal{T}(C\delta_x), g_y \rangle \neq 0$ and $\langle \mathcal{T}(C\delta_x), g_z \rangle \neq 0$. In particular, $(\mathcal{T}(C\delta_x)) \cdot g_y \not\equiv 0$ and $(\mathcal{T}(C\delta_x)) \cdot g_z \not\equiv 0$. However, by the disjointness of U_y and U_z, we have $g_y \cdot g_z \equiv 0$.

Assume $g_y = \mathcal{T} f_y$ and $g_z = \mathcal{T} f_z$ for some functions $f_y, f_z \in \mathcal{S}$. Then, by injectivity, we see that also $\delta_x \cdot f_y \not\equiv 0$ and $\delta_x \cdot f_z \not\equiv 0$, and therefore $f_y(x) \neq 0$ and $f_z(x) \neq 0$. This in turn implies that $(f_y \cdot f_z)(x) \neq 0$, which means in particular that $(f_y \cdot f_z) \not\equiv 0$, and by multiplication preservation and injectivity, $\mathcal{T} f_y \cdot \mathcal{T} f_z \not\equiv 0$, contradicting $g_y \cdot g_z \equiv 0$. This completes the proof of the third step. Moreover, using similar arguments one can easily show that $\mathrm{supp}(\mathcal{T}C\delta_x)$ does not depend on $C \neq 0$.

We define a function $u : \mathbb{R}^n \to \mathbb{R}^n$ by the following formula: $u(x) = \mathrm{supp}(\mathcal{T}\delta_x)$.

Step 4. At this step we show that if $f \in \mathcal{S}$ satisfies $f(x) = 1$, then $(\mathcal{T}f)(u(x)) = 1$, and if $f(x) = 0$, then $(\mathcal{T}f)(u(x)) = 0$.

Proof of Step 4. It is well known that the generalized functions which are supported at a single point u are only the functions which are sums of finitely many various derivatives of the generalized function δ_u; see for example [**GS2**], Section 4.5.

Thus, we deduce from Step 3 that for every x there exist an M and constants $\gamma_\alpha \in \mathbb{R}$ for $\alpha \in (\mathbb{N} \cup 0)^n$ with $|\alpha| \leq M$ such that

$$\mathcal{T}\delta_x = \gamma_0 \delta_{u(x)} + \sum_\alpha \gamma_\alpha \frac{\partial^\alpha}{\partial t^\alpha} \delta_{u(x)}$$

(where, as before, α is a multiindex, $|\alpha| = \sum_i \alpha_i$ and $\frac{\partial^\alpha}{\partial t^\alpha} = \frac{\partial^{|\alpha|}}{\prod \partial t_i^{\alpha_i}}$). Note that to be precise we must write $M(x)$ and $\gamma_\alpha(x)$ for each α, since these coefficients may, a priori, depend on the point x which we fixed. We may assume that for some α with $|\alpha| = M$ we have $\gamma_\alpha \neq 0$.

Consider a function $f \in \mathcal{S}$; then $f \cdot \delta_x = f(x)\delta_x$. Therefore, $(\mathcal{T}f) \cdot (\mathcal{T}\delta_x) = \mathcal{T}(f(x)\delta_x)$. Assume that $f(x) = 0$; then we have $(\mathcal{T}f) \cdot (\mathcal{T}\delta_x) \equiv 0$. Plugging in the formula for $\mathcal{T}\delta_x$ we see that in this case

$$(\mathcal{T}f) \cdot \left(\gamma_0 \delta_{u(x)} + \sum_\alpha \gamma_\alpha \frac{\partial^\alpha}{\partial t^\alpha} \delta_{u(x)} \right) \equiv 0.$$

We may use the formula for multiplication of a function by a derivative of the delta function, and get many equations. The equations corresponding to the αth derivatives with $|\alpha| = M$ are easily seen to be $(\mathcal{T}f)(u(x))\gamma_\alpha \frac{\partial^\alpha}{\partial t^\alpha}\delta_{u(x)} = 0$. Thus, since for one of them $\gamma_\alpha \neq 0$, we get $(\mathcal{T}f)(u(x)) = 0$.

Similarly, assume that $f(x) = 1$; then $(\mathcal{T}f) \cdot (\mathcal{T}\delta_x) = \mathcal{T}\delta_x$, and we may plug in the formula for $\mathcal{T}\delta_x$, getting the equation

$$(\mathcal{T}f) \cdot \left(\gamma_0 \delta_{u(x)} + \sum_\alpha \gamma_\alpha \frac{\partial^\alpha}{\partial t^\alpha} \delta_{u(x)} \right) = \gamma_0 \delta_{u(x)} + \sum_\alpha \gamma_\alpha \frac{\partial^\alpha}{\partial t^\alpha} \delta_{u(x)}.$$

Comparing the terms next to $\frac{\partial^\alpha}{\partial t^\alpha}\delta_{u(x)}$ we get $(\mathcal{T}f)(u(x)) = 1$ this time.

We remark that in exactly the same way, what one would get for a function $f \in \mathcal{S}$ with $f(x) = C$ is the formula $(\mathcal{T}f)(u(x)) = (\mathcal{T}C)(u(x))$. (This involves the fact, mentioned at the beginning, that evaluation of $\mathcal{T}C$ at a point has a meaning.) In other words, this step actually shows that the transform, on $\mathcal{S}$, is local: its value at $u(x)$ only depends on the value of f at x.

Step 5. We show here that the generalized function $\mathcal{T}\delta_x$ is some constant multiple of $\delta_{u(x)}$. We may thus write

$$\mathcal{T}\delta_x = c_x \delta_{u(x)}.$$

Proof of Step 5. From Step 3, we know that for every x there exist constants $\gamma_\alpha(x)$ and $M(x)$ such that

$$\mathcal{T}\delta_x = \gamma_0(x)\delta_{u(x)} + \sum_\alpha \gamma_\alpha(x) \frac{\partial^\alpha}{\partial t^\alpha} \delta_{u(x)},$$

and so that there exists some α with $|\alpha| = M$ and $\gamma_\alpha(x) \neq 0$, and with $\gamma_\alpha(x) = 0$ for $|\alpha| > M(x)$.

Let $f \in \mathcal{S}$, and let $g = \mathcal{T}f$. Assume that $f \in \mathcal{S}$ satisfies $f(x) = 1$. Step 4 implies that $g(u(x)) = 1$, and in fact we know that this implication is an equivalence, since all conditions are symmetric with respect to $\mathcal{T}$ and $\mathcal{T}^{-1}$, and therefore if $(\mathcal{T}f)(u(x)) = 1$ then $f(x) = 1$. We have

$$\mathcal{T}f \cdot \mathcal{T}\delta_x = \mathcal{T}\delta_x.$$

Thus we deduce that for any $g \in \mathcal{S}$ with $g(u(x)) = 1$ one has

$$g \cdot \mathcal{T}\delta_x = \mathcal{T}\delta_x.$$

Equivalently for any $g \in \mathcal{S}$ with $g(u(x)) = 0$ one has

$$g \cdot \left(\gamma_0(x)\delta_{u(x)} + \sum_\alpha \gamma_\alpha(x) \frac{\partial^\alpha}{\partial t^\alpha} \delta_{u(x)} \right) = 0.$$

Selecting specific functions g in $\mathcal{S}$ this readily implies that $\gamma_\alpha(x) = 0$ for $\alpha \neq 0$.

Step 6. At this step we note that $u : \mathbb{R}^n \to \mathbb{R}^n$ is invertible, and also that multiples of the delta functions at x are mapped by $\mathcal{T}$ to multiples of the delta function at $u(x)$.

Proof of Step 6. Indeed, first note that

$$\mathcal{T}(c\delta_x) = (\mathcal{T}c) \cdot (\mathcal{T}\delta_x) = (\mathcal{T}c)(u(x))C_x\delta_{u(x)},$$

so that multiples of the delta function δ_x are also mapped to multiples of $\delta_{u(x)}$.

To construct u^{-1}, simply note that all considerations above are also true for $\mathcal{T}^{-1}$, and we may define $v(y) = \operatorname{supp}(\mathcal{T}^{-1}\delta_y)$; then by the equations $\mathcal{T}^{-1}(\mathcal{T}\delta_x) = \delta_x$ and $\mathcal{T}(\mathcal{T}^{-1}\delta_y) = \delta_y$, we get $u(v(y)) = y$ and $v(u(x)) = x$ so that u is invertible.

Step 7. In this step we determine the general form of the transform, namely, we show that for any $f \in \mathcal{S}_1$ and $y = u(x)$,

$$(\mathcal{T}f)(y) = (\mathcal{T}(f(x)))(y).$$

(In fact, this was done already in the proof of Step 4.)

Proof of Step 7. Letting $y = u(x)$, for any $f \in \mathcal{S}$ we have

$$(\mathcal{T}f)(y) \cdot C_x\delta_y = \mathcal{T}(f \cdot \delta_x) = \mathcal{T}(f(x) \cdot \delta_x) = \mathcal{T}(f(x)) \cdot C_x\delta_y = \mathcal{T}(f(x))(y)C_x\delta_y.$$

Therefore, $(\mathcal{T}f)(y) = \mathcal{T}(f(x))(y)$, i.e.,

$$(\mathcal{T}f)(y) = \mathcal{T}(f(u^{-1}(y)))(y). \tag{1}$$

The same is true for functions $f \in \mathcal{S}_1$, since these can be multiplied by δ_x in exactly the same way, in view of the discussion at the beginning of the section.

Step 8. At this step we show that $u : \mathbb{R}^n \to \mathbb{R}^n$ is continuous and open.

Proof of Step 8. From Step 7 we see that for any $f \in \mathcal{S}$, if $f(x) = 1$ then $(\mathcal{T}f)(u(x)) = 1$, and if $f(x) = 0$ then $(\mathcal{T}f)(u(x)) = 0$ (actually we have already seen this at Step 4). Let A be a bounded open subset of $\mathbb{R}^n$; then we may take $f \in \mathcal{S}$ to equal 0 outside A and $\neq 0$ in A. Then for $\mathcal{T}f$, we have $(\mathcal{T}f) \neq 0$ exactly on $u(A)$, and we see that $u(A)$ is open as well. The same applies to u^{-1}, using $\mathcal{T}^{-1}$. Thus u is open. Symmetrically, u^{-1} is also open. Hence u, u^{-1} are continuous.

Step 9. Note that up to this point we have not used the fact that the functions are real-valued. We now restrict our attention to the case where the functions considered assume only real values. At this step we show that for every $x \in \mathbb{R}^n$ there is some $a(x) \in \mathbb{R}$ such that $(\mathcal{T}C)(x) = \operatorname{sgn}(C) \cdot |C|^{a(x)}$. Moreover, $a(x)$ is a C^∞ function of x, and is nowhere 0.

Proof of Step 9. Clearly, $\mathcal{T}C$ is multiplicative in C, so that for every x, $(\mathcal{T}C)(x)$ is multiplicative as well. We also know that as a function of C, $T_x(C) := (\mathcal{T}C)(x)$ is onto, because $\mathcal{T}$ is onto and because of formula (1).

Given a multiplicative function on $\mathbb{R}$, very little is needed in order to establish the existence of $a(x)$ as above. Indeed, one may consider the function $t_x(C) = \log |T_x(C)|$, which is additive, and one only needs to show that it is linear, which by the fact that T_x is onto and $T_x(1) = 1$, implies the formula with $a(x)$.

For example, if we show that T_x is monotone on $\mathbb{R}_+$, this is enough (a non-linear, additive function cannot be monotone). Assume by contradiction that there are $0 < C_1 < C_2$ such that $(\mathcal{T}C_1)(x) > (\mathcal{T}C_2)(x)$. In view of multiplicativity, this means that $(\mathcal{T}c)(x) > 1$ for $c = C_1/C_2 < 1$.

From Step 4 (applied to $\mathcal{T}^{-1}$) we know that this implies (since the constant function c is never equal to 1) that for all $y \in \mathbb{R}^n$ we have $(\mathcal{T}c)(y) > 1$.

Construct a function $f \in \mathcal{S}$ which equals c at x, equals 0 at some other x', and assumes values only between 0 and c (for example, take such a linear function on the interval $[x', x]$ and smooth it so as to make it to belong to $\mathcal{S}$ and have compact support).

Then by Step 4 again, $(\mathcal{T}f)(z) \neq 1$ for any z. However, by (1) we have

$$(\mathcal{T}f)(u(x)) = \mathcal{T}(f(x))(u(x)) = (\mathcal{T}c)(u(x)) > 1$$

and

$$(\mathcal{T}f)(u(x')) = \mathcal{T}(f(x'))(u(x')) = (\mathcal{T}0)(u(x')) = 0.$$

Thus, by the mean value theorem, there is some $z \in [u(x), u(x')]$ for which $(\mathcal{T}f)(z) = 1$, a contradiction.

We conclude that T_x is monotone increasing on $\mathbb{R}_+$, which (together with surjectivity and $T_x(1) = 1$) implies that it is of the form $T_x(C) = \operatorname{sgn}(C)|C|^{a(x)}$.

The fact that $a(x)$ is in C^∞ follows from the fact that $\mathcal{T}(C) = \operatorname{sgn}(C)|C|^{a(x)}$ is a C^∞ function of x for every C (for example, it suffices to take $C = 2$), which in turn follows from the fact that $C \in \mathcal{S}_1$. The fact that $a(x)$ is never equal to 0 follows from the fact that for every x, the function T_x is onto.

Step 10. We claim that $u(x) \in C^\infty$ as well.

Proof of Step 10. By Steps 7 and 9,

$$(\mathcal{T}f)(y) = \operatorname{sgn}(f(u^{-1}(y)))|f(u^{-1}(y))|^{a(y)} \tag{2}$$

for $f \in \mathcal{S}$. Fixing a point y_0, choose a function $f_i \in \mathcal{S}$ to be, in a neighborhood of y_0, equal to $f_i(x) = e^{\langle x, e_i \rangle}$. Then (2) becomes

$$(\mathcal{T}f_i)(y) = \exp[\langle u^{-1}(y), e_i \rangle a(y)] \in \mathcal{S}.$$

Hence taking the logarithm we get $\langle u^{-1}(y), e_i \rangle \cdot a(y) \in C^\infty(\mathbb{R}^n)$. Since $a \in C^\infty(\mathbb{R}^n)$ and $a \neq 0$, we deduce that $\langle u^{-1}(y), e_i \rangle \in C^\infty(\mathbb{R}^n)$ for any i. Hence $u^{-1} \in C^\infty$. Applying the same argument to $\mathcal{T}^{-1}$ we see that $u \in C^\infty$.

Step 11. We claim that $a(x) = 1$ for all x.

Proof of Step 11. If not, then for some x_0 either $a(x_0) < 1$ or $a(x_0) > 1$. We assume, without loss of generality, that $a(x_0) < 1$, since otherwise we would consider $\mathcal{T}^{-1}$ instead of $\mathcal{T}$, which is given by $\mathcal{T}^{-1}g = (g \circ u)^{1/a\circ u}$, so that the exponent is < 1 at the point $u^{-1}(x_0)$.

Since $a(x_0) = c < 1$ and a is continuous, there is a neighborhood U of x_0 such that $a|_U < c' < 1$. Consider a function $f \in \mathcal{S}$ passing through $(u^{-1}(x_0), 0)$, smoothed outside a small ball whose image under u is contained in U, on which it is linear. So, f is of the form $f(x) = \langle x - u^{-1}(x_0), \xi\rangle$ on this ball. By (2) we have

$$(\mathcal{T}f)(x) = |\langle u^{-1}(x) - u^{-1}(x_0), \xi\rangle|^{a(x)} \cdot \operatorname{sgn}(\langle u^{-1}(x) - u^{-1}(x_0), \xi\rangle).$$

If this function is differentiable at x_0, then by Step 10 so is $(\mathcal{T}f) \circ u$ at $y_0 = u^{-1}(x_0)$, which locally is simply

$$|\langle y - y_0, \xi\rangle|^{b(y)} \cdot \operatorname{sgn}(\langle y - y_0, \xi\rangle),$$

where $b(y) = a(u(y))$.

However, since $b(y_0) < 1$ is C^∞, it is easy to see that this function is *not* differentiable at y_0 (e.g. in the direction ξ). This is a contradiction, and we conclude that $a(x) \equiv 1$ for all x.

Thus, we conclude that

$$(\mathcal{T}f)(y) = f(u^{-1}(y)),$$

which is what we claimed. This completes the proof. □

4. Proof of Theorem 5

Now we return to considering complex-valued functions. We recall the theorem stated in the Introduction for complex-valued functions.

THEOREM 5. *Assume we are given a transform $\mathcal{T} : \mathcal{S} \to \mathcal{S}$ that admits a bijective extension $\mathcal{T}' : \mathcal{S}' \to \mathcal{S}'$ such that*

1. *For every $f \in \mathcal{S}$ and $g \in \mathcal{S}'$ we have $\mathcal{T}'(f * g) = (\mathcal{T}f) * (\mathcal{T}'g)$.*
2. *For every $f \in \mathcal{S}$ and $g \in \mathcal{S}'$ we have $\mathcal{T}'(f \cdot g) = (\mathcal{T}f) \cdot (\mathcal{T}'g)$.*

Then, for all $f \in \mathcal{S}$ we have $\mathcal{T}f = f \circ B$, or for all f we have $\mathcal{T}f = \overline{f \circ B}$, for some $B \in GL_n(\mathbb{R})$ with $|\det(B)| = 1$.

Before proving Theorem 5 let us prove two corollaries which provide a characterization of the Fourier transform.

THEOREM 4. *Assume we are given a transform $\mathcal{F} : \mathcal{S} \to \mathcal{S}$ that admits a bijective extension $\mathcal{F}' : \mathcal{S}' \to \mathcal{S}'$ such that*

1. *For every $f \in \mathcal{S}$ and $g \in \mathcal{S}'$ we have $\mathcal{F}(f * g) = (\mathcal{F}f) \cdot (\mathcal{F}g)$.*
2. *For every $f \in \mathcal{S}$ and $g \in \mathcal{S}'$ we have $\mathcal{F}(f \cdot g) = (\mathcal{F}f) * (\mathcal{F}g)$.*

Then, $\mathcal{F}$ is essentially the Fourier transform $\mathbb{F}$, that is, for some $B \in GL_n(\mathbb{R})$ with $|\det(B)| = 1$, we have either $\mathcal{F}(f) = \mathbb{F}(f \circ B)$ for all $f \in \mathcal{S}$ or $\mathcal{F}(f) = \overline{\mathbb{F}(f \circ B)}$ for all f.

PROOF OF THEOREM 4. Take $\mathcal{T} = \mathcal{F} \circ \mathbb{F}$; it satisfies the assumptions of Theorem 5. □

Now our main characterization of the Fourier transform (Theorem 2 in the Introduction) follows immediately from Theorem 4.

PROOF OF THEOREM 2. Indeed the condition $\mathcal{F}' \circ \mathcal{F}' = (-Id)^*$ implies that $\mathcal{F}'$ not only maps products to convolutions, but also vice versa: let $f' = \mathcal{F}f$ and $g' = \mathcal{F}'g$; then

$$\begin{aligned}\mathcal{F}'(f' * g') = \mathcal{F}'(\mathcal{F}f * \mathcal{F}'g) = \mathcal{F}'\mathcal{F}'(f \cdot g) = (-Id^*)(f \cdot g) \\ = (-Id^*)f \cdot (-Id^*)g = (\mathcal{F}f') \cdot (\mathcal{F}'g').\end{aligned}$$

□

PROOF OF THEOREM 5. We may assume without loss of generality that $\mathcal{T}$ is defined over the whole space $\mathcal{S}'$.

The proof is divided into several steps. We start with a short discussion, similar to that at the beginning of Section 3.

We already determined (at the beginning of Section 3, but now we deal with complex-valued functions) that there are certain subclasses $A \subset \mathcal{S}'$ which themselves consist of, say, continuous functions, and so evaluating them at a point is meaningful. We defined the class $\mathcal{S}_1 \subset \mathcal{S}'$ of all $\phi \in \mathcal{S}'$ such that for every $f \in \mathcal{S}$ we have $\phi \cdot f \in \mathcal{S}$, and moreover showed that for $\phi \in \mathcal{S}_1$ and $\psi \in \mathcal{S}'$ we have

$$\mathcal{T}(\psi \cdot \phi) = (\mathcal{T}\psi) \cdot (\mathcal{T}\phi),$$

where both sides are well defined.

Since in this section we will also use the convolution operator, let us repeat the discussion for the class $\mathcal{S}_2 \subset \mathcal{S}'$, which we define to be the set of all $\phi \in \mathcal{S}'$ such that for every $f \in \mathcal{S}$ we have $\phi * f \in \mathcal{S}$. Equivalently $\phi \in \mathcal{S}_2$ if and only if its Fourier transform $\mathbb{F}\phi$ belongs to $\mathcal{S}_1$. The delta-functions δ_x are an example of such functions (as are their various derivatives).

Note that for $\phi \in \mathcal{S}_2$ and $\psi \in \mathcal{S}'$, the convolution $\psi * \phi$ is well defined and belongs to $\mathcal{S}'$. Indeed, for $f \in \mathcal{S}$ we have that $\langle \psi * \phi, f \rangle = \langle \psi, (-Id)^*\phi * f \rangle$ is the action of a function in $\mathcal{S}'$ on a function in $\mathcal{S}$, which is well defined (indeed, if $\phi \in \mathcal{S}_2$ then the operator $\mathcal{S} \to \mathcal{S}$ given by $[f \mapsto (-Id)^*\phi * f]$ is continuous).

It is not difficult to see that if $\phi \in \mathcal{S}_2$ then also $\mathcal{T}\phi \in \mathcal{S}_2$, since for every $f \in \mathcal{S}$ there is some $g \in \mathcal{S}$ such that $\mathcal{T}g = f$ so that $(\mathcal{T}\phi) * f = \mathcal{T}(\phi * g) \in \mathcal{S}$.

Moreover, this implies that for $\phi \in \mathcal{S}_2$ and $\psi \in \mathcal{S}'$ we have $\mathcal{T}(\psi * \phi) = (\mathcal{T}\psi) * (\mathcal{T}\phi)$, where both sides are well defined: indeed, note that for any $g \in \mathcal{S}$, letting $f = \mathcal{T}^{-1}g$ we have

$$\mathcal{T}(\psi * \phi) * g = \mathcal{T}(\psi * \phi * f) = \mathcal{T}(\psi) * \mathcal{T}(\phi * f)$$

and likewise

$$(\mathcal{T}\psi * \mathcal{T}\phi) * g = \mathcal{T}(\psi) * \mathcal{T}(\phi * f).$$

Thus the convolution of these two with any function $g \in \mathcal{S}$ is the same (and well defined), which implies that they are the same element in $\mathcal{S}'$. We will use this observation mainly for the case where ϕ is a delta function.

Step 1. The following identities hold, despite the fact that a priori both functions in each identity belong to $\mathcal{S}' \setminus \mathcal{S}$. (For $C \in \mathbb{C}$ we also denote by $C \in \mathcal{S}'$ the constant function equaling C at every point, and for $x \in \mathbb{R}^n$ we denote by δ_x the usual delta function at x, satisfying $\int f\delta_x = f(x)$ for $f \in \mathcal{S}$.) We have

$$\begin{aligned}\mathcal{T}(C\delta_x) &= \mathcal{T}(\delta_x) \cdot \mathcal{T}(C), \\ \mathcal{T}(C_1 \cdot C_2) &= \mathcal{T}(C_1) \cdot \mathcal{T}(C_2), \\ \mathcal{T}(\delta_{x+y}) &= \mathcal{T}(\delta_x) * \mathcal{T}(\delta_y).\end{aligned}$$

Proof of Step 1. This follows from the discussion above and the discussion at the beginning of Section 3.

Step 2. Next we may determine the images of the three most elementary functions in $\mathcal{S}'$, the function δ_0 and the constant functions 0 and 1. We show that $\mathcal{T}\delta_0 = \delta_0$, $\mathcal{T}(1) = 1$, and $\mathcal{T}(0) = 0$.

Proof of Step 2. Indeed, for all $f \in \mathcal{S}'$, in view of the discussion above we have $\delta_0 * f = f$; hence $\mathcal{T}\delta_0 * g = g$ for all $g \in \mathcal{S}'$, in particular for $f = \delta_0$, and hence $\mathcal{T}\delta_0 = \delta_0$. Secondly, for all $f \in \mathcal{S}'$ we have $1 \cdot f = f$; hence $\mathcal{T}1 \cdot g = g$ for all $g \in \mathcal{S}'$, and hence (letting $g = 1$), $\mathcal{T}1 = 1$. Also, $0 \cdot f = 0$ for all $f \in \mathcal{S}'$, so $\mathcal{T}(0) \cdot g = \mathcal{T}(0)$ for all $g \in \mathcal{S}'$ (e.g., for $g = 0$) and so $\mathcal{T}(0) = 0$.

Step 3. At this step we claim that if the value of a function $f \in \mathcal{S}$ at 0 is either 0 or 1, then so is the value of the function $\mathcal{T}f$.

Proof of Step 3. Indeed, it is easy to check the formula $\delta_0 \cdot f = f(0)\delta_0$, and therefore we may apply the multiplication preservation (by Step 1) to get

$$\mathcal{T}\delta_0 \cdot \mathcal{T}f = \mathcal{T}(\delta_0 f(0)) = \mathcal{T}\delta_0 \cdot \mathcal{T}(f(0)),$$

which by Step 2 means that

$$(\mathcal{T}f)(0)\delta_0 = \delta_0 \cdot \mathcal{T}f = \delta_0 \cdot \mathcal{T}(f(0)) = \mathcal{T}(f(0))(0)\delta_0$$

and so

$$(\mathcal{T}f)(0) = (\mathcal{T}(f(0)))(0). \tag{3}$$

In particular, using Step 2, we see that if $f(0) = 1$ then $(\mathcal{T}f)(0) = 1$, and if $f(0) = 0$ then $(\mathcal{T}f)(0) = 0$.

Step 4. At this step we show that for $C \in \mathbb{C}$ the function $\mathcal{T}(C)$ is a constant function.

Proof of Step 4. Note that for any $f \in \mathcal{S}$ we have that $f * 1 = \int f$, and so $\mathcal{T}(\int f) = \mathcal{T}(f * 1) = (\mathcal{T}f) * 1 = \int(\mathcal{T}f)$. In particular, since any $C \in \mathbb{C}$ can be presented as $\int f$ for some $f \in \mathcal{S}$, we see that constants are mapped to constants.

Step 5. At this step we show that any constant multiple of the delta function at a point x is mapped to a function supported at one point (the definition of "support" was given in Definition 3.1 above).

Proof of Step 5. Indeed, assume that there are two different points $z, y \in \operatorname{supp}(\mathcal{T}(C\delta_x))$. Take disjoint neighborhoods of y and z, U_y and U_z.

By the definition of the support of a function in $\mathcal{S}'$, we may find two functions, $g_y, g_z \in \mathcal{S}$, supported on U_y and U_z respectively, which satisfy $\langle \mathcal{T}(C\delta_x), g_y\rangle \neq 0$, and $\langle \mathcal{T}(C\delta_x), g_z\rangle \neq 0$. In particular, $(\mathcal{T}(C\delta_x)) \cdot g_y \not\equiv 0$ and $(\mathcal{T}(C\delta_x)) \cdot g_z \not\equiv 0$. However, by the disjointness of U_y and U_z, we have $g_y \cdot g_z \equiv 0$.

Assume that $g_y = \mathcal{T}f_y$ and $g_z = \mathcal{T}f_z$ for some functions $f_y, f_z \in \mathcal{S}$. Then, by injectivity, we see that also $C\delta_x \cdot f_y \not\equiv 0$ and $C\delta_x \cdot f_z \not\equiv 0$, and therefore $f_y(x) \neq 0$ and $f_z(x) \neq 0$. This in turn implies that $(f_y \cdot f_z)(x) \neq 0$, which means in particular that $(f_y \cdot f_z) \not\equiv 0$, and by multiplication preservation and injectivity, $\mathcal{T}f_y \cdot \mathcal{T}f_z \not\equiv 0$, contradicting $g_y \cdot g_z \equiv 0$.

Step 6. We define a function $u : \mathbb{R}^n \to \mathbb{R}^n$ by the following formula: $u(x) = \operatorname{supp}(\mathcal{T}\delta_x)$. At this step we show that it is an additive function.

Proof of Step 6. Indeed, using Step 2 we have $\mathcal{T}(\delta_{x+y}) = \mathcal{T}(\delta_x) * \mathcal{T}(\delta_y)$, and thus

$$u(x) + u(y) = \operatorname{supp}(\mathcal{T}(\delta_x) * \mathcal{T}(\delta_y)) = \operatorname{supp}\mathcal{T}(\delta_x * \delta_y) = \operatorname{supp}\mathcal{T}(\delta_{x+y}) = u(x+y).$$

Step 7. At this step we show that in fact $\mathcal{T}\delta_x = \delta_{u(x)}$.

Proof of Step 7. It is well known that the generalized functions which are supported at a single point u are only the functions which are sums of finitely many various derivatives of the generalized function δ_u (as before, we refer to [**GS2**], Section 4.5).

Thus, we have by Step 5 that for every x there exists an M and constants $\gamma_\alpha \in \mathbb{C}$ for $\alpha \in (\mathbb{N} \cup 0)^n$ with $|\alpha| \le M$ such that

$$\mathcal{T}(\delta_x) = \gamma_0 \delta_{u(x)} + \sum_\alpha \gamma_\alpha \frac{\partial^\alpha}{\partial t^\alpha} \delta_{u(x)}$$

(where, as before, α is a multiindex, $|\alpha| = \sum_i \alpha_i$ and $\frac{\partial^\alpha}{\partial t^\alpha} = \frac{\partial^{|\alpha|}}{\prod \partial t_i^{\alpha_i}}$). We need to show that $\gamma_0 = 1$, and that for all $\alpha \neq 0$ we have $\gamma_\alpha = 0$. Note that to be precise we must write $\gamma_\alpha(x)$ for the coefficients since they may, a priori, depend on the point x which we fixed.

We start with the latter.

We look at δ_y, for $y = x/(M+1)$. Then the convolution of this function with itself $M+1$ times gives δ_x:

$$\overbrace{\delta_y * \delta_y * \cdots * \delta_y}^{M+1 \text{ times}} = \delta_x.$$

By the same reason as in Step 1 we may write

$$\overbrace{\mathcal{T}(\delta_y) * \mathcal{T}(\delta_y) * \cdots * \mathcal{T}(\delta_y)}^{M+1 \text{ times}} = \mathcal{T}(\delta_x).$$

By additivity of u we know that $u(y) = \operatorname{supp}(\mathcal{T}\delta_y) = u(x)/(M+1)$. Also, by the same reasoning as above, for some M' we have

$$\mathcal{T}(\delta_y) = \eta_0 \delta_{u(y)} + \sum_{1 \le |\alpha| \le M'} \eta_\alpha \frac{\partial^\alpha}{\partial t^\alpha} \delta_{u(y)}.$$

Using the easy fact that

$$\frac{\partial^\alpha}{\partial t^\alpha} \delta_z * \frac{\partial^{\alpha'}}{\partial t^{\alpha'}} \delta_w = \frac{\partial^{\alpha+\alpha'}}{\partial t^{\alpha+\alpha'}} \delta_{z+w}$$

we see that for the convolution of $\mathcal{T}\delta_y$ with itself $M+1$ times to equal $\mathcal{T}(\delta_x)$, the highest order derivative appearing with a non-zero coefficient must be 0. Indeed, if we let

$$M = \max\{|\alpha| : \eta_\alpha \neq 0\},$$

then there are finitely many different α with $|\alpha| = M$ and $\eta_\alpha \neq 0$. Consider the ones with the largest coefficient in the first coordinate, of those, take the ones with the largest second coordinate, an so on, until this is just one vector $\hat{\alpha}$. Then the derivative of order $(M+1)\hat{\alpha}$ is attained, in the convolution of $\mathcal{T}\delta_y$ with itself $(M+1)$ times, with the coefficient $(M+1)\eta_{\hat{\alpha}}$ only. Therefore the coefficients have no way of canceling out, despite the fact that there is no derivative of order larger than M in $\mathcal{T}\delta_x$. Therefore $\eta_{\hat{\alpha}} = 0$, in contradiction to the way is was chosen.

Therefore, $\mathcal{T}\delta_y = \eta_0 \delta_{u(y)}$. We conclude that in the expression for $\mathcal{T}\delta_x$ above, for all $\alpha \neq 0$ we have $\gamma_\alpha = 0$.

Finally, we need to explain that $\gamma_0 = 1$. Note that since $\delta_x \in \mathcal{S}_2$, we know that $(\mathcal{T}\delta_x) * (\mathcal{T}f)$ is well defined and is in $\mathcal{S}$ for any $f \in \mathcal{S}$. Of course, by the above this is none other than $(\gamma_0(x)\delta_{u(x)}) * (\mathcal{T}f) \in \mathcal{S}$. Hence we may convolve it with 1, and get

$$\gamma_0(x) \int (\mathcal{T}f) = \mathcal{T}(\delta_x * f * 1) = \mathcal{T}\Big(\int f\Big).$$

In particular, we see that $\gamma_0(x)$ does not depend on x, and so it equals $\gamma_0(0) = 1$.

Step 8. The mapping u defined above is 1-1, additive, onto, and $u(0) = 0$.

Proof of Step 8. We showed additivity at Step 6, the relation $u(0) = 0$ at Step 2, and 1-1 follows from Step 7 and injectivity. Next, notice that the conditions in the theorem are symmetric with respect to $\mathcal{T}$ and $\mathcal{T}^{-1}$, so that the same is true for $\mathcal{T}^{-1}$, which in particular shows that u is onto.

Step 9. The mapping $\beta : \mathbb{C} \to \mathbb{C}$ defined by $\beta(C) = \mathcal{T}(C)$ (where C on the right hand side is viewed as a constant function, and $\mathcal{T}(C)$, which by Step 4 is a constant function as well, is then taken as its constant value), is a multiplicative function, 1-1 and onto.

Proof of Step 9. Multiplicativity follows from the multiplicativity of $\mathcal{T}$ on constants as shown at Step 1. Injectivity follows from the injectivity of $\mathcal{T}$, and since the same is true for $\mathcal{T}^{-1}$, we see that β is onto.

Step 10. At this step we get the formal expression for the transform, which depends only on how $\mathcal{T}$ acts on constants and on delta-functions, namely, for u defined at Step 6 and β defined at Step 9, for any $f \in \mathcal{S}$ and $y \in \mathbb{R}^n$ we have

$$(\mathcal{T}f)(y) = \beta(f(u^{-1}(y))).$$

Proof of Step 10. Let $x = u^{-1}(y)$; then

$$\delta_x \cdot f = f(x) \cdot \delta_x,$$

so that, applying $\mathcal{T}$,

$$\delta_y \cdot (\mathcal{T}f)(y) = \mathcal{T}\delta_x \cdot \mathcal{T}f = \mathcal{T}(\delta_x \cdot f) = \delta_y \cdot \beta(f(x)).$$

Step 11. At this step we show that there exists some $B \in GL_n(\mathbb{R})$ such that $u(x) = Bx$.

Proof of Step 11. We know by Step 8 that u is additive and invertible, and we wish to show it is linear. It is a well-known elementary fact that this follows once we show that u is locally bounded. To show that u is locally bounded, we will use the assumption $\mathcal{T}\mathcal{S} = \mathcal{S}$. Indeed, look at any bounded region $A \subset \mathbb{R}^n$. Construct $f \in \mathcal{S}$ which is constant 1 on A, and identically 0 outside a large ball surrounding A. This can clearly be done. Next consider $\mathcal{T}f$. Then, by Step 10, $(\mathcal{T}f)(y) = 1$ for every y in $u(A)$, but since $\mathcal{T}f \in \mathcal{S}$, the set of points where $(\mathcal{T}f)(y) = 1$ must be bounded, and so $u(A)$ is bounded. A locally bounded and additive function $u : \mathbb{R}^n \to \mathbb{R}^n$ must be linear, and together with invertibility, the claim is proven.

Step 12. At this step we show that also β has a special form: $\beta(C) = \left(\frac{C}{|C|}\right)^k |C|^\alpha$ for some α with $\mathrm{Re}(\alpha) > 0$ and for $k \in \{\pm 1\}$.

Proof of Step 12. We know by Step 9 that β is multiplicative and invertible, and we wish to show it is of the special form given above. For this it suffices to check that β is continuous at every $C \in \mathbb{C}$.

We start with $C \in \mathbb{C}^*$ (clearly $\beta : \mathbb{C}^* \to \mathbb{C}^*$). Since we may write $\mathbb{C}^*$ as the direct product $\mathbb{C}^* = S^1 \times \mathbb{R}_{>0}$, and since from multiplicativity we have $\beta(re^{i\theta}) = \beta(r) \cdot \beta(e^{i\theta})$, it suffices to check that $\beta|_{S^1}$ is continuous at every θ and that $\beta|_{\mathbb{R}_{>0}}$ is continuous at every $r > 0$.

As before, for this the assumption $\mathcal{T}\mathcal{S} = \mathcal{S}$ suffices. Indeed, by Step 10 and Step 11 for every $f \in \mathcal{S}$ and $y \in \mathbb{R}^n$ we have $(\mathcal{T}f)(y) = \beta(f(B^{-1}y))$. Construct $f \in \mathcal{S}$ such that for $x = (a, 0, \ldots, 0)$ with, say, $a \in [-2\pi, 2\pi]$ one has $f(x) = e^{ia}$ and f has compact support. Then for $y = B(a, 0, \ldots, 0)$ with $a \in [-\pi, \pi]$ we have $(\mathcal{T}f)(y) = \beta(e^{ia})$, and so from the continuity of $\mathcal{T}f$ we deduce that $\beta|_{S^1}$ must be continuous. Similarly, for fixed but arbitrary $r_1 > r_0 > 0$ construct $f \in \mathcal{S}$ such that for $x = (a, 0, \ldots, 0)$ with $a \in [r_0, r_1]$ one has $f(x) = a$ and f has compact support. Then for $y = B(a, 0, \ldots, 0)$ with $a \in [r_0, r_1]$ we have $(\mathcal{T}f)(y) = \beta(a)$, and so from the continuity of $\mathcal{T}f$ we deduce that $\beta|_{[r_0, r_1]}$ must be continuous. Since r_0, r_1 were arbitrary, we get the continuity of $\beta|_{\mathbb{R}_{>0}}$.

In fact, it is clear that we have not used that $r_i > 0$, and so $\beta|_{\mathbb{R}}$ is continuous as well. If $C_n \to 0$ for $C_n \in \mathbb{C}$, then clearly we may write $\beta(C_n) = \beta(|C_n|)\beta(e^{i\theta_n}) \to 0$ since $\beta(e^{i\theta_n})$ is bounded and $|C_n| \to 0$. Thus, β is continuous also at 0, and the proof of this step is complete.

Step 13. At this step we complete the proof by showing that the restrictions $\alpha = 1$ for α described at Step 12, and $\det(B) = \pm 1$ for B described at Step 11, apply. Thus, for some $B \in GL_n$ with $\det(B) = \pm 1$,

$$(\mathcal{T}f)(y) = f(B^{-1}(y)) \quad \text{or} \quad (\mathcal{T}f)(y) = \overline{f(B^{-1}(y))}.$$

Proof of Step 13. We use the form of the transform given by Steps 10, 11, and 12: there exists some $B \in GL_n$, some α with $\mathrm{Re}(\alpha) > 0$ and $k \in \{\pm 1\}$ such that for $f \in \mathcal{S}$ one has

$$(\mathcal{T}f)(y) = \left(\frac{f(B^{-1}(y))}{|f(B^{-1}(y))|} \right)^k |f(B^{-1}(y))|^\alpha.$$

We will use the condition $\mathcal{T}(1 * f) = 1 * \mathcal{T}(f)$, namely $\mathcal{T}(\int f) = \int(\mathcal{T}f)$.

Let $\varepsilon > 0$ be arbitrary; a small $\delta > 0$ will be chosen later. Let $f \in \mathcal{S}$ be such that $f(x) = 1$ on $[-\frac{1}{2} + \delta, \frac{1}{2} - \delta]^n$, $f = 0$ outside $[-\frac{1}{2} - \delta, \frac{1}{2} + \delta]^n$, $f(x) \in [0, 1]$, and $\int f = 1$. Thus, by the above, $\int \mathcal{T}f = 1$ as well. Choose $\delta > 0$ small, so that the volume of $B([-\frac{1}{2} - \delta, \frac{1}{2} + \delta]^n \setminus [-\frac{1}{2} + \delta, \frac{1}{2} - \delta]^n)$ is smaller than ε.

It is easily confirmed that $\mathcal{T}f$ equals 1 on $B([-\frac{1}{2} + \delta, \frac{1}{2} - \delta]^n)$, equals 0 outside $B([-\frac{1}{2} - \delta, \frac{1}{2} + \delta]^n)$, and belongs to $[0, 1]$ everywhere, so that $|\int(\mathcal{T}f) - |\det(B)|| < \varepsilon$. Since $\int \mathcal{T}f = 1$, we get $|\det(B)| = 1$.

Next, let $f = e^{-c|x|^n}$, with c chosen so that $\int f = 1$. Then $\mathcal{T}(1 * f) = 1$ as well. Also, $(\mathcal{T}f)(y) = e^{-c\alpha|B^{-1}x|^n}$. Compare (using $|\det(B)| = 1$)

$$1 = \int e^{-c\alpha|B^{-1}x|^n} = (1/\alpha)$$

and get $\alpha = 1$. □

5. Some relevant history

In every result above, one of the main interests is that the linearity of the resulting transform is obtained from, say, multiplicativity, or from exchange of products with convolutions. If one *assumed* linearlity to begin with, the proofs would be much easier.

In a different setting, but similar spirit, *with the assumption of linearity* (and even more), there are several results of Lukacs. We quote one of them below, which involves also continuity; see [**L1**].

THEOREM 9 (Lukacs). *Suppose that $\alpha(s, F)$ is a bounded (meaning that the functions in the image are uniformly bounded) linear mapping of the space of distribution functions onto a set of continuous functions, and suppose that*

(i) $\alpha(s, F_1) \equiv \alpha(s, F_2)$ *if and only if* $F_1 \equiv F_2$.

(ii) *If* $F = F_1 * F_2$ *then* $\alpha(s, F) = \alpha(s, F_1)\alpha(s, F_2)$.

(iii) *Let F_n be a sequence of distribution functions. The necessary and sufficient condition for the weak convergence of F_n to some F is that* $\lim \alpha(s, F_n) = \alpha(s, F)$.

Then $\alpha(s, F) = \int_{-\infty}^{\infty} e^{ixA(s)}\, dF(x)$, where $A(s)$ is a continuous real-valued function such that the values assumed by $|A(s)|$ form a dense set in $[0, \infty)$.

COROLLARY 10 (Lukacs). *Suppose that the assumptions of the above theorem are satisfied, and that the mapping $\alpha(s, F)$ maps every degenerate distribution function to a non-negative definite function. Then $A(s) = Cs$ for some constant C.*

For definitions and details see [**L2**].

References

[AM1] S. Artstein-Avidan and V. Milman, *The concept of duality in convex analysis, and the characterization of the Legendre transform.* To appear in Annals of Mathematics.

[AM2] ———, *A characterization of the concept of duality.* Electronic Research Announcements in Mathematical Sciences, **14** (2007), 48–65.

[GS1] I. M. Gel′fand and G. E. Shilov, *Generalized functions*, vol. I. Properties and operations, Translated from the Russian by Eugene Saletan, Academic Press, New York–London, 1964.

[GS2] ———, *Generalized functions*, vol. II. Spaces of fundamental and generalized functions. Translated from the Russian by Morris D. Friedman, Amiel Feinstein and Christian P. Peltzer, Academic Press, New York–London, 1968.

[L1] E. Lukacs *A linear mapping of the space of distribution functions onto a set of bounded continuous functions*, Z. Wahrscheinlichkeitstheorie **3** (1964), 1–6.

[L2] ———, *Characteristic functions.* Griffin, London, 1960.

[S] E. M. Stein, *Harmonic analysis: Real-variable methods, orthogonality, and oscillatory integrals.* With the assistance of Timothy S. Murphy. Princeton Mathematical Series, 43. Monographs in Harmonic Analysis, III. Princeton University Press, Princeton, 1993.

[SS] E. M. Stein and R. Shakarchi, *Complex Analysis.* Princeton Lectures in Analysis, II. Princeton University Press, Princeton, 2003.

SCHOOL OF MATHEMATICAL SCIENCES, TEL AVIV UNIVERSITY, RAMAT AVIV, TEL AVIV 69978, ISRAEL

E-mail address: semyon@post.tau.ac.il

SCHOOL OF MATHEMATICAL SCIENCES, TEL AVIV UNIVERSITY, RAMAT AVIV, TEL AVIV 69978, ISRAEL

E-mail address: shiri@post.tau.ac.il

SCHOOL OF MATHEMATICAL SCIENCES, TEL AVIV UNIVERSITY, RAMAT AVIV, TEL AVIV 69978, ISRAEL

E-mail address: milman@post.tau.ac.il

Amer. Math. Soc. Transl.
(2) Vol. **226**, 2009

Geodesic Restrictions and L^p-Estimates for Eigenfunctions of Riemannian Surfaces

J. Bourgain

1. Introduction and preliminaries

Let (M, g) be a smooth compact Riemannian surface without boundary and Δ the Laplace operator associated to the metric g. Let $(\varphi_\lambda)_{\lambda \geq 0}$ be the eigenfunctions of Δ such that $-\Delta\varphi_\lambda = \lambda^2\varphi_\lambda$ and normalized in $L^2(M)$.

Recall first the general L^p-estimates ($p \geq 2$) established in [**So**].

Proposition 1.

$$\|\varphi_\lambda\|_p \leq C\lambda^{\delta(p)}, \tag{1.1}$$

where

$$\delta(p) = \begin{cases} \frac{1}{2}\left(\frac{1}{2} - \frac{1}{p}\right) & \text{if } 2 \leq p \leq 6, \\ \frac{1}{2} - \frac{2}{p} & \text{if } 6 \leq p \leq \infty. \end{cases} \tag{1.2}$$

These estimates are best possible as shown by the spherical harmonics of S^2. We also point out the paper [**S-Z**], which discusses when the above inequalities are saturated. In the particular case of hyperbolic surfaces, these inequalities may be improved by logarithmic factors in λ, while in fact much stronger conjectures have been made (that so far can only be approached in the arithmetic setting). The purpose of this note is to relate improvements upon the general moment inequalities from Proposition 1 with the issue of estimating the L^2-norm of geodesic restrictions of φ_λ.

Recall the recent work [**B-G-T**] (see also [**R**]) and

Proposition 2 ([**B-G-T**], Theorem 1). *Let $\gamma : [a, b] \to M$ be a smooth curve. Then*

$$\|\varphi_\lambda\|_{L^2(\gamma)} \leq C(1 + \lambda)^{\frac{1}{4}}\|\varphi_\lambda\|_{L^2(M)}. \tag{1.3}$$

Inequality (1.3) holds in particular for a geodesic restriction, which provides the worst case scenario (the example of the highest weight spherical harmonic concentrated on an equator shows that (1.3) is optimal).

2000 *Mathematics Subject Classification.* Primary 58J50; Secondary 58J37.
Key words and phrases. Eigenfunction, geodesic restriction.

Our purpose is to show that improving upon the general L^4-estimate, i.e.

$$\|\varphi_\lambda\|_{L^4(M)} \leq C\lambda^{\frac{1}{8}}, \tag{1.4}$$

is essentially equivalent to the improvement of the geodesic restriction estimate (1.3). More precisely, we have

PROPOSITION 3. *If $p \geq 2$ and $\gamma : [a, b] \to M$ is a geodesic, then*

$$\|\varphi_\lambda\|_{L^2(\gamma)} \leq C\lambda^{\frac{1}{2p}} \|\varphi_\lambda\|_{L^p(M)}. \tag{1.5}$$

Conversely,

PROPOSITION 4.

$$\|\varphi_\lambda\|_{L^4(M)} \leq C_\varepsilon \lambda^{\frac{1}{16}+\varepsilon} \max\{\|\varphi_\lambda\|_{L^2(\gamma)} : \gamma \textit{ geodesic}\}^{1/4} \tag{1.6}$$

with $\varepsilon > 0$ arbitrary.

Note that if we substitute the first inequality from (1.2) (i.e. for $2 \leq p \leq 6$) in (1.5), we recover (1.3). Conversely, (1.4) follows from (1.3) and (1.5) with $\varepsilon = 0$.

We also point out that all previous estimates remain valid for quasi-modes and result from a local analysis. A representation of φ_λ in local coordinates using an oscillatory integral is given by the following result (see [**B-G-T**], Theorem 4 and related references).

PROPOSITION 5. *Given $x_0 \in M$, there is a system of coordinates near x_0 such that for $x \in U = \{x : |x| < c\varepsilon\}$ and $\mathcal{X} \in \mathcal{S}(\mathbb{R})$, $\mathcal{X}(0) = 1$ we have*

$$\mathcal{X}(\sqrt{-\Delta} - \lambda)f(x) = \lambda^{\frac{1}{2}} \int e^{i\lambda\psi(x,y)} a(x, y, \lambda) f(y)\, dy + R_\lambda(f)(x), \tag{1.7}$$

where $\psi(x, y) = -d_g(x, y)$ is the geodesic distance, a is a non-negative smooth function supported by $\{|x| \leq c_0\varepsilon < c_1\varepsilon \leq |y| \leq c_2\varepsilon \ll 1\}$ $(c < c_0)$, and R_λ satisfies an estimate

$$\|R_\lambda(f)\|_\infty \leq C\|f\|_2. \tag{1.8}$$

Recall the following

LEMMA 1 (see Lemma 3.1 in [**B-G-T**]). *Using geodesic polar coordinates, write $y = \exp_0(r\omega)$ with $r > 0$, $\omega \in S^1$. Assuming $g(0) = Id$, we have*

$$(\partial_{x_1}\psi, \partial_{x_2}\psi)(0, y) = \omega. \tag{1.9}$$

The proofs of Propositions 3 and 4 use manipulations of oscillatory integral operators that are by now quite standard. In proving Proposition 4 we rely on an approach to Carleson–Sjölin type estimates [**C-S**], based on generalized Kakeya-type operators (in the spirit of [**B**] except that in the present 2-dimensional setting this more geometric point of view goes back at least to the work of A. Cordoba on Bochner–Riesz summation operators [**Cor**]). These Kakeya-type averaging operators depend on the phase function ψ in (1.6) and will naturally relate to geodesic restriction operators.

The extra λ^ε-factor in Proposition 4 may be unnecessary and comes from the variable coefficient analysis (it would not appear in the translation invariant setting).

In what follows, we will occasionally use symbols such as $\sim$ (resp. $\lesssim$) to indicate equivalence (resp. inequality) up to some constant.

2. Proof of Proposition 3

Using geodesic polar coordinates, we may assume that the geodesic γ is parametrized as $\gamma(t) = (t, 0)$, $|t| < c\varepsilon$. We need to evaluate

$$\Big\{ \int_{|t|<c\varepsilon} |\varphi_\lambda(\gamma(t)|^2 \, dt \Big\}^{\frac{1}{2}}. \tag{2.1}$$

Applying (1.7) with $f = \varphi_\lambda \equiv \varphi$, we are reduced to bound

$$\lambda^{\frac{1}{2}} \Big\{ \int \Big| \int e^{i\lambda\psi(\gamma(t),y)} a(y)\varphi(y)\, dy \Big|^2 dt \Big\}^{\frac{1}{2}}. \tag{2.2}$$

Proceeding by duality, let $\eta \in L^2[-c\varepsilon, c\varepsilon]$, $\|\eta\|_2 = 1$ and estimate

$$\Big| \iint e^{i\lambda\psi(\gamma(t),y)} a(y)\varphi(y)\eta(t)\, dy\, dt \Big| \le \Big\| \int e^{i\lambda\psi(\gamma(t),y)} \eta(t)\, dt \Big\|_{L^{p'}(dy)} \|\varphi\|_p. \tag{2.3}$$

To estimate the first factor, we decompose the y-range in dyadic regions according to the distance of y to the geodesic γ. Thus, for $k_* < k < \infty$, denote

$$\Omega_k = \{ y = (y_1, y_2) : |y_1| < c_2\varepsilon \text{ and } 2^{-k-1} < |y_2| \le 2^{-k} \}$$

and write, using Hölder's inequality,

$$\begin{aligned} &\Big\| \int e^{i\lambda\psi(\gamma(t),y)} \eta(t)\, dt \Big\|_{L^{p'}(dy)} \\ &\qquad = \Big(\sum_{k>k_*} \Big\| \int e^{i\lambda\psi(\gamma(t),y)} \eta(t)\, dt \Big\|^{p'}_{L^{p'}(\Omega_k)} \Big)^{\frac{1}{p'}} \\ &\qquad \le \Big(\sum_{k>k_*} 2^{-k(1-\frac{p'}{2})} \Big\| \int e^{i\lambda\psi(\gamma(t),y)} \eta(t)\, dt \Big\|^{p'}_{L^2(\Omega_k)} \Big)^{\frac{1}{p'}}. \end{aligned} \tag{2.4}$$

Next

$$\begin{aligned} &\Big\| \int e^{i\lambda\psi(\gamma(t),y)} \eta(t)\, dt \Big\|^2_{L^2(\Omega_k)} \\ &\qquad \le \iint \Big| \int_{\Omega_k} e^{i\lambda[\psi(\gamma(t),y) - \psi(\gamma(t'),y)]}\, dy \Big|\, |\eta(t)|\, |\eta(t')|\, dt\, dt'. \end{aligned} \tag{2.5}$$

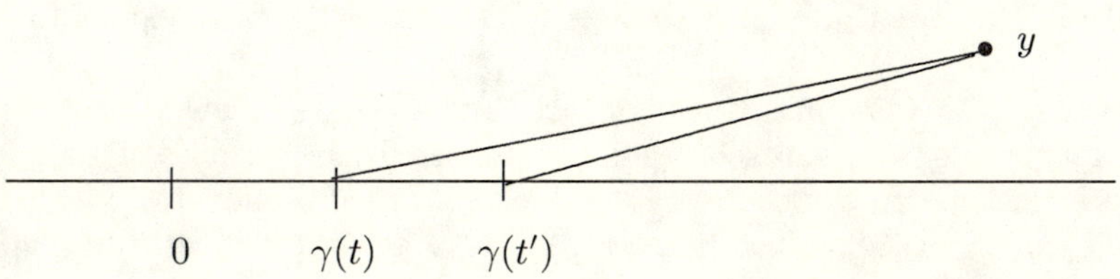

FIGURE 1

Applying Lemma 1 with 0 replaced by y, we see that

$$\begin{aligned} \nabla_y[\psi(\gamma(t), y) - \psi(\gamma(t'), y)] &= |\omega_y(\gamma(t)) - \omega_y(\gamma(t'))| \\ &\sim |\gamma(t) - \gamma(t')| \operatorname{dist}(y, \gamma) \\ &\sim 2^{-k} |t - t'|. \end{aligned} \tag{2.6}$$

It then follows from (2.5) and the trivial estimate that

$$\Big|\int_{\Omega_k} e^{i\lambda[\psi(\gamma(t),y)-\psi(\gamma(t'),y)]}\,dy\Big| \lesssim \min\Big(\frac{2^k}{\lambda|t-t'|}, 2^{-k}\Big). \tag{2.7}$$

Therefore

$$\begin{aligned}(2.5) &\lesssim \iint\limits_{|t-t'|>4^k\lambda^{-1}} \frac{2^k}{\lambda|t-t'|}|\eta(t)|\,|\eta(t')|\,dt\,dt' + 2^{-k}\iint\limits_{|t-t'|\le 4^k\lambda^{-1}} |\eta(t)|\,|\eta(t')|\,dt\,dt' \\ &\lesssim \frac{2^k}{\lambda}\log(2+4^{-k}\lambda)\|\eta\|_2^2 + 2^{-k}\min(1, 4^k\lambda^{-1})\|\eta\|_2^2.\end{aligned} \tag{2.8}$$

Substitution of (2.8) in (2.4) yields

$$\Big(\sum_{4^k<\lambda} 2^{-k(1-\frac{p'}{2})}2^{k\frac{p'}{2}}\lambda^{-\frac{p'}{2}}\log\frac{\lambda}{4^k}\Big)^{\frac{1}{p'}} + \Big(\sum_{4^k\ge\lambda} 2^{-k(1-\frac{p'}{2})}2^{-k\frac{p'}{2}}\Big)^{\frac{1}{p'}} \lesssim \lambda^{-\frac{1}{2p'}}$$

and hence

$$(2.2) \lesssim \lambda^{\frac{1}{2}-\frac{1}{2p'}} = \lambda^{\frac{1}{2p}},$$

which is (1.5).

3. Some background on oscillatory integral operators

We consider operators of the form

$$T_N f(x) = \int_I e^{iN\varphi(x,y)}a(x,y)f(y)\,dy, \tag{3.1}$$

where $a \in C_0^\infty(\mathbb{R}^{2d-1})$, and $\varphi \in C^\infty(\mathbb{R}^{2d-1})$ is a real-valued function satisfying the conditions for $(x,y) \in \operatorname{supp} a$,

$$\operatorname{rank}\frac{\partial^2\varphi}{\partial x\partial y} = d-1, \tag{3.2}$$

$$\frac{\partial}{\partial y}\Big\langle\frac{\partial\varphi}{\partial x}, t\Big\rangle = 0,\ t \in \mathbb{R}^d\setminus\{0\} \Rightarrow \det\Big(\frac{\partial^2}{\partial y^2}\Big\langle\frac{\partial\varphi}{\partial x}, t\Big\rangle\Big) \ne 0. \tag{3.3}$$

Condition (3.3) means that the map $y \mapsto \langle\frac{\partial\varphi}{\partial y}, t\rangle$ has only non-degenerate critical points.

Operators (3.1) were introduced in **[H]** and provide a generalization of the Fourier transform of measures carried by smooth hypersurfaces with curvature (Fourier restriction/extension theory) as well as an approach to Bochner–Riesz summation operators (note that here $(x,y) \in \mathbb{R}^d \times \mathbb{R}^d$ with phase function $|x-y|$ and the (3.1) format obtained by fixing one of the y-coordinates).

After coordinate change, φ may be given the form

$$\varphi(x,y) = x_1y_1 + \cdots + x_{d-1}y_{d-1} + x_d\langle Ay, y\rangle + O\big(|x|\,|y|(|x|^2+|y|^2)\big), \tag{3.4}$$

where A is a symmetric matrix, $\det A \ne 0$.

For $d=2$ (unlike in the higher-dimensional case) the mapping properties of T_N are well understood. It was shown in **[H]** that

$$\|T_N f\|_q \le C_\varphi N^{-\frac{2}{q}}\|f\|_r \tag{3.5}$$

holds, provided

$$q > 4 \quad \text{and} \quad \frac{3}{q} + \frac{1}{r} \le 1 \tag{3.6}$$

(which is the optimal range).

We will not use (3.5) but rather the ingredients of one of its proofs. The first step in the argument is a square-function inequality for the L^4_x-norm (depending on curvature and specific to the $d = 2$ case in this form). The second step is the use of 'variable-coefficient' Kakeya maximal operators associated to the system of curves

$$\Gamma_{y,\xi} = \big[\partial_y \varphi(x, y) = \xi\big] \tag{3.7}$$

(in our application, these curves are, by Lemma 1, the geodesics through the point corresponding to y). The reader will find an approach to Carleson–Sjölin estimates in a more general setting of Fourier integral separators in [**M-S-S**]. The short review below will be essentially self-contained and close to [**B**].

For $d = 2$, (3.4) becomes

$$\varphi(x_1, x_2, y) = x_1 y + x_2 y^2 + O\big(|x|\,|y|(|x|^2 + y^2)\big),$$

and applying another coordinate change in the x-variable, we obtain

$$\varphi(x_1, x_2, y) = x_1 y + x_2 y^2 + O(|y|^3 |x|). \tag{3.8}$$

Returning to (3.1), we start by partitioning the interval I where y ranges in subintervals I_α of size $|I_\alpha| \sim \frac{1}{\sqrt{N}}$ and write

$$Tf = \sum_{1 \le \alpha \le \alpha_*} T_\alpha f \qquad (\alpha_* \sim \sqrt{N}),$$

where

$$(T_\alpha f)(x) = \int_{I_\alpha} e^{iN\varphi(x,y)} f_\alpha(y) dy, \quad \text{with } f_\alpha = f \cdot 1_{I_\alpha}. \tag{3.9}$$

We first formulate the square function inequality.

LEMMA 2.

$$\|Tf\|_{L^4_{x,\mathrm{loc}}} \lesssim N^\varepsilon \Big\| \Big(\sum |T_\alpha f|^2\Big)^{\frac{1}{2}} \Big\|_{L^4_{x,\mathrm{loc}}} + N^{-C} \|f\|_2 \tag{3.10}$$

for all $\varepsilon > 0$, $C < \infty$.

Note that in the translation invariant case, we do have the clean square function inequality (cf. [**Cor**]). We did not make a serious attempt to remove the N^ε factor in our setting, though.

PROOF OF LEMMA 2. Upon writing

$$\|Tf\|_4^2 = \Big\| \sum_{\alpha,\beta} T_\alpha f \cdot \overline{T_\beta f} \Big\|_2,$$

the argument is based on almost orthogonality considerations. Write

$$(T_\alpha f \cdot \overline{T_\beta f})(x) = \iint e^{iN[\varphi(x,y_1) - \varphi(x,y_2)]} f_\alpha(y_1) \overline{f_\beta(y_2)}\, dy_1\, dy_2, \tag{3.11}$$

where by (3.8),

$$\varphi(x, y_1) - \varphi(x, y_2) = x_1(y_1 - y_2) + x_2(y_1 - y_2)(y_1 + y_2) + O\big(|x|\,|y_1 - y_2|\,|y|^2\big). \tag{3.12}$$

After a change of variables

$$\begin{cases} z_1 = y_1 - y_2, \\ z_2 = y_1 + y_2, \end{cases}$$

(3.12) gives

$$\phi(x, z_1, z_2) = x_1 z_1 + x_2 z_1 z_2 + O\big(|x|\,|z_1|\,|z|^2\big) \tag{3.13}$$

and

$$(3.11) = \iint_{(I_\alpha - I_\beta)\times(I_\alpha + I_\beta)} e^{iN\phi(x,z_1,z_2)} f_{\alpha\beta}(z_1, z_2)\, dz_1\, dz_2 = F_{\alpha,\beta}(x), \tag{3.14}$$

where

$$f_{\alpha\beta}(z_1, z_2) = f_\alpha\Big(\frac{z_1 + z_2}{2}\Big) f_\beta\Big(\frac{z_2 - z_1}{2}\Big).$$

Note that we may assume $|\alpha - \beta| \gtrsim 1$ (the other terms are obviously captured by the square function in (3.10)). Hence

$$|z_1| > \frac{1}{\sqrt{N}}. \tag{3.15}$$

We have to show that

$$\Big\|\sum F_{\alpha\beta}\Big\|_2^2 \lesssim N^\varepsilon \int \sum |F_{\alpha\beta}|^2 + N^{-C}\|f\|_2^4. \tag{3.16}$$

Inequality (3.16) clearly follows from an estimate

$$|\langle F_{\alpha\beta}, F_{\alpha'\beta'}\rangle| < K^{-C}\|f_\alpha\|_1 \cdot \|f_\beta\|_1 \cdot \|f_{\alpha'}\|_1 \cdot \|f_{\beta'}\|_1 \tag{3.17}$$

assuming

$$|\alpha - \alpha'| + |\beta - \beta'| > K, \tag{3.18}$$

which we apply with $K = N^\varepsilon$.

Assume (3.18). Writing

$$\langle F_{\alpha\beta}, F_{\alpha'\beta'}\rangle = \int_{(I_\alpha - I_\beta)\times(I_\alpha + I_\beta)} \int_{(I_{\alpha'} - I_{\beta'})\times(I_{\alpha'} + I_{\beta'})} \Big[\int e^{iN[\phi(x,z) - \phi(x,z')]} b(x)\, dx\Big] f_{\alpha\beta}(z) f_{\alpha'\beta'}(z')\, dz\, dz' \tag{3.19}$$

(where b is a localizing function) and performing the x-integration in (3.19), we will clearly obtain (3.17) from

(3.20)

$$\min\{|\nabla_x[\phi(x,z) - \phi(x,z')]| : x \in \operatorname{supp} b, z \in \operatorname{supp} f_{\alpha\beta}, z' \in \operatorname{supp} f_{\alpha'\beta'}\} \gtrsim \frac{K}{N}.$$

From (3.13), we have

$$\begin{aligned} &\phi(x,z) - \phi(x,z') \\ &\quad = x_1(z_1 - z_1') + x_2(z_1 z_2 - z_1' z_2') \\ &\qquad + O\Big(|x|\,|z_1 - z_1'|(|z|^2 + |z'|^2) + |x|\,|z_1|\,|z - z'|(|z| + |z'|)\Big) \end{aligned} \tag{3.21}$$

Assuming $|\nabla_x[\phi(x,z) - \phi(x,z')]| \lesssim \frac{K}{N}$, it follows from (3.21) that

$$\begin{cases} z_1 - z_1' + o(|z_1 - z_1'| + |z_1|\,|z_2 - z_2'|) < O(\frac{K}{N}), \\ z_1 z_2 - z_1' z_2' + o(|z_1 - z_1'| + |z_1|\,|z_2 - z_2'|) < O(\frac{K}{N}), \end{cases} \tag{3.22}$$

since $|z|, |z'| = o(1)$. Clearly (3.22) implies

$$|z_1 - z_1'| + |z_1|\, |z_2 - z_2'| < O\Big(\frac{K}{N}\Big), \tag{3.23}$$

and recalling (3.15),

$$|z_1 - z_1'| < O\Big(\frac{K}{N}\Big) \quad \text{and} \quad |z_2 - z_2'| < O\Big(\frac{K}{\sqrt{N}}\Big). \tag{3.24}$$

Therefore $|y_1 - y_1' - y_2 + y_2'| < O(\frac{K}{N})$, $|y_1 - y_1' + y_2 - y_1'| < O(\frac{K}{\sqrt{N}})$ implying $\operatorname{dist}(I_\alpha, I_{\alpha'}) \le |y_1 - y_1'| < O(\frac{K}{\sqrt{N}})$, $\operatorname{dist}(I_\beta, I_{\beta'}) \le |y_2 - y_2'| < O(\frac{K}{\sqrt{N}})$. But this contradicts (3.18).

This proves Lemma 2. □

Next, we need a more geometric interpretation of $T_\alpha f$ given by (3.9). Write

$$\begin{aligned} |(T_\alpha f)(x)| &= \Big| \int_{I_\alpha} e^{iN[\varphi(x,y) - \varphi(x,y_\alpha)]} f_\alpha(y)\, dy \Big| \\ &\sim \Big| \int_{I_\alpha} e^{iN(\partial_y \varphi)(x,y_\alpha)(y-y_\alpha)} f_\alpha(y)\, dy \Big| = \big| \widehat{f_\alpha^o}(N \cdot \partial_y \varphi(x, y_2)) \big|, \end{aligned} \tag{3.25}$$

where $y_\alpha \in I_\alpha$ is some fixed point and $f_\alpha^o(y) = f_\alpha(y + y_\alpha)$.

Let ζ be a smooth bump function satisfying $\zeta = 1$ on $[-1,1]$ and $\operatorname{supp} \zeta \subset [-2,2]$, and set $\zeta_{\frac{1}{\sqrt{N}}}(y) = \zeta(\sqrt{N} y)$. Hence $f_\alpha^o = f_\alpha^o \cdot \zeta_{\frac{1}{\sqrt{N}}}$ and $\widehat{f_\alpha^o} = \widehat{f_\alpha^o} * \widehat{\zeta_{\frac{1}{\sqrt{N}}}}$, implying by (3.25) that

$$\begin{aligned} |T_\alpha f(x)| &\le \int |\widehat{f_\alpha^o}(\xi)|\, |\widehat{\zeta_{\frac{1}{\sqrt{N}}}}(N \cdot \partial_y \varphi(x, y_\alpha) - \xi)|\, d\xi \\ &= \frac{1}{\sqrt{N}} \int |\widehat{f_\alpha^o}(\xi)| \Big| \widehat{\zeta}\Big(\sqrt{N}\big(\partial_y \varphi(x, y_\alpha) - \frac{\xi}{N}\big)\Big)\Big|\, d\xi. \end{aligned} \tag{3.26}$$

Hence, denoting

$$\rho_{y,\eta}^{(\frac{1}{\sqrt{N}})}(x) = \big| \widehat{\zeta}\big(\sqrt{N}(\partial_y \varphi(x,y) - \eta)\big)\big|, \tag{3.27}$$

also

$$|T_\alpha f(x)|^2 \le \frac{1}{\sqrt{N}} \int |\widehat{f}_\alpha(\xi)|^2 \rho_{y_\alpha, \frac{\xi}{N}}^{(\frac{1}{\sqrt{N}})}(x)\, d\xi. \tag{3.28}$$

Note that, roughly speaking, $\rho_{y,\eta}^{(\frac{1}{\sqrt{N}})}$ is the indicator function of a $\frac{1}{\sqrt{N}}$-neighborhood of the curve (3.7), i.e.

$$\Gamma_{y,\eta} = [\partial_y \varphi(x,y) = \eta]. \tag{3.29}$$

Also $|T_\alpha f|^2$ is a weighted average of $\rho_{y_\alpha,\eta}^{(\frac{1}{\sqrt{N}})}$ for varying η, with weight $\frac{1}{\sqrt{N}} \|f_\alpha\|_2^2$.

Returning to Lemma 2, it follows from the preceding that

LEMMA 3.

$$\begin{aligned} \|Tf\|_{L^4_{\text{loc}}}^4 &\lesssim N^{-1+\varepsilon} \sum_{\alpha,\beta} \|f\|_{L^2(I_\alpha)}^2 \|f\|_{L^2(I_\beta)}^2 \Big[\max_{\eta,\eta'} \int \rho_{y_\alpha,\eta}^{(\frac{1}{\sqrt{N}})} \cdot \rho_{y_\beta,\eta'}^{(\frac{1}{\sqrt{N}})} \Big] \\ &\quad + N^{-C} \|f\|_2^2. \end{aligned} \tag{3.30}$$

We will now apply the preceding to (1.7).

4. Proof of Proposition 4

In view of (1.7), we have

$$\int |\varphi_\lambda|^4 = \lambda \int |\varphi_\lambda(x)|^2 \Big| \int e^{i\lambda\psi(x,y)} a(x,y,\lambda)\varphi_\lambda(y)\, dy\Big|^2 dx + O(1). \tag{4.1}$$

Note that the phase function $\psi(x,y)$ is an $\mathbb{R}^2 \times \mathbb{R}^2$-function and the oscillatory integral in (4.1) over $y = (y_1, y_2) \in \mathbb{R}^2$, while in §3 the y-variable is 1-dimensional. To apply the discussion from §3, fix one of the variables y_2 (resp. y_1) and let $f(t) = \varphi_\lambda(t, y_2)$ $\big($or $f(t) = \varphi_\lambda(y_1, t)\big)$. We may then replace the first term of (4.1) by

$$\lambda \int |\varphi_\lambda(x)|^2 \Big| \int e^{i\lambda\varphi(x,t)} f(t)\, dt\Big|^2 dx, \tag{4.2}$$

where $\varphi(x,t) = \psi(x; t, y_2)$, say, and f may be assumed L^2-bounded. Recalling the restrictions on x, y in Proposition 5, we may clearly assume that $\operatorname{dist}(x, \mathbb{R}\times\{y_2\}) \gtrsim \varepsilon = O(1)$. Hence if x is an intersection point of geodesics $\Gamma_{t,\eta}$ and $\Gamma_{t',\eta'}$ through the respective points (t, y_2) and (t', y_2), then their angle of intersection is $\sim |t - t'|$.

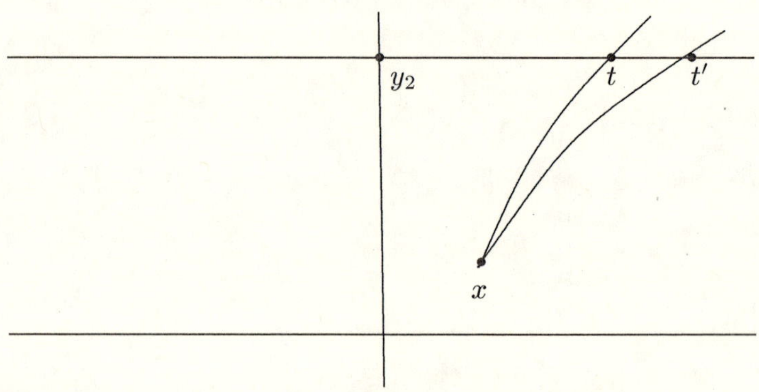

FIGURE 2

Here $\Gamma_{t,\eta}$ is defined as in (3.29), i.e.

$$\Gamma_{t,\eta} = [\partial_t \varphi(x,t) = \eta] \tag{4.3}$$

and is the geodesic through (t, y_2) parametrized by $\frac{x}{|x|} = (\eta, \cdot)$ in geodesic polar coordinates at (t, y_2). Using the notation from §3 with $N = \lambda$, we may therefore ensure that in (3.30)

$$\max_{\eta,\eta'} \int \rho_{t_\alpha,\eta}^{(\frac{1}{\sqrt{\lambda}})} \cdot \rho_{t_\beta,\eta'}^{(\frac{1}{\sqrt{\lambda}})} \lesssim \frac{1}{\sqrt{\lambda}|\alpha - \beta|}, \tag{4.4}$$

where the t_α are $\frac{1}{\sqrt{\lambda}}$-separated.

Returning to (4.2), write $f = \sum_\alpha f_\alpha$, $f_\alpha = f \cdot 1_{I_\alpha}$ as in §3 and denote, for $s \in \mathbb{Z}_+$,

$$\Omega_s = \{\alpha : \|f_\alpha\|_2^2 \sim 2^{-s}\}. \tag{4.5}$$

Estimate by Cauchy–Schwarz and (3.28),

$$\begin{aligned}\int |\varphi_\lambda|^2 \Big| \sum_{\alpha\in\Omega_s} \int e^{i\lambda\varphi(x,t)} f_\alpha(t)dt\Big|^2 \\ \le |\Omega_s| \sum_{\alpha\in\Omega_s} \int |\varphi_\lambda|^2 |T_\alpha f|^2 \\ \lesssim 2^s \sum_{\alpha\in\Omega_s} \frac{1}{\sqrt{\lambda}} \|f_\alpha\|_2^2 \max_\eta \Big\langle |\varphi_\lambda|^2, \rho_{t_\alpha,\eta}^{(\frac{1}{\sqrt{\lambda}})}\Big\rangle \lesssim \frac{2^s}{\lambda} B^2,\end{aligned} \tag{4.6}$$

where we denote

$$B = \max\big\{\|\varphi_\lambda\|_{L^2(\gamma)} : \gamma \text{ geodesic}\big\}. \tag{4.7}$$

On the other hand, application of Lemma 3 and (4.4) give

$$\begin{aligned}\int |\varphi_\lambda|^2 \Big|\sum_{\alpha\in\Omega_s} T_\alpha f\Big|^2 \le \|\varphi_\lambda\|_4^2 \Big\|\sum_{\alpha\in\Omega_s} T_\alpha f\Big\|_4^2 \\ \lesssim \Big[\lambda^{\varepsilon-1} \sum_{\alpha,\beta\in\Omega_s} \|f_\alpha\|_2^2 \|f_\beta\|_2^2 \frac{1}{\sqrt{\lambda}|\alpha-\beta|} + \lambda^{-C}\Big]^{\frac{1}{2}} \|\varphi_\lambda\|_4^2 \\ \lesssim 2^{-\frac{s}{2}} \lambda^{-\frac{3}{4}+\varepsilon} \|\varphi_\lambda\|_4^2.\end{aligned} \tag{4.8}$$

Summarizing, we get from (4.6), (4.8)

$$\begin{aligned}\|\varphi_\lambda\|_4^4 \lesssim \sum_s \min\{2^s B^2, 2^{-\frac{s}{2}} \lambda^{\frac{1}{4}+\varepsilon} \|\varphi_\lambda\|_4^2\} \\ \lesssim B^{\frac{2}{3}} \lambda^{\frac{1}{6}+\varepsilon} \|\varphi_\lambda\|_4^{\frac{4}{3}}\end{aligned} \tag{4.9}$$

and hence inequality (1.6).

This proves Proposition 4.

References

[B] J. Bourgain, *Some new estimates on oscillatory integrals.* Essays on Fourier analysis in honor of Elias M. Stein (Princeton, NJ, 1991), 83–112, Princeton Math. Ser., 42, Princeton Univ. Press, Princeton, NJ, 1995.

[B-G-T] N. Burq, P. Gérard and N. Tzvetkov, *Restrictions of the Laplace–Beltrami eigenfunctions to submanifolds*, Duke Math. J. **138** (2007), no. 3, 445–486.

[S-Z] C. D. Sogge and S. Zelditch, *Riemannian manifolds with maximal eigenfunction growth*, Duke Math. J. **114** (2002), 387–437.

[Cor] A. Córdoba, *A note on Bochner–Riesz operators*, Duke Math. J. **46** (1979), no. 3, 505–511.

[C-S] L. Carleson and P. Sjölin, *Oscillatory integrals and a multiplier problem for the disc*, Studia Math. **44** (1972), 287–299.

[H] L. Hörmander, *Oscillatory integrals and multipliers on FL^p*, Ark. Mat. **11** (1973), 1–11.

[M-S-S] G. Mockenhaupt, A. Seeger, and C. Sogge, *Local smoothing of Fourier integral operators and Carleson–Sjölin estimates*, J. Amer. Math. Soc. **6** (1993), no. 1, 65–130.

[R] A. Reznikov, *Norms of geodesic restrictions for eigenfunctions on hyperbolic surfaces and representation theory*, Preprint (2004) arXiv:math.AP/0403437.

[So] C. D. Sogge, *Concerning the L^p-norm of spectral clusters for second-order elliptic operators on compact manifolds*, J. Funct. Anal. **77** (1988), no 1, 123–138.

School of Mathematics, Institute for Advanced Study, Princeton, NJ 08540, USA
E-mail address: bourgain@ias.edu

Amer. Math. Soc. Transl.
(2) Vol. **226**, 2009

A Blaschke-Type Condition for Analytic and Subharmonic Functions and Application to Contraction Operators

S. Favorov and L. Golinskii

To V. P. Havin on the occasion of his 75th birthday

ABSTRACT. Let E be a closed set on the unit circle. We find a Blaschke-type condition, optimal in a sense of the order, on the Riesz measure of a subharmonic function v in the unit disk with a certain growth in the direction of E. In the special case where E is a finite set and $v = \log |f|$ with an analytic function f, our result agrees with a recent result of A. Borichev, L. Golinskii, and S. Kupin. An application to contractions close to unitary operators in Hilbert space is given.

1. Introduction

In [**1**] the authors study the zero sets for functions analytic in the unit disk $\mathbb{D}$ that grow exponentially fast near a finite set E of points on the unit circle $\mathbb{T}$. Here is the main result of [**1**]. As usual, $x_+ = \max\{x, 0\}$.

THEOREM BGK. *Let $E \subset \mathbb{T}$ be a finite set, $f \in \mathcal{A}(\mathbb{D})$ an analytic function in $\mathbb{D}$, $|f(0)| = 1$, and*

$$|f(z)| \le \exp\Big(\frac{D}{\rho^q(z,E)}\Big) \tag{1}$$

with $D, q \ge 0$, $\rho(z,E) = \operatorname{dist}(z,E)$. Let Z_f be the zero set of f, where each zero is counted according to its multiplicity. Then for any $\varepsilon > 0$,

$$\sum_{z\in Z_f} (1-|z|)\,\rho(z,E)^{(q-1+\varepsilon)_+} \le C(\varepsilon, q, E)\, D. \tag{2}$$

Note that both (1) and (2) make sense for arbitrary (infinite) subsets E on $\mathbb{T}$, so the question arises naturally as to whether the finiteness of E, which is an essential ingredient in the proof of Theorem BGK, can be relaxed. We should also

2000 *Mathematics Subject Classification.* Primary 30D50; Secondary 31A05, 47B10.
Key words and phrases. Analytic function in the unit disk, subharmonic function in the unit disk, Riesz measure, Blaschke-type estimate, discrete spectrum, contraction operator.

mention that in the case of $E = \mathbb{T}$, $\rho(z, E) = 1 - |z|$ we come to the well-known Blaschke-type condition for analytic functions in $\mathbb{D}$ with radial growth. The theory of such functions goes back to M. M. Djrbashian [**4**]; see also W. Hayman and B. Korenblum [**10**], F. A. Shamoyan [**16**], A. M. Jerbashian [**12**].

Our investigation shows that the natural setting for the above problem is the class of subharmonic functions v and their Riesz measures (generalized Laplacians) $\mu = (1/2\pi)\triangle v$ rather than analytic functions and their zero sets. It turns out that arbitrary closed sets E can be involved, and the results are optimal. Note that in the case of subharmonic functions of the form $\log|f|$ with $f \in \mathcal{A}(\mathbb{D})$, the Riesz measure is a discrete and integer-valued measure supported on Z_f, and $\mu\{z\}$ equals the multiplicity of the zero of f at z.

To formulate our result we need the following quantitative characteristic of the "sparseness" of a closed set E in terms of its t-neighborhood

$$E_t := \{\zeta \in \mathbb{T} : \rho(\zeta, E) < t\}. \tag{3}$$

Put

$$I(\alpha, E) := \int_0^2 \frac{\nu_E(s)}{s^{\alpha+1}}\, ds \le +\infty, \qquad \nu_E(t) = |E_t|, \quad \alpha \in \mathbb{R}. \tag{4}$$

Here and in what follows we denote by $|A|$ the normalized Lebesgue measure of a set $A \subset \mathbb{T}$. It is not hard to see that $I(\alpha, E) < \infty$ for any E and an arbitrary negative α. In the case where $\nu_E(t) = O(t^\beta)$, $\beta > 0$, one has $I(\alpha, E) < \infty$ for $\alpha < \beta$. Examples of sets E and evaluations of their characteristic I are given in Section 2.

The main result of our paper is the following

THEOREM 1. *Let $E \subset \mathbb{T}$ be a closed set. Let v be a subharmonic function in $\mathbb{D}$, $v \not\equiv -\infty$, μ its Riesz measure, and let*

$$v(z) \le \frac{1}{\rho^q(z, E)}$$

for all $z \in \mathbb{D}$ and some $q > 0$.

(i) *If $I(q, E) < \infty$, then*

$$\int_{\mathbb{D}} (1 - |\lambda|)\, d\mu(\lambda) < \infty. \tag{5}$$

Moreover, if $v(0) = 0$, then

$$\int_{\mathbb{D}} (1 - |\lambda|)\, d\mu(\lambda) \le 1 + q2^q I(E, q). \tag{6}$$

(ii) *If $I(q, E) = \infty$ and $I(\alpha, E) < \infty$ for some $\alpha < q$, then*

$$\int_{\mathbb{D}} (1 - |\lambda|)\, (\rho(\lambda, E))^{q-\alpha}\, d\mu(\lambda) < \infty. \tag{7}$$

Moreover, if $v(0) = 0$, then

$$\int_{\mathbb{D}} (1 - |\lambda|)\, (\rho(\lambda, E))^{q-\alpha}\, d\mu(\lambda) < 36(2^{-\alpha} + q(90)^{q-\alpha}\, I(\alpha, E)). \tag{8}$$

The result turns out to be optimal in the following sense.

THEOREM 2. *Let $E \subset \mathbb{T}$ be a closed set so that $I(\alpha, E) = +\infty$ for some $\alpha \geq 0$. Then for the subharmonic function $v_0(z) = \rho^{-q}(z, E)$ with the Riesz measure μ_0, and $q \geq \alpha$, we have*

$$\int_{\mathbb{D}} (1 - |\lambda|) \, (\rho(\lambda, E))^{q-\alpha} \, d\mu_0(\lambda) = +\infty. \tag{9}$$

As a consequence we obtain

THEOREM 3. *Let $E \subset \mathbb{T}$ be a closed set, and let $I(\alpha, E) < \infty$ for some $\alpha \in \mathbb{R}$. Suppose $f \in \mathcal{A}(\mathbb{D})$, and*

$$|f(z)| \leq \exp\Big(\frac{D}{\rho^q(z, E)}\Big), \quad z \in \mathbb{D}. \tag{10}$$

Then

$$\sum_{z \in Z_f} (1 - |z|) \, \rho(z, E)^{(q-\alpha)_+} < \infty. \tag{11}$$

Moreover, if $|f(0)| = 1$, then

$$\sum_{z \in Z_f} (1 - |z|) \, \rho(z, E)^{(q-\alpha)_+} \leq C(q, \alpha, E) \, D,$$

where $C(q, \alpha, E)$ does not depend on f. In the case of $q > \alpha$ one can take $C(q, \alpha, E) = 36(2^{-\alpha} + q(90)^{q-\alpha} \, I(\alpha, E))$.

The classical Blaschke condition arises for $E = \mathbb{T}$, $\alpha = -q < 0$ and $q \to 0$.

The first statement of Theorem 3 is a certain uniqueness theorem: under the assumption (10), the divergence of (11) yields $f \equiv 0$.

We proceed as follows. In Section 2 the main results are proved. In Section 3 an application to the spectral theory of contractions in Hilbert space is discussed.

2. Proof of the main results

There is a simple way to compute the function ν_E defined in (4), in terms of the complementary arcs of E. Let

$$\mathbb{T} \setminus E = \bigcup_j \gamma_j, \qquad |\gamma_j| \downarrow 0.$$

Then

$$\nu_E(t) = \sum_{j=N+1}^{\infty} |\gamma_j| + \frac{2N}{\pi} \arcsin\frac{t}{2} + |E|,$$

where $N = N(t)$ is taken from

$$|\gamma_{N+1}| \leq \frac{2}{\pi} \arcsin\frac{t}{2} < |\gamma_N|.$$

So the exponent α in (4) for which $I(\alpha, E) < \infty$ can be easily determined. For instance, one can take any $\alpha < 1$ for

$$E = \left\{ e^{i\varphi_n} : \varphi_n = \sum_{k=n}^{\infty} \frac{1}{2^k} \right\} \cup \{1\},$$

any $\alpha < 1 - \frac{1}{\gamma}$ works for

$$E = \left\{ e^{i\varphi_n} : \varphi_n = c \sum_{k=n}^{\infty} \frac{1}{k^\gamma}, \ \gamma > 1 \right\} \cup \{1\}, \quad c^{-1} = \sum_{k=1}^{\infty} \frac{1}{k^\gamma},$$

and for

$$E = \left\{ e^{i\varphi_n} : \varphi_n = c \sum_{k=n}^{\infty} \frac{1}{k \log^2 k} \right\} \cup \{1\}, \quad c^{-1} = \sum_{k=2}^{\infty} \frac{1}{k \log^2 k},$$

$I(\alpha, E) < \infty$ for any $\alpha < 0$. For the generalized Cantor set $\mathcal{C}_\beta$ (in the case of standard Cantor set, $\beta = 1/3$) one can take any $\alpha < 1 - d(\beta)$, where

$$d(\beta) = \frac{\log 2}{\log 2 - \log(1 - \beta)}$$

is the Hausdorff dimension of $\mathcal{C}_\beta$ (see, e.g., [**6**]).

Let us list some elementary properties of ν_E:

(1) ν_E is a continuous and strictly monotone increasing function on $[0, t_0(E)]$ and $\nu_E(t) = 1$ for $t_0(E) \le t < \infty$;
(2) $\nu_E(0+) = |E|$;
(3) $\nu_E(t) \ge |E| + t/\pi$ for each closed $E \ne \mathbb{T}$ and $0 < t \le t_1(E)$;
(4) $\nu_E(t) = O(t)$ as $t \to 0$ if and only if E is a finite set.

Our argument relies upon some basic results of potential theory in the complex plane (see [**15**, Chapters 3, 4]). Let v be a subharmonic function on $\mathbb{D}$, $v \not\equiv -\infty$. The Riesz measure $\mu = (1/2\pi)\Delta v$ is a positive Radon measure on $\mathbb{D}$. If v has a harmonic majorant on a subdomain $\Omega \subseteq \mathbb{D}$, then the following representation holds:

$$v(z) = u(z) - \int_\Omega G_\Omega(z, \lambda)\, d\mu(\lambda), \quad z \in \Omega, \tag{12}$$

where u is the least harmonic majorant in Ω, and $G_\Omega(z, \lambda)$ is Green's function for Ω, that is,

$$G_\Omega(z, \lambda) = \log(1/|z - \lambda|) - h(z, \lambda), \tag{13}$$

where $h(z, \cdot)$ is a harmonic function in Ω such that $h(z, \lambda) = \log(1/|z - \lambda|)$ for $\lambda \in \partial\Omega$. Note that if $\Omega \ne \mathbb{D}$ and $0 \in \Omega$, then $h(0, \lambda) > 0$ for $\lambda \in \Omega$.

Proof of Theorem 1. We begin with the proof of (6). It easily follows from $I(q, E) < \infty$, $q > 0$, and (4) that now $\nu_E(0+) = |E| = 0$. Put $\rho(z) = \rho(z, E)$.

The monotone permutation theorem (also known as the "layer cake representation"; see, e.g., [**14**, Theorem 1.13]),

$$\int_X f^r(x)\, d\sigma(x) = r \int_0^\infty y^{r-1} \sigma(\{x : f(x) > y\})\, dy, \tag{14}$$

is of importance in our argument. Here $r > 0$, (X, σ) is a measure space, and $f \ge 0$ is a measurable function on X. We have (see (4))

$$\int_\mathbb{T} \frac{dm(\zeta)}{\rho^q(\zeta)} = q \int_0^\infty y^{q-1} \left| \left\{ \zeta : \rho(\zeta) < \frac{1}{y} \right\} \right| dy = 2^{-q} + qI(q, E) < \infty. \tag{15}$$

Therefore the Poisson integral

$$U(z) = \int_\mathbb{T} \frac{1 - |z|^2}{|\zeta - z|^2} \frac{dm}{\rho^q(\zeta)}, \quad z \in \mathbb{D},$$

is a well-defined harmonic function. Let us prove that v admits a harmonic majorant on the whole disk $\mathbb{D}$. A simple geometrical inequality shows that for all $|z| \le 1$ and $0 < \tau < 1$ one has $\rho(z) \le 2\rho(\tau z)$. The function $v_0 = \rho^{-q}$ is obviously subharmonic in $\mathbb{D}$ since $v_0(z) = \sup_{\zeta\in E}(|z-\zeta|^{-q})$, and continuous at any point $\zeta \in \mathbb{T}\setminus E$. By the well-known property of the Poisson integral,

$$\lim_{z\to\zeta} U(z) = v_0(\zeta) \ge 2^{-q} v_0(\tau\zeta), \qquad \zeta \in \mathbb{T}\setminus E, \quad 0<\tau<1.$$

As long as $\zeta \in E$, one has $\lim_{z\to\zeta} U(z) = +\infty$. But $v_0(\tau z)$ is a *bounded* subharmonic function in $\mathbb{D}$, so by the Maximum Principle, $U(z) \ge 2^{-q} v_0(\tau z)$, $z \in \mathbb{D}$. It remains only to let $\tau \to 1$ to make sure that $2^q U$ is a desired harmonic majorant for v. Representation (12) now takes the form

$$v(z) = u(z) - \int_{\mathbb{D}} \log\left|\frac{1-\bar\lambda z}{z-\lambda}\right| d\mu(\lambda), \quad z \in \mathbb{D},$$

where $u \le U$ in $\mathbb{D}$. For $z=0$ we come to

$$\int_{\mathbb{D}} (1-|\lambda|)\, d\mu(\lambda) \le \int_{\mathbb{D}} \log\frac{1}{|\lambda|}\, d\mu(\lambda) = u(0) \le 2^q U(0) = 2^q \|\rho^{-q}\|_{L^1(\mathbb{T})},$$

and (6) follows from (15).

The case when $I(q,E)=\infty$ but $I(\alpha,E)<\infty$ for some $\alpha<q$ is much subtler.

For a fixed $t\in(0,1)$ denote by $\Omega=\Omega_t$ the connected component of the open subset $\{z\in\mathbb{D} : \rho(z)>t\}$ that contains the origin, so Ω is a non-empty subdomain of $\mathbb{D}$. Put

$$\Gamma := \{z\in\mathbb{D} : \rho(z)=t\}, \qquad E_t^c = \mathbb{T}\setminus E_t,$$

where E_t is defined in (3). It is easy to check that E_t is a finite union of disjoint open arcs, so E_t^c is a finite union of disjoint closed arcs. The boundary $\partial\Omega$ is contained in the closure of $E_t^c\cup\Gamma$. The function v is bounded from above in Ω, so (12) holds for v in Ω, with the least harmonic majorant u.

Let V be a harmonic function in $\mathbb{D}$ such that

$$V(\zeta) = \begin{cases} t^{-q}, & \zeta\in E_t, \\ \rho^{-q}(\zeta), & \zeta\in E_t^c. \end{cases} \tag{16}$$

Note that V is continuous in the closed unit disk. For $0<t<1$ and each $z\in\Gamma$ there is $\zeta'\in E$ such that $|z-\zeta'|=t$. Let $\omega(\lambda,\gamma_z,\mathbb{D})$ be the harmonic measure of the arc $\gamma_z=\{\zeta\in\mathbb{T} : |\zeta-\zeta'|\le t\}$. It easily follows from the explicit expression (see, e.g., [**7**, Chapter 1]) that $\omega(\lambda,\gamma_z,\mathbb{D})\ge 1/6$ for $|\lambda-\zeta'|=t$, $\lambda\in\mathbb{D}$. Since $\gamma_z\subset E_t$, we get $V(z)\ge t^{-q}\omega(z,\gamma_z,\mathbb{D})\ge t^{-q}/6$. Therefore, we have

$$\limsup_{z\to\zeta} v(z) \le 6V(\zeta), \qquad \forall\zeta\in\partial\Omega.$$

By the Maximum Principle,

$$v(z)\le 6V(z) \quad \text{and so} \quad u(z)\le 6V(z), \qquad z\in\Omega.$$

Let us prove (8). Since $v(0)=0$, (12) for $z=0$ can be written as

$$\int_\Omega G_\Omega(0,\lambda)\, d\mu(\lambda) = u(0) \le 6V(0) = \frac{6\nu_E(t)}{t^q} + 6\int_{E_t^c} \frac{dm(\zeta)}{\rho^q(\zeta)}. \tag{17}$$

We proceed with another application of (14) to obtain

$$\begin{aligned}\int_{E_t^c}\frac{dm(\zeta)}{\rho^q(\zeta)}&=q\int_0^\infty y^{q-1}\,|\{\zeta\in\mathbb{T}:\,t<\rho(\zeta)<1/y\}|\,dy\\&=2^{-q}-\frac{\nu_E(t)}{t^q}+q\int_t^2\frac{\nu_E(s)}{s^{q+1}}\,ds\,.\end{aligned}\tag{18}$$

Hence, the following bound holds:

$$\int_\Omega G_\Omega(0,\lambda)\,d\mu(\lambda)\le 6\cdot 2^{-q}+6q\int_t^2\frac{\nu_E(s)}{s^{q+1}}\,ds\,.\tag{19}$$

Our next step is to estimate the Green function $G_\Omega(0,\cdot)$ in a smaller domain Ω_τ with $\tau=(6\pi+3)t$. We now assume that $t\in(0,\frac{1}{6\pi+3})$, and so $\tau\in(0,1)$. For $z\in\Gamma$ there is $\zeta'\in E$ such that $|z-\zeta'|=t$, so $t\ge 1-|z|$, and hence

$$\log\frac{1}{|z|}\le\log\frac{1}{1-t}\le\frac{3t}{2}\,.$$

The harmonic function $h(0,\cdot)$ from (13) does not exceed $3t/2$ on Γ and equals zero on E_t^c, so

$$h(0,\lambda)\le\frac{3t}{2}\,,\qquad \lambda\in\Omega.\tag{20}$$

We will distinguish two situations for λ. Let $|\lambda|\le 1-2t$. Then by (20)

$$G_\Omega(0,\lambda)=\log\frac{1}{|\lambda|}-h(0,\lambda)\ge 1-|\lambda|-\frac{3t}{2}\ge\frac{1-|\lambda|}{4}.$$

Let $|\lambda|>1-2t$. This is where the restriction $\lambda\in\Omega_\tau$ is essential. Let $\omega(\lambda,E_t,\mathbb{D})$ be the harmonic measure of the set E_t. We put $g(\lambda)=9t\omega(\lambda,E_t,\mathbb{D})$. The same argument as above implies $\omega(\lambda,E_t,\mathbb{D})\ge 1/6$ for $\lambda\in\Gamma$, so

$$h(0,\lambda)\le g(\lambda),\ \lambda\in\partial\Omega\implies h(0,\lambda)\le g(\lambda),\ \lambda\in\Omega,$$

and

$$G_\Omega(0,\lambda)\ge 1-|\lambda|-g(\lambda).$$

We will find the upper bound for $g(\lambda)$ in Ω_τ. For $\lambda=|\lambda|e^{i\theta}$ write

$$g(\lambda)=9t\int_{E_t}\frac{1-|\lambda|^2}{|\zeta-\lambda|^2}\,dm(\zeta)=\frac{9t(1-|\lambda|^2)}{2\pi}\int_{e^{i\varphi}\in E_t}\frac{d\varphi}{(1-|\lambda|)^2+4|\lambda|\sin^2\frac{\varphi-\theta}{2}}\,.$$

Since $\rho(\lambda)\ge\tau$ and $1-|\lambda|<2t$, we have

$$\begin{aligned}\rho(e^{i\theta})&=|e^{i\theta}-\zeta_1|=|\lambda-\zeta_1+e^{i\theta}-\lambda|\\&\ge(6\pi+3)t-(1-|\lambda|)>(6\pi+1)t\end{aligned}$$

for some $\zeta_1\in E$. Next, for $e^{i\varphi}\in E_t$ there is $\zeta_2\in E$ with $|e^{i\varphi}-\zeta_2|=\rho(e^{i\varphi})\le t$, so

$$|\theta-\varphi|\ge 2\left|\sin\frac{\theta-\varphi}{2}\right|=|e^{i\theta}-\zeta_2+\zeta_2-e^{i\varphi}|\ge\rho(e^{i\theta})-t\ge 6\pi t.$$

Therefore, as $|\lambda|>1-2t>9/10$, we come to

$$g(\lambda)\le\frac{5\pi t(1-|\lambda|)}{2}\int_{6\pi t\le|\varphi-\theta|\le\pi}\frac{d\varphi}{(\varphi-\theta)^2}<\frac{5(1-|\lambda|)}{6}.$$

Finally, the following lower bound holds for all $\lambda \in \Omega_\tau$:

$$G_\Omega(0,\lambda) \ge \frac{1-|\lambda|}{6},$$

and we can continue (19) as

$$\int_{\Omega_\tau} (1-|\lambda|)\, d\mu(\lambda) \le 36 \left(2^{-q} + q \int_{\tau/(6\pi+3)}^{2} \frac{\nu_E(s)}{s^{q+1}}\, ds \right). \tag{21}$$

The latter holds for each $\tau \in (0,1)$.

Define $\tilde\rho := \min\{\rho, 1\}$. It is easily checked that

$$\tilde\rho(\lambda) \le \rho(\lambda) \le 2\tilde\rho(\lambda) \quad \forall \lambda \in \mathbb{D}, \tag{22}$$

and $\{\lambda : \rho(\lambda) > \tau\} = \{\lambda : \tilde\rho(\lambda) > \tau\}$ for $0 < \tau < 1$. Formula (14), applied to the measure $d\sigma = (1-|\lambda|)d\mu$ and function $\tilde\rho \le 1$, gives

$$\int_{\mathbb{D}} (\tilde\rho(\lambda))^{q-\alpha} d\sigma(\lambda) = (q-\alpha) \int_0^1 \tau^{q-\alpha-1}\, d\tau \int_{\{\rho>\tau\}} d\sigma(\lambda).$$

A simple geometric inequality shows that for all $z \in \mathbb{D}$ and $0 < \tau < 1$ one has $\rho(z) \le 2\rho(\tau z)$. Hence if $\rho(z) > \tau$ then $\rho(\tau z) > \tau/2$, so the whole interval $[0,z]$ belongs to the set $\{\rho > \tau/2\}$. The latter means that $\{\rho > \tau\} \subset \Omega_{\tau/2}$. Hence by using (22) we obtain

$$\int_{\mathbb{D}} (\rho(\lambda))^{q-\alpha} d\sigma(\lambda) \le 2^{q-\alpha}(q-\alpha) \int_0^1 \tau^{q-\alpha-1}\, d\tau \int_{\Omega_{\tau/2}} (1-|\lambda|)\, d\mu,$$

and (21) leads to (8), as claimed.

To prove (5), (7) in the case where $v(0) > -\infty$, one can apply (6) (respectively, (8)) to the function $v_1(z) = (v(z) - v(0))(1 + 2^q|v(0)|)^{-1}$. Indeed, $v_1(z) \le \rho^{-q}(z)$ for all $z \in \mathbb{D}$, and the Riesz measure of v_1 coincides with the Riesz measure of v up to a constant factor (which depends on $v(0)$).

If $v(0) = -\infty$, take the harmonic function h in the disk $\{|z| < 1/2\}$ such that $v = h$ for $|z| = 1/2$, and put

$$v_1(z) = \begin{cases} \max(v(z), h(z)) & \text{for } |z| < 1/2, \\ v(z) & \text{for } |z| \ge 1/2. \end{cases}$$

Clearly, $v_1(z) \le v(z) \le \rho^{-q}(z, E)$. Moreover, v_1 is subharmonic in $\mathbb{D}$ (see, e.g., [**15**, Theorem 2.4.5]), and the restriction of its Riesz measure μ_1 to the set $\{z \in \mathbb{D} : |z| > 1/2\}$ is the same as that for μ. Therefore, the integrals in (5), (7) taken for μ and μ_1 differ by a bounded term. Since $v_1(0) \ne \infty$, the result follows. The proof is complete. □

PROOF OF THEOREM 2. We invoke again the harmonic function V defined in (16). It is clear that $V = v_0$ on E_t^c, and $V \le t^{-q} = v_0$ on Γ. Denote by u_0 the least harmonic majorant of v_0 on $\Omega = \Omega_t$, which is defined above. We have

$$V(\zeta) \le v_0(\zeta) \le u_0(\zeta),\ \zeta \in \partial\Omega \implies V(z) \le u_0(z),\ z \in \Omega.$$

Since $h(0,\cdot) \ge 0$ and $v_0(0) = 1$, we obtain from (12) that, for $z = 0$,

$$V(0) \le u_0(0) \le \int_\Omega \log\frac{1}{|\lambda|}\, d\mu_0(\lambda) + 1. \tag{23}$$

For $V(0)$ we see (cf. (17)–(18)) that

$$V(0) = 2^{-q} + q\int_t^2 \frac{\nu_E(s)}{s^{q+1}}\,ds\,. \tag{24}$$

Therefore, in the case of $q = \alpha$ we have

$$\int_{\mathbb{D}} \log\frac{1}{|\lambda|}\,d\mu_0(\lambda) = +\infty.$$

Now let $q > \alpha$. We apply (14) with $d\sigma = \log|\lambda|^{-1}\,d\mu_0$, keeping (23)–(24) in mind:

$$\begin{aligned}\int_{\mathbb{D}} (\rho(\lambda, E))^{q-\alpha} d\sigma(\lambda) &= (q-\alpha)\int_0^2 y^{q-\alpha-1}\sigma(\{\lambda : \rho > y\})\,dy \\ &\ge (q-\alpha)\int_0^1 y^{q-\alpha-1}\,dy \int_{\Omega_y} \log\frac{1}{|\lambda|}\,d\mu_0(\lambda) \\ &\ge (q-\alpha)\int_0^1 y^{q-\alpha-1}\left\{2^{-q} - 1 + q\int_y^1 \frac{\nu_E(s)}{s^{q+1}}\,ds\right\}dy \\ &= q\int_0^1 \frac{\nu_E(s)}{s^{\alpha+1}}\,ds + 2^{-q} - 1,\end{aligned}$$

and hence

$$\int_{\mathbb{D}} \log\frac{1}{|\lambda|}(\rho(\lambda, E))^{q-\alpha}\,d\mu_0(\lambda) = +\infty.$$

To go over to (9) we note that in view of (12) and (13) the function ρ^{-q} in the disk $\{z : |z| < 1/2\}$ differs from $\int_{|\lambda|<1/2} \log(1/|\lambda - z|)\,d\mu_0(\lambda)$ by a harmonic function. Hence, the latter integral is finite at $z = 0$, and (9) follows. The proof is complete. □

The following example shows that the result in Theorem 3 is optimal as well.

EXAMPLE. Let $z_n = 1 - 1/(n+1)$, $n = 1, 2, \ldots$. Consider an infinite canonical product (see, e.g., [**11**, p. 132])

$$f(z) := \prod_{n=1}^{\infty} P\left(\frac{1-z_n^2}{1-z_n z}\right), \qquad P(z) = (1-z)\,e^z.$$

It follows from the Taylor expansion of P that $|1 - P(z)| < 4|z|^2$ for $|z| \le 2$. Since $|1 - z_n^2| < 2|1 - z_n z|$ for $z \in \mathbb{D}$, we get

$$\log|f(z)| \le \sum_{n=1}^{\infty}\left|1 - P\left(\frac{1-z_n^2}{1-z_n z}\right)\right| < 4\sum_{n=1}^{\infty}\left|\frac{1-z_n^2}{1-z_n z}\right|^2. \tag{25}$$

Furthermore, for $z \in \mathbb{D}$ we have

$$\left|\frac{1-z_n^2}{1-z_n z}\right| \le \frac{2}{|n(1-z)+1|} \le \frac{2\sqrt{2}}{n|1-z|+1}.$$

Hence for $n(z) = [|1-z|^{-1}] + 1 \in \mathbb{N}$ the right-hand side of (25) does not exceed

$$32\,n(z) + \frac{32}{|1-z|^2}\sum_{n=n(z)+1}^{\infty}\frac{1}{n^2} \le \frac{64}{|1-z|} + \frac{32}{n(z)|1-z|^2} \le \frac{96}{|1-z|}.$$

So f satisfies (10) with $E = \{1\}$, $q = 1$, and the choice $\alpha = 1 - \varepsilon$ is optimal.

The growth of canonical products with positive zeros was studied in [**8**].

3. Discrete spectrum of contraction operators

The relationship between the spectral theory of non-selfadjoint operators and the theory of analytic functions via perturbation determinants is well established due to the works of I. Gohberg, M. Krein, V. Matsaev, and others. Some variants of this approach were used recently in [**2, 3**].

The results of Section 2, in particular Theorem 3, can be applied to operator theory in the same fashion as it is done in [**1, 2, 3**].

Let U be a unitary operator in a Hilbert space $\mathcal{H}$, and $\sigma(U) = E$ its spectrum. The resolvent $R_U(z) = (U - z)^{-1}$ is an operator-valued function analytic in $\mathbb{D}$, and (see, e.g., [**13**, Section V.3.8])

$$\|R_U(z)\| = \rho^{-1}(z, \sigma(U)), \qquad z \in \mathbb{D}.$$

A bounded linear operator T on $\mathcal{H}$ is called a *contraction* if $\|T\| \le 1$, where $\|\cdot\|$ is the standard operator norm. Assume that $T - U$ is a compact operator. If $E \ne \mathbb{T}$, then by a general result of perturbation theory, $\sigma(T) = E \cup \sigma_d(T)$, where the discrete component of the spectrum $\sigma_d(T) \subset \overline{\mathbb{D}} \setminus E$ is an at most countable set with all possible accumulation points in E, and each $z_k \in \sigma_d(T)$ is an eigenvalue of finite algebraic multiplicity.

Our goal here is to obtain some quantitative bound for the rate of convergence of z_k to E in the case where $T - U \in \mathcal{S}_q$, $q > 0$, the Schatten–von Neumann classes of compact operators, i.e., the sequence of eigenvalues of the operator $|T - U|$ belongs to ℓ^q. The norm in $\mathcal{S}_q$ is denoted by $\|\cdot\|_q$, and the standard operator norm $\|\cdot\|$ is denoted by $\|\cdot\|_\infty$. The family $\{\mathcal{S}_q\}$ is nested: $\mathcal{S}_{q_1} \subset \mathcal{S}_{q_2}$ for $0 < q_1 < q_2 \le \infty$, and $\|A\|_{q_2} \le \|A\|_{q_1}$. If $A \in \mathcal{S}_q$, $q \ge 1$, and S is a bounded linear operator, then $AS \in \mathcal{S}_q$ and $\|AS\|_q \le \|A\|_q \|S\|$. An extensive information on the subject is Gohberg–Krein [**9**], Simon [**17**], and Dunford–Schwartz [**5**, Chapter XI.9].

The key analytic tool is the so-called renormalized determinant $\det_n(I + A)$, defined for $A \in \mathcal{S}_n$ and $n = 1, 2, \dots$. In particular, we need the following continuity property (cf. [**17**, Theorem 9.2, (c)]):

$$|\det{}_n(I + A) - \det{}_n(I + B)| \le \|A - B\|_n \exp\left(C_n(1 + \|A\|_n^n + \|B\|_n^n)\right), \tag{26}$$

where C_n is a constant that depends only on n. If $T - U \in \mathcal{S}_n$, then the perturbation determinant

$$u_n(z) := \det{}_n(T - z)(U - z)^{-1} = \det{}_n\left(I + (T - U)(U - z)^{-1}\right)$$

is a well-defined and analytic function in $\mathbb{D}$, and its zero set in $\mathbb{D}$ is $Z_{u_n} = \sigma_d(T) \cap \mathbb{D}$, with the order of each zero equal to the algebraic multiplicity of the eigenvalue.

THEOREM 4. *Let U be a unitary operator on $\mathcal{H}$, $\sigma(U) = E \ne \mathbb{T}$, and $I(\alpha, E) < \infty$ in (4) for some $\alpha \in \mathbb{R}$. Let a contraction T be a $\mathcal{S}_q$-perturbation of U: $T - U \in \mathcal{S}_q$, $q > 0$. Then*

$$\sum_{z \in \sigma_d(T)} (1 - |z|)\rho(z, E)^{(q-\alpha)_+} < \infty. \tag{27}$$

Moreover, there is a constant $\delta = \delta_q$, $0 < \delta_q < 1$, depending only on q and such that for $\|T - U\|_q < \delta_q$ we have

$$\sum_{z \in \sigma_d(T)} (1 - |z|)\rho(z, E)^{(q-\alpha)_+} \le C(q, E) \max(\|T - U\|_q, \|T - U\|_q^q). \tag{28}$$

REMARK. Obviously, the eigenvalues of T with $|z| = 1$ do not enter the left hand side, so we claim nothing about this part of the discrete spectrum. However, this part is missing as long as T is a completely non-unitary contraction.

PROOF. Take a positive integer n with $q \le n < q+1$, and so $T - U \in \mathcal{S}_n$. By [**5**, Lemma XI.9.22, (d)],

$$\begin{aligned} |u_n(z)| &\le \exp\left(c_q\|(T-U)(U-z)^{-1}\|_q^q\right) \\ &\le \exp\left(c_q\|T-U\|_q^q\,\|(U-z)^{-1}\|^q\right) \\ &\le \exp\left(c_q\|T-U\|_q^q\rho^{-q}(z,E)\right). \end{aligned}$$

An application of the first statement of Theorem 3 proves (27).

Throughout the rest of the proof, c_q stands for various positive constants which depend only on q. Since $\|T-U\| < 1$, we see that T is invertible, and

$$u_n(0) = \det{}_n\left(I + (T-U)U^{-1}\right) \ne 0.$$

Put $f(z) = u_n(z)/u_n(0)$, $|f(0)| = 1$, so that

$$\log|f(z)| \le \frac{c_q\|T-U\|_q^q}{\rho^q(z,E)} + \log\frac{1}{|u_n(0)|}.$$

The bound (26) with $A = (T-U)U^{-1}$, $B = 0$ gives (in view of the monotonicity of norms):

$$|u_n(0) - 1| \le \|T-U\|_q \exp\left(C_q'(1 + \|T-U\|_q^n)\right) < \frac{1}{2},$$

as soon as δ_q is small enough. Hence

$$\log\frac{1}{|u_n(0)|} \le \log 4\,|u_n(0) - 1| \le c_q\,\|T-U\|_q.$$

An application of the second statement of Theorem 3 gives (28). □

References

[1] A. Borichev, L. Golinskii, and S. Kupin, *A Blaschke-type condition and its application to complex Jacobi matrices*, preprint `arXive:0712.0407v2`.

[2] M. Demuth and G. Katriel, *Eigenvalue inequalities in terms of Schatten norm bounds on differences of semigroups, and application to Schrödinger operators*, Ann. Henri Poincaré **9** (2008), 817–834.

[3] M. Demuth, M. Hansmann, and G. Katriel, *On the distribution of eigenvalues of non-selfadjoint operators*, preprint `arXive:0802.2468v1`.

[4] M. M. Djrbashian, *Theory of factorization and boundary properties of functions meromorphic in the disk*, Proceedings of the ICM, Vancouver, B.C., 1974, vol. 2, pp. 197–202, Canadian Math. Congress, Montreal, Que., 1975.

[5] N. Dunford and J. T. Schwartz, *Linear operators.* Part II: Spectral theory. Self adjoint operators in Hilbert space, John Wiley and Sons, New York–London, 1963.

[6] K. Falconer, *Fractal geometry. mathematical foundations and applications.* John Wiley and Sons, Chichester, New York, 1990.

[7] J. Garnett, *Bounded analytic functions.* Graduate Texts in Mathematics, vol. 236. Springer, New York, 2007.

[8] M. Girnyk *On asymptotic properties of certain canonical products*, Sib. Math.J. **15** (1974), no. 5, 1036–1048.

[9] I. Gohberg and M. Krein, *Introduction to the theory of linear nonselfadjoint operators.* Translations of Mathematical Monographs, vol. 18, Amer. Math. Soc., Providence, RI, 1969.

[10] W. K. Hayman and B. Korenblum, *A critical growth rate for functions regular in a disk*, Michigan Math. J. **27** (1980), 21–30.

[11] H. Hedenmalm, B. Korenblum, and K. Zhu, *Theory of Bergman spaces.* Graduate Texts in Mathematics, vol. 199. Springer-Verlag, New York, 2000.

[12] A. M. Jerbashian, *On the theory of weighted classes of area integrable regular functions*, Complex Variables **50** (2005), 155–183.

[13] T. Kato, *Perturbation theory for linear operators.* Springer-Verlag, Berlin, 1966.

[14] E. Lieb and M. Loss, *Analysis.* Graduate Studies in Mathematics, vol. 14, Amer. Math. Soc., Providence, RI, 1997.

[15] T. Ransford, *Potential theory in the complex plane.* London Math. Soc. Student Texts, vol. 28, Cambridge University Press, 1995.

[16] F. A. Shamoyan, *Zeros of functions analytic in the disk and growing near the boundary*, Izv. Akad. Nauk Armyan. SSR Ser. Mat. **18** (1983), no. 1, 15–27. (Russian)

[17] B. Simon, *Trace ideals and their applications.* Mathematical Surveys and Monographs, vol. 120. Amer. Math. Soc., Providence, RI, 2005.

MATHEMATICAL SCHOOL, KHARKOV NATIONAL UNIVERSITY, SWOBODY SQUARE 4, 61077 KHARKOV, UKRAINE

E-mail address: `Sergey.Ju.Favorov@univer.kharkov.ua`

MATHEMATICS DIVISION, INSTITUTE FOR LOW TEMPERATURE PHYSICS AND ENGINEERING, 47 LENIN AVENUE, 61103 KHARKOV, UKRAINE

E-mail address: `leonid.golinskii@gmail.com`

Amer. Math. Soc. Transl.
(2) Vol. **226**, 2009

New Estimates for the Length of the Erdős–Herzog–Piranian Lemniscate

Alexander Fryntov and Fedor Nazarov

ABSTRACT. Let $p(z)$ be a monic polynomial of a fixed degree $n \geq 1$. Consider the lemniscate

$$L_p := \{z : |p(z)| = 1\}.$$

Let $|L_p|$ be the length of L_p. Erdős, Herzog, and Piranian conjectured that

$$|L_p| \leq |L_{p_0}| = 2n + O(1), \tag{1}$$

where $p_0(z) = z^n - 1$. Despite the efforts of many people, this conjecture still remains unresolved. The goal of the note is to present a new approach to the problem that allows one to show that $|L_p| \leq |L_{p_0}|$ if p is sufficiently close to p_0, and to prove the asymptotic estimate $|L_p| \leq 2n + o(n)$ as $n \to \infty$ for all monic polynomials p.

1. Introduction. In 1958 Erdős, Herzog, and Piranian ([**1**], Problem 12) asked whether the polynomial $p_0(z) = z^n - 1$ has the maximum length of the lemniscate $L = L_p = \{z \in \mathbb{C} : |p(z)| = 1\}$ among all monic polynomials $p(z) = z^n + \cdots$ of degree n. After 50 years, this conjecture still remains unresolved. The first upper bound $|L| \leq 4\pi n$ was obtained by Dolzhenko in 1960 in his thesis [**2**] and published in 1963 [**3**]. Meanwhile, in 1961, Pommerenke [**4**] published a much worse estimate $74n^2$, which became known much wider than Dolzhenko's result. Apparently unaware of Dolzhenko's work, Borwein [**5**] published the estimate $|L| \leq 8e\pi n$ in 1995. The first real improvement came in 1999 when Eremenko and Hayman [**6**] proved the conjecture for $n = 2$, showed that all critical points of the extremal polynomial must lie on the lemniscate, and obtained the estimate $|L| \leq 9.173n$ for all n. In 2007, this upper bound was superceded by the estimate $|L| \leq 2\pi n$ proved by Danchenko [**7**], which remained the best published upper bound by the time of this writing. Several more papers devoted to or motivated by the lemniscate problem were published (see [**8**], [**9**], and [**10**], for instance).

The goal of this note is to present a new approach to the problem based on an explicit formula for the length. Unfortunately, we have not been able to get a full solution either, but at least we managed to show that $|L_p|$ attains a local maximum when $p = p_0$ and to obtain an upper bound of the form $2n + o(n)$.

2. Acknowledgements. This work was done when the first author was visiting University of Wisconsin, Madison. Our special thanks go to Andreas Seeger,

---

2000 *Mathematics Subject Classification.* Primary 30C10.
Key words and phrases. Erdős–Herzog–Piranian conjecture, lemniscate length.

whose generous support made this visit possible. We are also grateful to Alexandre Eremenko, Mikhail Sodin, and Alexander Volberg for valuable discussions. The second author was partially supported by the NSF grant DMS0501067.

3. Notation. Throughout this paper, we denote by p a monic polynomial, by n its degree, by η an arbitrary root of p, by ζ an arbitrary root of p', and by ξ an arbitrary root of pp'. This notation will be used without any further comments; so $\sum_\zeta$, say, will always mean the sum over all roots of p' counted with their multiplicities, etc.

As usual, $\mathbb{C}$ and $\mathbb{R}$ stand for the sets of complex and real numbers respectively. We will also denote by D_r the disk $\{z \in \mathbb{C} : |z| < r\}$ and by T_r its boundary circle $\{z \in \mathbb{C} : |z| = r\}$. We denote by $d(F, z)$ the distance from the point $z \in \mathbb{C}$ to the set $F \subset \mathbb{C}$.

If f is a complex-valued smooth function defined on an open set $\Omega \subset \mathbb{C}$, we shall treat it as a function of complex variable $z = x + iy$ and use the complex notation for the partial derivatives

$$\partial f = f_z = \frac{1}{2}(f_x - if_y), \qquad \bar{\partial} f = f_{\bar{z}} = \frac{1}{2}(f_x + if_y),$$

and for the differential forms

$$dz = dx + idy, \qquad d\bar{z} = dx - idy.$$

In this notation,

$$d\bar{z} \wedge dz = 2i\, dx \wedge dy$$

and

$$df = f_z dz + f_{\bar{z}} d\bar{z} = \partial f\, dz + \bar{\partial} f\, d\bar{z}.$$

All other notation conventions will be introduced at the moment of their first appearance.

4. The Stokes formula. Let E be a bounded open subset of the complex plane $\mathbb{C}$. Suppose that the boundary of E consists of finitely many smooth Jordan arcs. Let ω be a differential 1-form on $\mathbb{C}$ with locally bounded coefficients that are smooth outside a finite set K, and suppose that the differential $d\omega$ (which makes sense everywhere in $\mathbb{C} \setminus K$) has locally integrable coefficients with respect to the area measure. (We shall call such forms quasismooth.) It is not hard to check that the classical Stokes formula

$$\int_{\partial E} \omega = \iint_E d\omega \tag{2}$$

remains valid for quasismooth differential 1-forms.

Now let $p = p(z)$ be a monic polynomial of a fixed degree $n \geq 2$ (the case $n = 1$ is rather trivial, so we will not consider it here). With any such polynomial, we associate the sets

$$L = L_p = \{z \in \mathbb{C} : |p(z)| = 1\}, \quad E = E_p = \{z \in \mathbb{C} : |p(z)| < 1\}.$$

Let K be a finite set containing the critical points of p, and let $s = s(z)$ be a smooth complex-valued function defined on $\mathbb{C} \setminus K$. Suppose that s coincides with the outward unit normal vector to ∂E on $\partial E \setminus K$. Then

$$|L| = \int_L \frac{dz}{is} = \frac{1}{i} \int_L \bar{s}\, dz. \tag{3}$$

If the differential 1-form $\omega = \bar{s}\,dz$ is quasismooth, then the Stokes formula yields

$$i|L| = \iint_E d\omega\,.$$

Evaluating $d\omega$ we have

$$d\omega = (\partial\bar{s}\,dz + \bar\partial\bar{s}\,d\bar{z}) \wedge dz = \bar\partial\bar{s}\,d\bar{z} \wedge dz = 2i\,\overline{\partial s}\,dx \wedge dy\,.$$

Therefore,

$$|L| = 2\iint_E \overline{\partial s}\,dx \wedge dy\,.$$

Since the quantity on the left hand side is real, we conclude that also

$$|L| = 2\iint_E \partial s\,dA = 2\,\mathrm{Re}\iint_E \partial s\,dA\,, \tag{4}$$

where A is the plane area measure.

The outward unit normal vector to ∂E is the normalized gradient of the function $\log|p|$. Computing it, we get

$$s = \frac{\overline{\varphi}}{|\varphi|} = \frac{|\varphi|}{\varphi}, \quad \text{where} \quad \varphi = \frac{p'}{p}$$

on L_p. We have a great deal of freedom extending s from L_p to the entire complex plane. The most obvious extension is given by the right hand side of the last formula, which makes sense everywhere except the zeroes and the critical points of p. This way we get a form $\omega = \bar{s}\,dz$ such that $\int_{L_p}\omega = i|L_p|$ and for every other piecewise smooth curve γ, we have $\left|\int_\gamma \omega\right| \le |\gamma|$. Yet another direct computation shows that

$$2\partial s = -\frac{|\varphi|}{\varphi}\frac{\varphi'}{\varphi} = \frac{|\varphi|}{\varphi}\Big(\sum_\eta \frac{1}{z-\eta} - \sum_\zeta \frac{1}{z-\zeta}\Big),$$

where the first sum is taken over all roots η of p and the second sum is taken over all roots ζ of p' (counted with multiplicities in both cases). It is not hard to see from this that ω is quasismooth and for every bounded open set G, one has

$$\Big|\iint_G d\omega\Big| \le \iint_G \Big(\sum_\xi \frac{1}{|z-\xi|}\Big)\,dA(z),$$

where the sum is taken over all roots ξ of pp'. Our first task will be to use this extension to show that $|L_p| \le |L_{p_0}|$ for all p sufficiently close to $p_0(z) = z^n - 1$.

5. A simple lemniscate problem.

LEMMA 1. *Let $p(z) = z^n + a_2 z^{n-2} + a_3 z^{n-3} + \cdots + a_n$ satisfy $a_n \in \mathbb{R}$ and $\max_{2\le k\le n}|a_k| = 1$. Let $\widetilde{L}_p = \{z \in \mathbb{C} : \mathrm{Re}\, p(z) = 0\}$. Then for every $r \ge 2$, one has*

$$|\widetilde{L}_p \cap D_r| \le 2nr - c_n,$$

where $c_n > 0$ depends on n only.

PROOF. Note first that the intersection $\widetilde{L}_p \cap T_\rho$ consists of at least $2n$ points for every $\rho \ge 2$. Indeed, $\mathrm{Re}\, z^n$ has n positive maxima interlaced with n negative minima of size ρ^n on T_ρ. All other terms together can contribute no more than $\rho^{n-2} + \rho^{n-3} + \cdots + 1 < \rho^n$, so the signs of $\mathrm{Re}\, p(z)$ at those maxima and minima are the same as those of $\mathrm{Re}\, z^n$. Thus, we have at least $2n$ sign changes, and therefore at least $2n$ roots for $\mathrm{Re}\, p(z)$ on T_ρ.

Now consider a big sphere S of radius R touching the complex plane at the origin (we think of the complex plane as a horizontal plane, with the sphere above it). Let τ be the central projection of the plane to the lower hemisphere of S from the center of the sphere. Then, by the well-known Poincaré formula,

$$|\tau(\widetilde{L}_p)| = \pi R \int_S \#(\mathcal{E}_\theta \cap \tau(\widetilde{L}_p))\, d\mu(\theta),$$

where μ is the surface measure on S normalized by the condition $\mu(S) = 1$, and $\mathcal{E}_\theta = \{x \in S : \langle x, \theta\rangle = 0\}$ is the great circle of S orthogonal to θ. Since every horizontal circle on the lower hemisphere lying outside $\tau(D_r)$ intersects $\tau(\widetilde{L}_p)$ in at least $2n$ points, we conclude that

$$|\tau(\widetilde{L}_p \setminus D_r)| \geq \pi R n - 2nR \arctan \frac{r}{R}.$$

Combining this inequality with the previous identity, we see that

$$|\tau(\widetilde{L}_p \cap D_r)| \leq 2nR \arctan \frac{r}{R} - \pi R \int_S [n - \#(\mathcal{E}_\theta \cap \tau(\widetilde{L}_p))]\, d\mu(\theta).$$

Since $\tau^{-1}(\mathcal{E}_\theta)$ is a line on the complex plane $\mathbb{C}$, we conclude that for almost all θ, one has

$$\#(\mathcal{E}_\theta \cap \tau(\widetilde{L}_p)) \leq n,$$

so the integrand is non-negative almost everywhere.

Now let us pass to the limit as $R \to \infty$. The left hand side tends to $|\widetilde{L}_p \cap D_r|$. The first term on the right hand side tends to $2nr$. Finally, the integral can be rewritten as

$$\int_{\mathbb{C}} [n - \#(\widetilde{L}_p \cap \Gamma_w)] \left(1 + \frac{|w|^2}{R^2}\right)^{-3/2} \frac{dA(w)}{2|w|},$$

where Γ_w is the line passing through $w \in \mathbb{C}$ in the direction perpendicular to w (this is a simple exercise in the change of variable theorem), and therefore it tends to

$$\int_{\mathbb{C}} [n - \#(\widetilde{L}_p \cap \Gamma_w)] \frac{dA(w)}{2|w|}$$

in view of the monotone convergence theorem.

Thus it remains to show that this integral is bounded away from 0 on the class of polynomials under consideration. We shall show that it is true even for the smaller integral $\int_{D_1}$.

The first observation is that for a fixed monic polynomial p the number of the intersections of $\widetilde{L}_p$ with Γ_w is a stable quantity (meaning that it does not change under small perturbations of the coefficients of p) for almost all w. This, together with the dominated convergence theorem, allows us to conclude that our integral is a continuous function of the coefficients $a_2, \ldots, a_n$ running over a compact set. Thus, it will suffice to show that this integral never vanishes.

To this end, fix p and let j be the least index for which $a_j \neq 0$. Consider the line $\Gamma(\alpha) = \{te^{i\alpha} : t \in \mathbb{R}\}$. We have

$$\operatorname{Re} p(te^{i\alpha}) = t^n \operatorname{Re} e^{in\alpha} + t^{n-j} \operatorname{Re}[a_j e^{i(n-j)\alpha}] + \cdots = q(t).$$

Note now that $\operatorname{Re}[a_j e^{i(n-j)\alpha}]$ preserves sign on intervals of length $\frac{\pi}{n-j}$ and that $\operatorname{Re} e^{in\alpha}$ changes sign on every interval whose length exceeds $\frac{\pi}{n}$. Thus, we can find α such that $\operatorname{Re}[a_j e^{i(n-j)\alpha}]$ and $\operatorname{Re} e^{in\alpha}$ are both non-zero and have the same sign.

If $q(t)$ had n real roots, every derivative of it would have only real roots. But the $(n-j)$-th derivative is a polynomial of the form $\kappa t^j + \lambda$ where $\kappa, \lambda \neq 0$ have the same sign, and such a polynomial has at least one non-real root for all $j \geq 2$. Thus $q(t)$ must have non-real roots as well. Since the existence of non-real roots is a stable property of polynomials, we conclude that for all w sufficiently close to 0 whose argument is sufficiently close to $\alpha + \frac{\pi}{2}$, the line Γ_w intersects $\widetilde{L}_p$ at strictly less than n points, which is more than enough to ensure the strict positivity of the integral in question. □

A simple rescaling argument yields the following result. Let $p(z) = z^n + a_2 z^{n-2} + a_3 z^{n-3} + \cdots + a_n$ satisfy $a_n \in \mathbb{R}$, $\max_{2\leq k\leq n} |a_k|^{1/k} = a > 0$. Let $\widetilde{L}_p = \{z \in \mathbb{C} : \operatorname{Re} p(z) = 0\}$. Then for every $r \geq 2a$, one has

$$|\widetilde{L}_p \cap D_r| \leq 2nr - c_n a$$

It is worth noting for what follows that under these assumptions, the roots ζ of p' lie in D_{2a}.

6. The first variation estimate. Suppose p is a monic polynomial that is close to p_0. Using an appropriate shift of the argument, we can ensure that the coefficient at z^{n-1} equals 0. After that is done, we can make the free term real by using an appropriate rotation. Hence we can assume without loss of generality that $p = p_0 + q$, where $q(z) = \sum_{k=2}^n a_k z^{n-k}$, the coefficients a_k are small, and $a_n \in \mathbb{R}$.

Let

$$a = \max_k |a_k|^{1/k}.$$

Fix $r \in (4a, \frac{1}{4})$ to be chosen later and consider the parts of the domains E_p and E_{p_0} lying inside and outside T_r separately. We shall start with the comparison of $|L_p \setminus D_r|$ and $|L_{p_0} \setminus D_r|$.

Using the same differential 1-form ω as before, we get

$$i|L_p \setminus D_r| + \int_{E_p \cap T_r} \omega = \iint_{E_p \setminus D_r} d\omega$$

and

$$\int_{L_{p_0} \setminus D_r} \omega + \int_{E_{p_0} \cap T_r} \omega = \iint_{E_{p_0} \setminus D_r} d\omega.$$

Let G be the symmetric difference of E_p and E_{p_0}. Since

$$\left| \int_\gamma \omega \right| \leq |\gamma|$$

for every piecewise smooth curve γ, we immediately conclude from the above formulae that

$$|L_p \setminus D_r| \leq |L_{p_0} \setminus D_r| + |G \cap T_r| + \iint_{G \setminus D_r} \sum_\xi \frac{1}{|z-\xi|} \, dA(z).$$

Now let us note that for every $\rho \in (4a, \frac{1}{4})$, the circle of radius ρ^n centered at 1 is transversal to the unit circle. Since z^n travels over that circle at constant speed, simple geometric considerations show that $||p_0(z)| - 1| \geq c\rho^{n-1} \min_j |z - z_j|$ if $z \in T_\rho$, where the z_j are the $2n$ solutions of the system $|p(z_0)| = 1$, $|z| = \rho$, and $c > 0$ is some absolute constant. Since $|q| \leq 2a^2\rho^{n-2}$ on T_ρ, we immediately conclude that $G \cap T_\rho$ is contained in the union of $2n$ arcs of length $Ca^2\rho^{-1}$, i.e., that

$$|G \cap T_\rho| \le Ca^2\rho^{-1} \tag{5}$$

with some absolute positive constant C depending on n only. In particular,

$$|G \cap T_r| \le Ca^2 r^{-1}.$$

To estimate the double integral over $G \setminus D_r$, note that since p_0' does not vanish outside $D_{1/4}$, and since $|q| \le 2^{n+1}a^2$ in D_2, we can use the regular perturbation theory to conclude that $G_1 = G \setminus D_{1/4}$ is contained in the Ca^2-neighborhood of L_{p_0}, so $A(G_1) \le Ca^2$. Also, when a is small enough, the distance from G_1 to every root ξ of pp' is bounded from below by some constant c depending on n only. Thus,

$$\iint_{G_1} \sum_\xi \frac{1}{|z-\xi|}\, dA(z) \le Ca^2.$$

The other part $G_2 = G \cap D_{1/4} \setminus D_r$ is more interesting. If η is a root of p, then we still have $d(G_2, \eta) \ge c$, so the integral of $\frac{1}{|z-\eta|}$ over G_2 does not exceed $CA(G_2)$. In view of (5), we have

$$A(G_2) \le Ca^2 \int_r^{1/4} \frac{d\rho}{\rho} \le Ca^2 \log\frac{1}{r}.$$

If ζ is a root of p', then $|\zeta| \le 2a$, and therefore

$$\frac{1}{|z-\zeta|} \le \frac{2}{|z|}$$

on G_2. Thus, the integral of $\frac{1}{|z-\zeta|}$ over G_2 is bounded by

$$Ca^2 \int_r^\infty \frac{d\rho}{\rho^2} = Ca^2 r^{-1}.$$

Bringing all these estimates together, we finally conclude that

$$|L_p \setminus D_r| \le |L_{p_0} \setminus D_r| + Ca^2 r^{-1}.$$

Now it is time to compare $|L_p \cap D_r|$ with $|L_{p_0} \cap D_r|$. For the latter, we will use the trivial lower bound $|L_{p_0} \cap D_r| \ge 2nr$. To estimate the former, consider the polynomial $f(z) = 1 + p(z) = z^n + q(z)$, the lemniscate

$$\widetilde{L} = \{z : \operatorname{Re} f = 0\},$$

and the region $F_p = \{z : \operatorname{Re} f(z) > 0\}$. Obviously, $E_p \subset F_p$. The key observation is that the difference $H = F_p \setminus E_p$ is rather small in the disk D_r. Indeed, since the inequality $|p| < 1$ is equivalent to $2\operatorname{Re} f - |f|^2 > 0$ and $|f| \le 2r^n$ in D_r, we conclude that $H \cap D_r \subset \{z : |\operatorname{Re} f(z)| \le 2r^{2n}\}$. Arguing as above, we see that

$$|L_p \cap D_r| \le |\widetilde{L} \cap D_r| + |H \cap T_r| + \iint_H \Big(\sum_\xi \frac{1}{|z-\xi|}\Big)\, dA(z).$$

The length $|H \cap T_r|$ can be easily estimated from above by Cr^{n+1}. Indeed, the condition $|\operatorname{Re} f| \le 2r^{2n}$ implies $|\operatorname{Re} z^n| \le 2(r^{2n} + a^2 r^{n-2}) \le \frac{r^n}{2}$. Hence, the speed with which Re f changes when z travels over T_r with unit speed across H is at least

$$\frac{nr^{n-1}}{2}-\sum_{k=2}^{n}(n-k)a^k r^{n-k-1}\geq\frac{nr^{n-1}}{4},$$

and the above estimate immediately follows.

To estimate the area integral, note that $\iint_H \frac{dA(z)}{|z-\xi|}$ has the geometric meaning of the average (over the angle the lines make with the positive axis) length of the cross-sections of H by the lines passing through ξ. These cross-sections can be easily controlled by the Remez theorem (see [**12**], Theorem 5.1.1), which asserts that the measure of the set where the polynomial $at^n+\cdots$ with real coefficients is less than ε in absolute value is at most $4(\varepsilon/|a|)^{1/n}$. Applying this estimate to the polynomial Re $f(\xi+e^{i\alpha}t)$, we conclude that for every ξ,

$$\iint_{H\cap D_r}\frac{dA(z)}{|z-\xi|}\leq 4\cdot 2^{1/n}r^2\int_0^\pi\frac{d\alpha}{|\cos n\alpha|^{1/n}}=Cr^2\,.$$

Finally, according to Lemma 1,

$$|\widetilde{L}\cap D_r|\leq 2nr-ca\,.$$

Bringing all the estimates together, we obtain

$$|L_p|\leq|L_{p_0}|-ca+C(r^2+a^2r^{-1}),$$

and it remains to put $r=a^{2/3}$ and choose a sufficiently small to get the desired result.

7. The simplest upper bound for the lemniscate length. We start with using the above extension to show that

$$|L|\leq 2\pi(2n-1)\,.$$

We have

$$|L|=-\iint_E\frac{|\varphi|}{\varphi}\frac{\varphi'}{\varphi}\,dA\leq\iint_E\left|\frac{\varphi'}{\varphi}\right|dA\,.$$

Recall now that

$$-\frac{\varphi'}{\varphi}=\sum_\eta\frac{1}{z-\eta}-\sum_\zeta\frac{1}{z-\zeta}.$$

Thus

$$|L|\leq\sum_\xi\iint_E\frac{1}{|z-\xi|}\,dA\,.$$

Since the logarithmic capacity of E is equal to 1, Pólya's theorem (see [**11**], Theorem 5.3.5) implies that its area does not exceed π. Thus,

$$|L|\leq(2n-1)\iint_{\mathbb{D}}\frac{dA}{|z|}=2\pi(2n-1)\,.$$

Note, by the way, that if p has multiple roots, then every root of p of multiplicity m appears m times in $\sum_\eta$ and $m-1$ times in $\sum_\zeta$. If we take this cancellation into account, then the last estimate can be improved to $2\pi(2k-1)$, where k is the number of distinct roots of p.

8. An improved upper bound. To obtain a better estimate, we consider a different extension of s, which differs from the one we considered before by the factor $|p|$ (this factor is identically 1 on L). So, we define

$$s = |p|\frac{|\varphi|}{\varphi} = \frac{p|p'|}{p'} = \frac{|p'|}{\varphi}.$$

In this case

$$2\partial s = \frac{p|p'|}{p'}\Big(\frac{2p'}{p} - \frac{p''}{p'}\Big) = 2|p'| - \frac{|p\varphi|}{\varphi}\psi = |p'| - \frac{|p\varphi|}{\varphi}\frac{\varphi'}{\varphi},$$

where

$$\psi := \frac{p''}{p'} = \sum_{\zeta} \frac{1}{z-\zeta}.$$

The lemniscate length can now be evaluated as follows:

$$|L| = 2\iint_E |p'|\,dA - \iint_E \frac{|p\varphi|}{\varphi}\psi\,dA. \tag{6}$$

Since p maps E onto the unit disk and covers each point of the unit disk n times, we have

$$\iint_E |p'|^2\,dA = \pi n.$$

The Cauchy inequality yields

$$\iint_E |p'|\,dA \le \pi\sqrt{n}.$$

On the other hand,

$$\Big|\iint_E \frac{|p\varphi|}{\varphi}\psi\,dA\Big| \le \iint_E |\psi|\,dA \le \sum_{\zeta}\iint_E \frac{dA}{|z-\zeta|} \le (n-1)\iint_{\mathbb{D}} \frac{1}{|z|}\,dA = 2\pi(n-1).$$

Combining these estimates, we obtain the upper bound

$$|L| \le 2\pi(n-1+\sqrt{n}),$$

which is only marginally worse than Danchenko's estimate $2\pi n$.

9. Asymptotic estimate. In this section we obtain an asymptotic estimate of the lemniscate length of a monic polynomial p of degree n as n approaches infinity.

Using the same extension as in the previous section, we get the inequality

$$|L| \le \pi\sqrt{n} + J,$$

where

$$J = -\operatorname{Re}\iint_E \frac{|\varphi p|}{\varphi}\frac{\varphi'}{\varphi}\,dA.$$

In this section we will estimate the integral J more accurately to get the best possible asymptotic estimate for $|L|$.

Take $\delta \in (0, \frac{1}{4})$. Let $E_\delta \subset E$ be the set of all points $z \in E$ satisfying

$$\Big|\frac{\varphi'}{\varphi}\Big| \ge 2\delta n + \sum_{\xi}\frac{8\delta}{|z-\xi|} \tag{7}$$

and

$$|z-\xi| \ge \frac{2\delta}{\sqrt{n}} \qquad \text{for all } \xi. \tag{8}$$

We start with estimating the integral over $E \setminus E_\delta$. The set $E \setminus E_\delta$ is the union of the set B_1 where condition (7) is broken and the set B_2 where condition (8) is broken. Since the area of the latter set is at most $4\pi\delta^2$, we have

$$\text{(9)} \qquad \iint_{B_1} \left|\frac{\varphi'}{\varphi}\right| dA \leq 2\delta n A(B_1) + \sum_{\xi} 8\delta \iint_{B_1} \frac{dA(z)}{|z-\xi|} \leq 2\pi n\delta + 16\pi(2n-1)\delta,$$

$$\text{(10)} \qquad \iint_{B_2} \left|\frac{\varphi'}{\varphi}\right| dA \leq \sum_{\xi} 4\delta \iint_{B_2} \frac{dA(z)}{|z-\xi|} \leq 4\pi(2n-1)\delta\,,$$

so

$$\iint_{E\setminus E_\delta} \left|\frac{\varphi'}{\varphi}\right| dA \leq 2\pi n\delta + 16\pi(2n-1)\delta + 4\pi\delta(2n-1) \leq 42\pi\delta n$$

and hence,

$$J \leq 42\pi\delta n + \iint_{E_\delta} |p'|\, dA + J_\delta,$$

where

$$J_\delta = -\operatorname{Re} \iint_{E_\delta} \frac{|\varphi p|}{\varphi} \psi\, dA\,.$$

Using the estimate

$$\iint_{E} |p'|\, dA \leq \pi\sqrt{n}$$

once more, we see that it is enough to estimate J_δ.

We rewrite J_δ using the explicit expression for ψ:

$$\text{(11)} \qquad J_\delta = \sum_{\zeta} \operatorname{Re} \iint_{\delta} \frac{|p\varphi|}{\varphi} \frac{1}{\zeta - z}\, dA\,.$$

Now we partition the complex plane into equal squares with side length

$$\ell = \frac{\delta^2}{\sqrt{n}}$$

and define the set F as the minimal union of these squares that contains E_δ; that is, F includes all the squares that intersect E_δ. For each square Q in F, let $\widetilde{Q}$ be the twice larger square with the same center. Finally, let

$$\widetilde{F} = \bigcup_{Q:Q\subset F} \widetilde{Q}\,.$$

Note that for every root ξ of pp',

$$\text{(12)} \qquad d(\widetilde{F}, \xi) > \frac{2\delta}{\sqrt{n}} - \frac{4\delta^2}{\sqrt{n}} > \frac{\delta}{\sqrt{n}}\,.$$

Since, for every function Φ,

$$\iint_{E_\delta} \operatorname{Re}\left(|p|\,\Phi\right) dA \leq \iint_{E_\delta} \operatorname{Re}_+ \Phi\, dA \leq \iint_{F} \operatorname{Re}_+ \Phi\, dA\,,$$

we obtain

$$\text{(13)} \qquad J_\delta \leq \sum_{\zeta} \iint_{F} \operatorname{Re}_+ \Big(\frac{|\varphi|}{\varphi} \frac{1}{\zeta - z}\Big)\, dA(z)\,.$$

Now let

$$I_\zeta(Q) := \iint_Q \mathrm{Re}_+ \frac{|\varphi|}{(\zeta - z)\varphi}\, dA(z)\,, \tag{14}$$

where Q is a square of the set F and ζ is a root of p'. Let w be any point in Q. Then

$$I_\zeta(Q) \le \iint_Q \mathrm{Re}_+ \Big(\frac{|\varphi|}{(\zeta - z)\varphi} - \frac{|\varphi|}{(\zeta - w)\varphi}\Big)\, dA(z) + \frac{1}{|\zeta - w|}\iint_Q \mathrm{Re}_+\,\theta\, dA\,,$$

where

$$\theta = \frac{|(\zeta - w)|}{(\zeta - w)}\,\frac{|\varphi|}{\varphi}\,.$$

Denote the first term by $I_\zeta^{(1)}(Q)$ and the second, by $I_\zeta^{(2)}(Q)$.

We have

$$I_\zeta^{(1)}(Q) \le \iint_Q \frac{|w - z|}{|\zeta - z|\,|\zeta - w|}\, dA(z) \le \frac{2\ell}{|\zeta - w|}\iint_Q \frac{1}{|z - \zeta|}\, dA\,.$$

Since $\ell = \delta^2/\sqrt{n}$ and $|w - \zeta| \ge \delta/\sqrt{n}$ (see (12)), we have

$$I_\zeta^{(1)}(Q) \le 2\delta \iint_Q \frac{1}{|\zeta - z|}\, dA(z)\,. \tag{15}$$

To estimate $I_\zeta^{(2)}(Q)$, note that

$$\mathrm{Re}_+\,\theta = \frac{1}{\pi} + \sum_{k \ne 0} a_k\, \theta^k\,, \tag{16}$$

where a_k are the Fourier coefficients of the function $\mathrm{Re}_+\, z$ on the unit circle. To estimate the sum over $k \ne 0$, we need

LEMMA 2. *Let Q be a square with side length* 1, *and let u be a real harmonic function in the twice larger square $\widetilde{Q}$ with the same center, such that*

$$|\bar\partial u| > R$$

everywhere in $\widetilde{Q}$. Then

$$\Big|\iint_Q e^{iu}\, dA\Big| \le \frac{4}{R}\,.$$

PROOF. The function $f = 1/\partial u$ is analytic in the square $\widetilde{Q}$ and $|f| < 1/R$. By the Cauchy theorem $|f'| \le 2/R$ in Q. Since u is a real function, we have $\bar\partial u = 1/\bar f$, and the Stokes formula yields

$$\int_{\partial Q} \bar f e^{iu} dz = \iint_Q \bar\partial(\bar f e^{iu})\, d\bar z \wedge dz = \iint_Q (\bar f' e^{iu} + i e^{iu})\, d\bar z \wedge dz\,.$$

Thus

$$\Big|\iint_Q e^{iu}\, dA\Big| \le \frac{1}{2}\Big[\int_{\partial Q} |\bar f|\,|dz| + 2\iint_Q |\bar f'|\, dA\Big] \le \frac{4}{R}\,.$$

We are done. □

A simple scaling argument shows that if the side length of Q is not 1 but ℓ, then the estimate changes to

$$\Big|\iint_Q e^{iu}\, dA\Big| \le \frac{4}{R\ell} A(Q)\,.$$

We apply this lemma to estimate

$$\iint_Q \theta^k \, dA .$$

Take $w \in Q$ to satisfy

$$\frac{\ell^2}{|\zeta - w|} = \frac{A(Q)}{|\zeta - w|} = \iint_Q \frac{1}{|\zeta - z|}\, dA .$$

By Lemma 2,

$$\frac{1}{|\zeta - w|} \iint_Q \theta^k \, dA \le \frac{4}{R\ell |k|} \iint_Q \frac{1}{|\zeta - z|}\, dA ,$$

where

$$R = \inf_{z \in \widetilde{Q}} |\partial u| .$$

It remains to estimate R from below. We have

$$|\partial u| = \frac{1}{2} \cdot \left| \frac{\varphi'}{\varphi} \right| .$$

Take $z_0 \in E_\delta \cap Q$. According to (7) we have

$$|\partial u(z_0)| \ge \delta n + \sum_\xi \frac{4\delta}{|\xi - z_0|} .$$

Since $d(\widetilde{Q}, \zeta) \ge \delta/\sqrt{n}$, $|z - z_0| \le 4\ell$, and $\delta \le \frac{1}{4}$, we get

$$\frac{1}{2} \le \frac{|\xi - z_0|}{|\xi - z|} \le 2 .$$

Note that

$$\left(\frac{\varphi'}{\varphi} \right)' = \sum_\eta \frac{1}{(z - \eta)^2} - \sum_\zeta \frac{1}{(z - \zeta)^2} .$$

Hence

$$\left| \frac{1}{2} \left(\frac{\varphi'}{\varphi} \right)' \right| \le \sum_\xi \frac{1}{2|z - \xi|^2} \le \frac{\sqrt{n}}{\delta} \sum_\xi \frac{1}{2|z - \xi|} \le \frac{\sqrt{n}}{\delta} \sum_\xi \frac{1}{|z_0 - \xi|} .$$

Finally, for $z \in \widetilde{Q}$ we have

$$|\partial u(z)| \ge |\partial u(z_0)| - 4\ell \max_{\widetilde{Q}} \left| \frac{1}{2} \left(\frac{\varphi'}{\varphi} \right)' \right| \ge \delta n,$$

and therefore, in view of (16),

$$I_\zeta^{(2)}(Q) \le \frac{4a}{\delta^3 \sqrt{n}} \iint_Q \frac{1}{|\zeta - z|}\, dA + \frac{1}{\pi} \iint_Q \frac{1}{|\zeta - z|}\, dA , \tag{17}$$

where

$$a = \sum_{k \ne 0} \frac{|a_k|}{|k|} .$$

Summing the inequalities (17) and (15) over all squares $Q \in \widetilde{F}$ and then over ζ we get

$$J_\delta \le \left(\frac{1}{\pi} + 2\delta + \frac{4a}{\delta^3 \sqrt{n}} \right) \iint_F \sum_\zeta \frac{1}{|\zeta - z|}\, dA . \tag{18}$$

To finish the estimate we need to estimate the logarithmic capacity of the set F. To this end, we estimate the logarithmic derivative φ of p first. We have

$$|\varphi| \le \sum_{\eta} \frac{1}{|\eta - z|} \le \frac{n\sqrt{n}}{\delta}$$

on F (see (12)). Since $d(E, z) \le 2\ell$ for every point $z \in F$, we have

$$\log|p| \le 2\ell \max_F |\varphi| \le \frac{2\ell n\sqrt{n}}{\delta} = 2\delta n\,.$$

Thus,

$$|p| \le e^{2\delta n}\,.$$

The last inequality means that the logarithmic capacity of F is at most $e^{2\delta}$, and therefore its area is at most $\pi e^{4\delta}$. Thus,

$$|L| \le 42\pi\delta n + 2\pi\sqrt{n} + e^{2\delta}\left(\frac{1}{\pi} + 2\delta + \frac{4a}{\delta^3\sqrt{n}}\right)2\pi(n-1)$$

for every $\delta > 0$. Now it is clear that the optimal choice of δ is about $n^{-1/8}$, which results in the estimate

$$|L| \le 2n + O(n^{7/8})\,.$$

We have no doubt that the power $7/8$ can be substantially improved, although to bring it below $1/2$ seems quite a challenging problem.

References

[1] P. Erdős, F. Herzog, and G. Piranian, *Metric properties of polynomials*, J. Anal. Math. 6 (1958), 125–148. Also available at `http://www.math-inst.hu/~p_erdos/1958-05.pdf`

[2] E. P. Dolzhenko, *Differentiability properties of functions and certain questions of approximation theory*, Ph.D. (Candidate of Sciences degree) Thesis, Moscow State University, Moscow 1960. (Russian)

[3] ______, *Some metric properties of algebraic hypersurfaces*, Izv. Akad. Nauk SSSR Ser. Mat. 27:2 (1963), 241–252. (Russian)

[4] Ch. Pommerenke, *On metric properties of complex polynomials*, Michigan Math. J. 8:2 (1961), 97–115.

[5] P. Borwein, *The arc length of the lemniscate* $|p(z)| = 1$, Proc. Amer. Math. Soc. 123:3 (1995), 797–799.

[6] A. Eremenko and W. Hayman, *On the length of lemniscates*, Michigan Math. J. 46:2 (1999), 409–415. Also available at `http://arxiv.org/abs/0805.2295`

[7] V. I. Danchenko, *The lengths of lemniscates. Variations of rational functions*, Mat. Sb. 198 (2007), no. 8, 51–58; English transl. in Sb. Math. 198 (2007), no. 7-8, 1111–1117.

[8] O. S. Kuznetsova and V. G. Tkachev, *Length functions of lemniscates*, Manuscripta Math. 112 (2003), no. 4, 519–538. Also available at `http://arxiv.org/abs/math/0306327`

[9] G. A. Barsegian, *Gamma-lines of polynomials and a problem by Erdős–Herzog–Piranian*, Topics in analysis and its applications, 119–122, NATO Sci. Ser. II Math. Phys. Chem., 147, Kluwer Acad. Publ., Dordrecht, 2004.

[10] C. Wang and L. Peng, *The arc length of the lemniscate* $|w^n + c| = 1$, Rocky Mountain J. Math. 36:1 (2006), 337–347.

[11] T. Ransford, *Potential theory in the complex plane*, London Mathematical Society Student Texts, 28. Cambridge University Press, Cambridge, 1995.

[12] P. Borwein and T. Erdélyi, *Polynomials and polynomial inequalities*, Graduate Texts in Mathematics, 161. Springer-Verlag, New York, 1995.

Department of Mathematics, University of Wisconsin-Madison, 480 Lincoln Drive, Madison, WI 53706, USA

E-mail address: `nazarov@math.wisc.edu`

Amer. Math. Soc. Transl.
(2) Vol. **226**, 2009

Compact Approximation by Bounded Functions and Functions Continuous up to the Boundary

Paul M. Gauthier and Mark S. Melnikov

ABSTRACT. A characterization is given for domains on which each holomorphic function can be approximated by bounded ones. An analogous result is obtained for approximation by functions which moreover can be continuously extended to the boundary. Related questions in potential theory are also considered.

1. $H^\infty(\Omega)$ approximation

As usual, we denote by $\mathbb{C}$ the open complex plane and by $\overline{\mathbb{C}}$ the closed complex plane $\mathbb{C} \cup \{\infty\}$. For a subset $E \subset \overline{\mathbb{C}}$, we denote by ∂E the boundary of E in $\overline{\mathbb{C}}$. The analytic capacity and the continuous analytic capacity (see [**2**]) of E are denoted respectively by $\gamma(E)$ and $\alpha(E)$. For an open subset $\Omega \subset \overline{\mathbb{C}}$, we denote by $H(\Omega)$ the space of holomorphic functions on Ω endowed with the topology of uniform convergence on compact subsets. As usual, $H^\infty(\Omega)$ denotes the subspace of bounded holomorphic functions on Ω, and $A(\Omega)$ denotes the subspace of holomorphic functions on Ω which continuously extend to $\overline{\Omega}$.

The need for a criterion of the denseness of $H^\infty(\Omega)$ in $H(\Omega)$ appeared first in Corollary 3.7 in [**3**], where only special cases were obtained. Raymond Mortini posed this problem during the 2007 function theory workshop in Oberwolfach; published in [**5**]. In this note, we give a characterization of domains Ω for which $H^\infty(\Omega)$ is dense in $H(\Omega)$. Our characterization is employed in [**4**]. We also give a characterization for density of the space $A(\Omega)$ and consider analogous questions for superharmonic and harmonic functions.

THEOREM 1. *Let Ω be an open subset of the closed complex plane $\overline{\mathbb{C}}$. The following conditons are equivalent*:

(i) *$H^\infty(\Omega)$ is dense in $H(\Omega)$*;

2000 *Mathematics Subject Classification.* Primary 30E10.

Key words and phrases. Analytic capacity, complex approximation.

This research was done while the first author was visiting the Centre de Recerca Matemàtica in Bellaterra, Spain, and was supported by MEC (Spain) and NSERC (Canada). The second author was supported by grants nos. MTM2004-00519 and 2001SGR00431.

(ii) *for each open set* $U \subset \overline{\mathbb{C}}$, *with* $\partial U \subset \Omega$,

(1) $$\text{if } U \setminus \Omega \neq \emptyset, \text{ then } \gamma(U \setminus \Omega) > 0;$$

(iii) *for each open set* $U \subset \overline{\mathbb{C}}$, *condition* (1) *holds.*

PROOF. Without loss of generality, we shall assume that $\infty \in \Omega$.

Suppose (iii) is not satisfied. Then, there is an open set U such that $U \setminus \Omega \neq \emptyset$, but $\gamma(U \setminus \Omega) = 0$. Since $U \setminus \Omega$ is closed in U and the analytic capacity of any continuum is positive, it follows that $U \setminus \Omega$ is totally disconnected, and so there is a bounded open set $V \subset U$ such that $V \setminus \Omega \neq \emptyset$ and $\partial V \subset \Omega$. Of course, $\gamma(V \setminus \Omega) = 0$, so (ii) fails. We have shown that (ii) $\Longrightarrow$ (iii) and obviously (iii) $\Longrightarrow$ (ii).

Now, suppose (ii) fails, but (i) holds. Then, there is a V as in the previous paragraph. Choose a point $a \in V \setminus \Omega$ and consider the function $(z-a)^{-1}$. By assumption, for each n, there is a function $g_n \in H^\infty(\Omega)$ such that

$$\max_{z \in \partial V} |g_n(z) - (z-a)^{-1}| < \frac{1}{n}.$$

Thus,

$$\max_{z \in \partial V} |(z-a)g_n(z) - 1| < \frac{\delta}{n},$$

where δ is the diameter of V. Since $\gamma(V \setminus \Omega) = 0$, each function g_n extends to a holomorphic function on V, which we continue to denote by g_n. By the maximum principle,

$$\max_{z \in V} |(z-a)g_n(z) - 1| < \frac{\delta}{n},$$

and in particular, for $z = a$, we have $1 < \delta/n$. Of course, for $n > \delta$ this is a contradiction. Thus, (i) $\Longrightarrow$ (ii).

It remains to show that (iii) $\Longrightarrow$ (i). Suppose Ω satisfies (iii) and f is holomorphic on Ω. We wish to approximate f on an arbitrary compact subset $K \subset \Omega$. We may assume that $\infty \in K$ and K has only finitely many complementary components, V_j, $j = 1, \ldots, n$, each of which meets the complement of Ω. For every $j = 1, \ldots, n$, fix $a_j \in V_j \setminus \Omega$. Each V_j is bounded and (of course) each a_j is finite. The rational functions with poles at the points a_j are dense in the holomorphic functions on K. Thus, we may assume that f itself is such a rational function. From the partial fractions decomposition, $f = \sum f_j$, where f_j is a rational function whose only pole is at a_j. Thus, it is sufficient to approximate each f_j on $\overline{\mathbb{C}} \setminus V_j$ by a function $\phi_j \in H^\infty(V_j \setminus E_j)$, where $E_j = V_j \setminus \Omega$. To approximate f_j it is sufficient to approximate $(z - a_j)^{-1}$.

Fix j and simplify the notation by setting $a = a_j$, $V = V_j$, $E = E_j$. It is sufficient to show that the function $(z-a)^{-1}$ can be uniformly approximated on $\overline{\mathbb{C}} \setminus V$ by a function $\phi \in H^\infty(\overline{\mathbb{C}} \setminus E)$.

Let R be the distance of E to the complement of V. For $0 < r < R$, set $E_r = \overline{B}(a,r) \cap E \supset B(a,r) \setminus \Omega$ and $\gamma_r = \gamma(E_r)$. Then $E_r \subset V$ and $\gamma_r > 0$ by condition (iii), since $a \in B(a,r) \setminus \Omega$. From the definition of capacity, there is a function ϕ_r which is holomorphic and bounded by 1 outside of a compact subset of E_r, takes the value 0 at ∞, and satisfies $|\phi_r'(\infty)| > \gamma(E_r)/2$. Then, in a neighborhood of ∞,

$$\phi_r(z) = \frac{\phi_r'(\infty)}{z-a} + \sum_{k=2}^{\infty} \frac{a_k}{(z-a)^k}.$$

The following estimate is elementary (see [**2**, p. 201]). For $k \geq 2$,

$$|a_k| \leq ek\gamma_r r^{k-1}. \tag{2}$$

From (2), we have

$$\left|\frac{1}{z-a} - \frac{\phi_r(z)}{\phi_r'(\infty)}\right| \leq \sum_{k=2}^{\infty} \frac{2|a_k|}{\gamma_r|z-a|^k} \leq \frac{2e}{R}\sum_{k=2}^{\infty} k\left(\frac{r}{R}\right)^{k-1} = \epsilon(r),$$

where

$$\epsilon(r) = \frac{2e}{R}\left(\frac{1}{(1-r/R)^2} - 1\right).$$

As $r \to 0$ we have that $\epsilon(r) \to 0$. Thus, (iii) $\Longrightarrow$ (i).

This completes the proof of the theorem. □

2. $A(\Omega)$ approximation

For an open set $\Omega \subset \overline{\mathbb{C}}$, let $A(\Omega)$ denote the class of functions holomorphic in Ω and having continuous extensions to $\overline{\Omega}$. We consider the analogous problem of characterizing those open sets Ω for which $A(\Omega)$ is dense in $H(\Omega)$, in the topology of uniform convergence on compact subsets.

THEOREM 2. *Let Ω be an open subset of the closed complex plane $\overline{\mathbb{C}}$. The following conditions are equivalent*:

(i) *$A(\Omega)$ is dense in $H(\Omega)$*;

(ii) *for each open set $U \subset \overline{\mathbb{C}}$, with $\partial U \subset \Omega$,*

$$\text{if } U \setminus \Omega \neq \emptyset, \text{ then } \alpha(U \setminus \Omega) > 0. \tag{3}$$

PROOF. In considering approximation by bounded functions, we assumed that $\infty \in \Omega$. Here, we shall make the opposite normalization, namely, we assume that $\infty \notin \Omega$.

The proof that (i) $\Longrightarrow$ (ii) is an obvious variant of the corresponding proof in the previous theorem.

The proof that (ii) $\Longrightarrow$ (i) is also similar to the corresponding proof in the previous theorem. The problem can again be reduced to the following. Suppose E is a compact set with the property that if U is open and $E \cap U \neq \emptyset$, then $\alpha(E \cap U) > 0$. It is sufficient to show that, if V is an open neighborhood of E, then there is a point $a \in E$ such that each rational function, whose only pole is at a, can be approximated uniformly on $\overline{\mathbb{C}} \setminus V$ by a function $\phi \in A(\overline{\mathbb{C}} \setminus E)$. In the proof of the previous theorem, we were able to take an arbitrary point $a \in E$. Here, we must be more careful.

Since $\alpha(E \cap V) > 0$, and E is compact, we may choose a point $a \in E$ such that $\alpha_r > 0$, for each $r > 0$, where $E_r = \overline{B}(a,r) \cap E$ and $\alpha_r = \alpha(E_r)$. From the definition of continuous analytic capacity, there is a continuous function ϕ_r bounded by 1, which is holomorphic outside a compact subset of E_r, takes the value 0 at ∞, and satisfies $|\phi_r'(\infty)| > \alpha(E_r)/2$. The rest of the proof that (ii) $\Longrightarrow$ (i) is the same as the proof that (iii) $\Longrightarrow$ (i) in the previous theorem. □

3. Related questions in potential theory

The following result, recently obtained by Sylvain Roy, is a potential-theoretic analog of our results. Let $SH(\Omega)$ denote the class of superharmonic functions on an open set Ω in $\mathbb{R}^d$. For a compact subset $K \subset \Omega$, let ω_i, $i = 1, \ldots, m$, denote the bounded components of $\mathbb{R}^d \setminus K$ which are not contained in Ω and let ω_{m+1} be the unbounded component of $\mathbb{R}^d \setminus K$.

The following is a portion of Proposition 6.3 of Syvain Roy in [**6**].

THEOREM 3 (Roy [**6**]). *Let Ω be a Greenian open subset of $\mathbb{R}^d$ $(d \geq 2)$ and let K be a compact subset of Ω. The following conditions are equivalent:*

(i) *for each $u \in SH(\Omega)$, there exists a lower bounded function $v \in SH(\Omega)$ such that $v = u$ on K;*

(ii) *$\omega_i \setminus \Omega$ is non-polar, for $i = 1, \ldots, m$ if $d \geq 3$, and for $i = 1, \ldots, m+1$ if $d = 2$.*

We remark that it is natural to restrict our attention to Greenian open sets since it is precisely such open sets which admit non-constant lower bounded superharmonic functions. Let $c(E)$ denote the capacity of a subset $E \subset \mathbb{R}^d$ and let $\overline{\mathbb{R}}^d$ be the one-point compactification of $\mathbb{R}^d$. In the terminology of the present paper the preceding theorem of Roy may be reformulated as follows.

THEOREM 4. *Let Ω be a Greenian open subset of $\mathbb{R}^d$ $(d \geq 2)$. The following conditions are equivalent:*

(i) *for each $u \in SH(\Omega)$, there is a decreasing sequence $\{v_n\}_{n \in \mathbb{N}}$ of lower bounded subharmonic functions in $SH(\Omega)$, which is eventually equal to u on each compact subset of Ω;*

(ii) *for each bounded open set U in $\mathbb{R}^d$ with $\partial U \subset \Omega$, and also each open set U in $\overline{\mathbb{R}}^d$ with $\partial U \subset \Omega$ if $d = 2$,*

$$\text{if } U \setminus \Omega \neq \emptyset, \text{ then } c(U \setminus \Omega) > 0. \tag{4}$$

PROOF. Suppose first that condition (ii) holds. Let K be a compact subset of Ω. Let ω_i, $i = 1, \ldots, m$, denote the bounded components of $\mathbb{R}^d \setminus K$ which are not contained in Ω, and let ω_{m+1} be the unbounded component of $\mathbb{R}^d \setminus K$. Then, $\omega_i \setminus \Omega \neq \emptyset$, for $i = 1, \ldots, m$ if $d \geq 3$, and for $i = 1, \ldots, m+1$ if $d = 2$. Thus, by condition (ii), $c(\omega_i \setminus \Omega) > 0$, for $i = 1, \ldots, m$ if $d \geq 3$, and for $i = 1, \ldots, m+1$ if $d = 2$. Consequently, condition (ii) of Theorem 3 is satisfied. By Theorem 3, we have a lower bounded superharmonic function v_K on Ω which agrees with u on K. Let $K_j \nearrow \Omega$ be an exhaustion of Ω by compact sets. Setting

$$v_j = \min\{v_{K_1}, v_{K_2}, \ldots, v_{K_j}\},$$

we have (i).

Now suppose tha condition (i) holds. Let U be any bounded open set in $\mathbb{R}^d$ for which $\partial U \subset \Omega$ and $U \setminus \Omega \neq \emptyset$. Let $K = \partial U$ and $\omega = U$. Then ω is a bounded component of $\mathbb{R}^d \setminus K$ which is not contained in Ω. Since we suppose that (i) holds, we have *a fortiori* condition (i) of Theorem 3. Thus, by Theorem 3, $\omega \setminus \Omega$ is not polar. Hence, $c(U \setminus \Omega) > 0$. Similarly, if $d = 2$ and U is an open set in $\overline{\mathbb{R}}^d$ with $\partial U \subset \Omega$, it follows from Theorem 3 that $c(U \setminus \Omega) > 0$. Thus, we have (ii). □

Let us turn to the analogous problem for harmonic functions. For an open subset $\Omega \subset \mathbb{R}^d$ (and also an open subset of $\overline{\mathbb{R}}^2 = \overline{\mathbb{C}}$, when $d = 2$), we denote by $h(\Omega)$

the space of harmonic functions on Ω endowed with the topology of uniform convergence on compact subsets. Moreover, $h^\infty(\Omega)$ denotes the subspace of bounded harmonic functions on Ω

THEOREM 5. *Let Ω be a Greenian open subset of $\mathbb{R}^d$ $(d \geq 2)$. Consider the following conditions:*

(i) *$h^\infty(\Omega)$ is dense in $h(\Omega)$;*

(ii) *for each bounded open set U in $\mathbb{R}^d$ with $\partial U \subset \Omega$, and also each open set U in $\overline{\mathbb{R}}^2$ with $\partial U \subset \Omega$ if $d = 2$,*

$$\text{if } U \setminus \Omega \neq \emptyset, \text{ then } c(U \setminus \Omega) > 0. \tag{5}$$

Condition (i) *implies condition* (ii).

PROOF. Suppose (ii) fails, but (i) holds. Suppose first that $d \geq 3$. Then there is a bounded open set U with $\partial U \subset \Omega$ such that $U \setminus \Omega \neq \emptyset$ but $c(U \setminus \Omega) = 0$. Choose a point $a \in U \setminus \Omega$. Let K be a compact neighborhood of ∂U contained in Ω. By assumption, there is a sequence of functions $g_n \in h^\infty(\Omega)$ such that

$$\max_{x \in K} \left| g_n(x) - \|x - a\|^{2-n} \right| \to 0, \quad n \to \infty.$$

Since $c(U \setminus \Omega) = 0$, each function g_n extends to a harmonic function on U, which we continue to denote by g_n. By the maximum principle, $\{g_n\}$ is a uniformly Cauchy sequence on $\overline{U}$ which thus converges uniformly to a function $g \in C(\overline{U}) \cap h^\infty(U)$. Now the sequence $\{g_n(x)\}$ converges uniformly to $\|x - a\|^{2-n}$ on K. Choose a point $b \in K$ such that $\|b - a\| = \min \|x - a\|$ for $x \in K$. Then,

$$\|b - a\|^{2-n} > \max_{x \in \partial U} \|x - a\|^{2-n}.$$

From the uniform convergence of the sequence $\{g_n\}$ on K, it follows that $\|g(b)\| > \max \|g(x)\|$ for $x \in \partial U$, which contradicts the maximum principle for g. Hence (ii) implies (i). □

We thank Stephen Gardiner for providing us with the following example which shows that in the preceding theorem, the reverse implication does not hold. Suppose Ω is a Denjoy domain in $\mathbb{R}^d$. This means that $\mathbb{R}^d \setminus \Omega \subset \mathbb{R}^{d-1}$. It is well known (see, for example, [**7**]) that every bounded harmonic function on a half-space is the Poisson integral of its non-tangential boundary values. It follows that if $u \in h^\infty(\Omega)$, where Ω is a Denjoy domain satisfying

$$\lambda_{d-1}(\mathbb{R}^d \setminus \Omega) = 0, \tag{6}$$

where λ_{d-1} is Lebesgue measure in $\mathbb{R}^{d-1}$, then u is symmetric with respect to $\mathbb{R}^{d-1}$. However, there exist functions $u \in h(\Omega)$, which are positive in $\Omega^+ = \{x \in \Omega : x_d > 0\}$ and negative in $\Omega^- = \{x \in \Omega : x_d < 0\}$. For example, we may take $u(x) = x_d$. Of course, such a function cannot be approximated by a function which is symmetric. Thus, $h^\infty(\Omega)$ is not dense in $h(\Omega)$. To obtain an example of a domain Ω which satisfies (ii) but not (i), it is hence sufficient to give an example of a Denjoy domain satisfying (ii) and (6).

Let $E \subset \mathbb{R}^d$. Following [**1**], we denote by $k(E)$ the kernel of E:

$$k(E) = \{x \in \mathbb{R}^d : U \cap E \text{ is non-polar for all neighbourhoods } U \text{ of } x\}.$$

In [**1**], we find that, $\check{\Omega} \setminus \Omega$ is polar, where $\check{\Omega} = \mathbb{R}^d \setminus k(\mathbb{R}^d \setminus \Omega)$. Now, let E_o be a compact non-polar subset of $\mathbb{R}^d$ which lies in $\mathbb{R}^{d-1}$ and for which $\lambda_{d-1}(E_o) = 0$.

Let $\Omega_o = \mathbb{R}^d \setminus E_o$. If $\check{\Omega}_o = \mathbb{R}^d$, it follows that $\mathbb{R}^d \setminus \Omega_o$ is polar. That is, E_o is polar, which contradicts the choice of E_o. Hence, $\check{\Omega}_o \neq \mathbb{R}^d$. In other words, $k(E_o) = k(\mathbb{R}^d \setminus \Omega_o) \neq \emptyset$. Now, set $E = k(E_o)$ and consider the Denjoy domain $\Omega = \mathbb{R}^d \setminus E$. Then Ω satisfies condition (ii). Moreover, $\lambda_{d-1}(E) \leq \lambda_{d-1}(E_o) = 0$. Hence, Ω is a Denjoy domain satisfying both (ii) and (6), and from the previous paragraph we see that $h^\infty(\Omega)$ is not dense in $h(\Omega)$. Thus, condition (ii) does not imply condition (i).

References

[1] D. H. Armitage, S. J. Gardiner and I. Netuka, *Separation of points by classes of harmonic functions*, Math. Proc. Camb. Philos. Soc. **113** (1993), no. 3, 561–571.

[2] T. W. Gamelin, *Uniform Algebras*, Prentice-Hall, Englewood Cliffs, NJ, 1969.

[3] P. Gorkin, F. León-Saavedra, and R. Mortini, *Bounded universal functions in one and several complex variables*, Math. Z. **258** (2008), 745–762.

[4] K.-G. Grosse-Erdmann and R. Mortini, *Universal functions for composition operators with non-automorphic symbol*, to appear in J. Anal. Math.

[5] R. Mortini, *X-universal functions*, in Workshop: Normal families and complex dynamics, organizers Steinmetz, Rippon, Zalcman. Oberwolfach Rep. **4** (2007), no. 1, 515–516.

[6] S. Roy, *Extreme Jensen measures*, Ark. Mat. **46** (2008), 153–182.

[7] E. M. Stein and G. Weiss, *Introduction to Fourier Analysis*. Princeton Univerisity Press, 1971.

Département de Mathématiques et de Statistique, Université de Montréal, H3C 3J7, Canada

E-mail address: gauthier@dms.umontreal.ca

Departament de Matemàtiques, Universitat Autònoma de Barcelona, Spain

E-mail address: markmel@mat.uab.es

Amer. Math. Soc. Transl.
(2) Vol. **226**, 2009

Un Théorème de Helson pour des Séries de Walsh

Jean-Pierre Kahane

A Victor Havin, en témoignage d'une longue et fidèle amitié, et de nos bonnes conversations en français

Henry Helson [**1**] a établi en 1954 le théorème suivant : *Si les sommes partielles d'une série trigonométrique sont positives, les coefficients tendent vers zéro* . C'est une des perles de la théorie des séries trigonométriques, encore mise en valeur par le fait que l'hypothèse n'entraîne pas que la série trigonométrique est une série de Fourier–Lebesgue (Katznelson 1965 [**2**]).

Nous allons montrer l'analogue des théorèmes de Helson et de Katznelson pour les séries de Walsh, avec un complément au théorème de Katznelson qui s'étend aux séries trigonométriques.

THÉORÈME I. *Si les sommes partielles d'une série de Walsh sont positives, les coefficients tendent vers zéro.*

THÉORÈME II. *Soit ψ une application croissante de $\mathbb{R}^+$ sur $\mathbb{R}^+$, telle que $\lim_{x\to 0}(\psi(x)/x^2) = 0$. Alors*

a) il existe une mesure de probabilité singulière μ sur $[0,1]$ dont les sommes partielles de la série de Walsh sont positives, et dont les coefficients de Fourier–Walsh, $\widehat{\mu}(n)$, vérifient $\sum_0^\infty \psi(|\widehat{\mu}(n)|) < \infty$;

b) le même énoncé vaut pour les séries et les coefficients de Fourier au sens usuel.

Précisons les notations pour les séries de Walsh. Au lieu de $[0,1]$, il est commode de les définir sur le groupe multiplicatif $\{-1,1\}^{\mathbb{N}^*}$, que nous désignerons par $\mathbb{D}$. Les fonctions coordonnées r_1, r_2, ... sont les fonctions de Rademacher, et elles engendrent par multiplication les fonctions de Walsh w_n ($n = 0, 1, 2, \ldots$) qui sont les caractères de $\mathbb{D}$. On ordonne ainsi les $w_n = 1$, $w_1 = r_1$, $w_2 = r_2$, $w_3 = r_1 r_2$, $w_4 = r_3$ etc., c'est-à-dire

$$w_n = \prod r_j^{\alpha_j}, \quad \alpha_j = 0 \text{ ou } 1 \quad \text{et} \quad \sum \alpha_j < \infty . \tag{1}$$

Aux entiers n on associe ainsi les suites finies (α_j) de 0 et de 1, et l'ordre croissant des n est l'ordre lexicographique inverse des mots (α_j).

2000 *Mathematics Subject Classification.* Primary 42C10.
Key words and phrases. Helson, Fourier-Walsh.

Une série de Walsh est une série formelle de la forme $\sum_0^\infty c_n w_n$, où les coefficients c_n sont réels ou complexes. Nous nous bornerons aux c_n réels. Les fonctions de Walsh opérant par multiplication sur les séries de Walsh, nous aurons besoin du lemme suivant.

LEMME. *Soit S une série de Walsh et w_m une fonction de Walsh. Ecrivons $w_m S$ sous la forme*

$$w_m S = \sum_0^\infty d_n w_n \,.$$

Alors, pour chaque k, la somme partielle d'ordre 2^k, $\sum_0^{2^k-1} d_n w_n$, est le produit par w_m d'une différence de sommes partielles de la série S.

PREUVE. La série $\sum_0^\infty d_n w_n w_m$ n'est autre que la série S dont on a modifié l'ordre des termes. Reste à montrer que les $w_n w_m$ $(n = 0, 1, \ldots, 2^k - 1)$ constituent, à l'ordre près, une suite de 2^k fonctions de Walsh consécutives. Pour cela, écrivons les w_n sous la forme (1) (avec ici $\alpha_j = 0$ ou 1 pour $j \le k$ et $\alpha_j = 0$ pour $j > k$) et $w_m = \prod r_j^{\beta_j}$ $(\beta_i = 0$ ou $1)$. Les $w_n w_m$ s'écrivent $\prod r_j^{\gamma_j}$ avec $\gamma_j = 0$ ou 1 et $\gamma_j = \alpha_j + \beta_j$ modulo 1. Ainsi $(\gamma_j)_{j=1,2,\ldots,k}$ parcourt $\{0,\, 1\}^k$ quand $(\alpha_j)_{j=1,2,\ldots,k}$ parcourt $\{0,\, 1\}^k$, tandis que $(\gamma_j)_{j>k} = (\beta_j)_{j>k}$. Donc les indices des $w_n w_m$, qui s'écrivent $\sum_1^k + \sum_{k+1}^\infty \gamma_j 2^{j-1}$, parcourent un segment des entiers de longueur 2^k. □

A toute série de Walsh S est associée une martingale dyadique constituée par ses sommes partielles d'ordre 2^k $(k = 0,\, 1,\, 2,\, \ldots)$

$$M_k = M_k(S) = \sum_0^{2^k-1} c_n w_n \,,$$

et on obtient ainsi toutes les martingales dyadiques définies sur $\mathbb{D}$. Rappelons des propriétés des martingales dyadiques dont nous nous servirons.

P1. S est la série de Fourier–Walsh d'une mesure de Radon réelle sur $\mathbb{D}$, c'est-à-dire $c_n = \int w_n \, d\mu$, $\mu \in \mathcal{M}(\mathbb{D})$, si et seulement si les M_k sont bornés dans $L^1(\mathbb{D})$. Dans ce cas, les M_k tendent presque partout sur $\mathbb{D}$ vers la densité de la partie absolument continue de μ, et les M_k tendent vers μ dans $\mathcal{M}(\mathbb{D})$ au sens faible, comme formes linéaires sur $C(\mathbb{D})$.

P2. S est la série de Fourier–Walsh d'une fonction réelle intégrable sur $\mathbb{D}$, soit $c_n = \int f w_n$, $f \in L^1(\mathbb{D})$, si et seulement si les M_k sont uniformément intégrables. Dans ce cas, les M_k tendent vers f presque partout et dans $L^1(\mathbb{D})$.

P3. S est la série de Fourier–Walsh d'une mesure positive $\mu \in \mathcal{M}^+(\mathbb{D})$ si et seulement si les M_k sont positives, $M_k \ge 0$.

Ici comme dans la suite, positif signifie ≥ 0.

PREUVE DU THÉORÈME I. Supposons les sommes partielles de la série S positives, et de plus $c_0 = 1$. Alors S est la série de Fourier–Walsh d'une mesure de probabilité $\mu \in \mathcal{M}_1^+(\mathbb{D})$, soit $c_n = \widehat{\mu}(n)$, et

$$M_k = M_k(r_1, r_2, \ldots, r_k) = \sum_0^{2^k-1} \widehat{\mu}(n) w_n \,.$$

Ecrivons

$$M_{k+1} = M_k + r_{k+1} N_k\,, \quad N_k = N_k(r_1, r_2, \dots, r_k),$$

$$N_k = \sum_0^{2^k-1} \widehat{N}_k(m) w_m,$$

$$N_k^* = \sup_{0 \le n < 2k} \Big| \sum_0^n \widehat{N}_k(m) w_m \Big|.$$

L'hypothèse que les sommes partielles de S soient positives se traduit par

$$N_k^* \le M_k \qquad (k = 0, 1, \dots). \tag{2}$$

Supposons que les $\widehat{\mu}(n)$ ne tendent pas vers 0, c'est-à-dire qu'il existe un $a > 0$, une suite strictement croissante d'entiers k_j, et des entiers $n_j \in [2^{k_j}, 2^{k_j+1}[$ tels que $|\widehat{\mu}(n_j)| \ge a$, et tentons d'établir une contradiction.

Supposons d'abord $n_j = 2^{k_j}$. Les N_{k_i} sont bornés dans $L^1(\mathbb{D})$ et $|\widehat{N}_{k_j}(0)| \ge a$ puisque $\widehat{N}_{k_i}(0) = \widehat{\mu}(2^{k_j})$. Quitte à remplacer la suite (k_j) par une sous-suite, nous pouvons supposer que les N_{k_j} convergent faiblement vers une mesure $\nu \in \mathcal{M}(\mathbb{D})$. Ainsi

$$\widehat{\nu}(n) = \lim_{i \to \infty} \widehat{N}_{k_i}(n)$$

et en particulier $\widehat{\nu}(0) \ne 0$.

Les sommes

$$\nu_k = \sum_0^{2^k-1} \widehat{\nu}(n) w_n \qquad (k = 0, 1, 2, \dots)$$

forment une martingale dyadique bornée dans $L^1(\mathbb{D})$ (propriété P1) et, pour chaque k,

$$\nu_k = \lim_{j \to \infty} \sum_0^{2^{k-1}} \widehat{N}_{k_i}(n) w_n\,.$$

L'hypothèse de positivité, sous la forme (2), entraîne

$$\Big| \sum_0^{2^k-1} \widehat{N}_{k_i}(n) w_n \Big| \le M_{k_i} \tag{3}$$

lorsque $k_j \ge k$, donc $|\nu_k| \le M_{k_i}$.

Or les M_{k_i} convergent presque partout vers une $f \in L^1$, la densité de la partie absolument continue de μ (propriété P1). Donc

$$|\nu_k| \le f\,.$$

Cela entraîne que les ν_k sont uniformément intégrables, donc convergent dans $L^1(\mathbb{D})$ (propriété P2), donc que ν est absolument continue.

D'autre part, si l'on décompose μ en sa partie absolument continue et sa partie singulière, $\mu = \mu^a + \mu^s$, on peut écrire également $M_k = M_k^a + M_k^s$ et $N_k = N_k^a + N_k^s$. Les M_k^a convergent vers f dans $L^1(\mathbb{D})$, donc les N_k^a tendent vers 0 dans $L^1(\mathbb{D})$, donc ν est la limite faible des $N_{k_j}^s$. Or, pour tout $\varepsilon > 0$ et tout k,

$$(N_k^s > \varepsilon) \subset \Big(M_k^s > \frac{\varepsilon}{2}\Big) \cup \Big(M_{k+1}^1 > \frac{\varepsilon}{2}\Big)$$

et la mesure du second membre tend vers 0 quand $k \to \infty$. Donc ν est singulière. La contradiction est établie dans le cas particulier $n_j = 2^{k_i}$.

Passons au cas général. On considère maintenant les

$$N'_{k_j} = w_{m_j} N_{k_j}\,, \qquad m_j = n_j - 2^{k_j}\,.$$

Ainsi $\widehat{N}'_{k_j}(0) = \widehat{N}_{k_j}(m_j) = \widehat{\mu}(n_j)$. Comme ci-dessus, quitte à restreindre la suite (n_j), les N'_{k_j} convergent faiblement vers une mesure ν' singulière, non nulle puisque $\widehat{\nu}'(0) \neq 0$. Pour montrer que ν' est absolument continue, le point crucial est l'analogue de (3) que l'on obtient en appliquant le lemme à N_{k_j} (pour S) et w_{m_j} (pour w_m). On obtient ainsi, pour $k \geq k_j$,

$$\left| \sum_0^{2^k-1} \widehat{N}'_{k_j}(n) w_n \right| \leq 2M_{k_j}\,, \tag{4}$$

ce qui permet d'achever la démonstration que ν' est absolument continue. La contradiction est ainsi établie dans le cas général, et cela achève la preuve du théorème I. □

REMARQUE. Cette preuve est calquée sur celle de Helson. Comme ici, Helson met en évidence, sous l'hypothèse que les coefficients ne tendent pas vers zéro, une mesure ν non nulle qui est à la fois singulière et absolument continue. Mais la démonstration de la continuité absolue est différente. Helson utilise un théorème de Frédéric et Marcel Riesz qui appartient à la théorie des fonctions analytiques. On utilise ici la théorie des martingales. L'emploi en parallèle des martingales et des fonctions analytiques est classique en analyse harmonique depuis les théorèmes de Paley et de Littlewood–Paley.

Une autre différence entre les preuves est l'utilisation de l'hypothèse. La positivité des sommes partielles entraîne que les sommes partielles sont bornées dans L^1, et Helson n'utilise rien d'autre. Au contraire, nous avons utilisé de manière essentielle une autre conséquence de la positivité, la formule (2) (équivalente à la positivité des sommes partielles).

PREUVE DU THÉORÈME II. *Partie a).* On construit la mesure μ comme produit infini

$$\mu = \prod (1 + X_k)$$

où les X_k sont de polynômes de Walsh indépendants à valeurs dans l'intervalle $[-1, 1]$ et de valeur moyenne nulle : $|X_k| \leq 1$ et $EX_k = 0$. Posons $EX_k^2 = \sigma_k^2$. Lorsque $\sum \sigma_k^2 < \infty$, le produit infini converge dans $L^2(\mathbb{D})$, donc μ est absolument continue. Lorsque $\sum \sigma_k^2 = \infty$, μ est une mesure de probabilité singulière.

Voici une démonstration rapide de ce dernier fait, tirée de [**3**]. Supposons $\sum \sigma_k^2 = \infty$. Les $\frac{X_k}{\sigma_k}$ forment un système orthonormal dans $L^2(\mathbb{D}, \lambda)$, où λ désigne la mesure de Haar sur $\mathbb{D}$. On vérifie que les $\frac{X_k}{\sigma_k} - \sigma_k$ sont de valeur moyenne nulle et deux à deux orthogonales dans $L^2(\mathbb{D}, \mu)$:

$$E_\mu \Big(\frac{X_k}{\sigma_k} - \sigma_k \Big) = E_\lambda \Big(\Big(\frac{X_k}{\sigma_\kappa} - \sigma_k \Big)(1 + X_k) \Big) = 0$$

et, pour $k \neq k'$,

$$\begin{aligned}E_\mu\Big(\Big(\frac{X_k}{\sigma_k}-\sigma_k\Big)\Big(\frac{X_{k'}}{\sigma_{k'}}-\sigma_{k'}\Big)\Big) &= E_\lambda\Big(\Big(\frac{X_k}{\sigma_k}-\sigma_k\Big)\Big(\frac{X_{k'}}{\sigma_{k'}}-\sigma_{k'}\Big)(1+X_k)(1+X_{k'})\Big)\\ &= E_\lambda\Big(\Big(\frac{X_k}{\sigma_k}-\sigma_k\Big)(1+X_k)\Big)E_\lambda\Big(\Big(\frac{X_{k'}}{\sigma_{k'}}-\sigma_{k'}\Big)(1+X_{k'})\Big)=0.\end{aligned}$$

De plus,

$$\begin{aligned}E_\mu\Big(\Big(\frac{X_k}{\sigma_k}-\sigma_k\Big)^2\Big) &= E_\lambda\Big(\Big(\frac{X_k}{\sigma_k}-\sigma_k\Big)^2(1+X_k)\Big)\\ &\leq 2E_\lambda\Big(\frac{X_k}{\sigma_k}-\sigma_k\Big)^2 = 2(1+\sigma_k^2)\leq 4\,.\end{aligned}$$

Pour toute suite $(b_k)\in\ell^2$, les série $\sum b_k\frac{X_k}{\sigma_k}$ et $\sum b_k\big(\frac{X_k}{\sigma_k}-\sigma_k\big)$ convergent respectivement dans $L^2(\mathbb{D},\lambda)$ et dans $L^2(\mathbb{D},\mu)$. Si μ n'était pas orthogonale à λ, il existerait un point de $\mathbb{D}$ et une suite d'entiers n_i tels que les sommes partielles d'ordre n_i des deux séries convergent en ce point. Par différence, les sommes partielles d'ordre n_i de la série $\sum b_k\,\sigma_k$ convergeraient. Or on peut choisir les $b_k>0$, $(b_k)\in\ell^2$, de façon que la série $\sum b_k\sigma_k$ diverge. Donc $\mu\perp\lambda$.

Remarquons que l'hypothèse d'indépendance des X_k peut être remplacée par une condition d'orthogonalité forte, à savoir

$$\int\prod X_k^{\alpha_k}=0$$

pour toutes les suites (α_k) constituées de 0, 1 et 2, et finies ($\sum\alpha_k<\infty$), contenant au moins un 1 et au plus deux 2. C'est sous cette forme que la méthode a été introduite et utilisée par Peyrière dans l'étude de la singularité mutuelle des produits de Riesz [**3**].

On imposera donc la condition $\sum\sigma_k^2=\infty$, et il s'agit maintenant de construire les X_k de façon que les sommes partielles de la série de Fourier–Walsh de μ soient positives, et que les coefficients vérifient la majoration $|\widehat{\mu}(n)|\leq\psi(n)$. Pour cela, il est commode de disposer de polynômes de Walsh de la forme suivante :

$$\varphi=\varphi(r_1,r_2,\ldots,r_\ell)=\sum_1^{2^\ell}\varepsilon_n w_n\,,\quad \varepsilon_n=\pm1,$$

pour lesquels

$$\|\varphi\|_v=\sup_{1\leq p\leq 2^\ell}\|\sum_1^p\varepsilon_n w_n\|_\infty<C\,2^{\ell/2}=C\|\varphi\|_2\,,$$

C étant une constante absolue. Voici une construction classique de tels polynômes : on pose $P_0=Q_0=1$, $P_{\ell+1}=P_\ell+r_{\ell+1}Q_\ell$, $Q_{\ell+1}=P_\ell-r_{\ell+1}Q_\ell$ ($\ell=0,1,\ldots$), on vérifie que $P_\ell^2+Q_\ell^2=2^{\ell+1}$, d'où

$$\|P_\ell\|_v\leq 2^{\ell/2}+2^{(\ell-1)/2}+\cdots\leq 2^{\ell/2}\frac{\sqrt2}{\sqrt2-1}$$

et on pose $\varphi(r_1,r_2,\ldots,r_\ell)=r_1P_\ell$. C'est la construction donnée dans le cas trigonométrique par Harold Shapiro puis par Walter Rudin, et la suite (ε_r) s'appelle la suite de Rudin–Shapiro.

Etant donné un ensemble d'entiers positifs J, de cardinal 2^ℓ, on désignera par $\varphi((r_j),\, j \in J)$ le polynôme de Walsh obtenu à partir de $\varphi(r_1, r_2, \ldots, r_\ell)$ en substituant à $r_1, r_2, \ldots, r_\ell$ les r_j, $j \in J$, dans l'ordre croissant des j. La construction de μ va dépendre essentiellement du choix d'une suite très rapidement croissante d'entiers ℓ_k, que nous ferons plus tard. Pour chaque k, soit J_k un ensemble d'entiers, de cardinal 2^{ℓ_k}, situé à droite de J_{k-1} : $\inf J_k > \sup J_{k-1}$. Posons

$$\begin{aligned} a_k &= \frac{1}{2C}\, 2^{-\ell_k/2}, \\ X_k &= a_k\, \varphi((r_j),\, j \in J_k)\,. \end{aligned}$$

Explicitons les normes de X_k dans L^2, dans U (maximum des valeurs absolues des sommes partielles), dans A (somme des valeurs absolues des coefficients) et dans PM (sup des valeurs absolues des coefficients) :

$$\begin{aligned} \|X_k\|_2 &= \frac{1}{2C}\,, \\ \|X_k\|_U &< \frac{1}{2}\,, \\ \|X_k\|_A &= \frac{1}{2C}\, 2^{\ell_k/2}\,, \\ \|X_k\|_{PM} &= \frac{1}{2C}\, 2^{-\ell_k/2}\,. \end{aligned}$$

Comme les J_k sont disjoints, les X_k sont bien indépendants, et on a bien $EX_k = 0$ et $\sum \sigma_k^2 = \infty$.

Posons

$$\Pi_k = (1 + X_1)(1 + X_2) \cdots (1 + X_k)\,.$$

Une somme partielle de la série de Fourier–Walsh de μ dont l'ordre est compris entre $\sup J_j$ et $\sup J_{j+1}$ est la somme de Π_k et d'une somme partielle de $\Pi_k X_{k+1}$, ce que nous écrivons

$$S_.(\mu) = S_.(X_{k+1}) = \Pi_k + S_{..}(\Pi_k X_{k+1})\,.$$

Cette dernière somme partielle se décompose à son tour en

$$S_{..}(\Pi_k X_{k+1}) = \Pi_k S_{...}(X_{k+1}) + S_{....}(\Pi_k) \times \text{un coefficient de } X_{k+1}\,.$$

Donc

$$\begin{aligned} S_.(X_{k+1}) &\geq \Pi_k - \Pi_k \|X_{k+1}\|_U - S_{....}(\Pi_k)\|X_{k+1}\|_{PM} \\ &\geq \frac{1}{2}\Pi_k - \frac{1}{2C} 2^{-(\ell_{k+1}/2)} \|\Pi_k\|_A\,. \end{aligned}$$

Imposons la condition que, pour tout k,

$$\frac{1}{2C}\, 2^{-(\ell_{k+1}/2)} \|\Pi_k\|_A \leq \frac{1}{4} \inf \Pi_k\,. \tag{5}$$

Il en résulte que $S_.(X_{k+1}) \geq \frac{1}{4}\Pi_k$, donc $S_.(\mu) \geq 0$.

Ecrivons $\psi(x) = x^2\, \varepsilon(x)$. Alors

$$\sum \varphi(\widehat{\mu}(n)) \leq \sum_k \|\Pi_k\, X_{k+1}\|_2^2\, \varepsilon(\|\Pi_k\, X_{k+1}\|_{PM})\,.$$

Or

$$\|\Pi_k X_{k+1}\|_2^2 \leq \frac{1}{4C^2}\|\Pi_k\|_A^2$$

et

$$\varepsilon(\|\Pi_k X_{k+1}\|)_{PM} \leq \varepsilon(\|X_{k+1}\|_{PM}) = \varepsilon\Big(\frac{1}{2C}2^{-(\ell_{k+1}/2)}\Big).$$

Si, outre (5), on impose à la suite (ℓ_k) la condition

$$\sum \|\Pi_k\|_A^2 \varepsilon\Big(\frac{1}{2C}\,2^{-(\ell_{k+1}/2)}\Big) < \infty\,, \tag{6}$$

ce qui est possible, la conclusion du théorème est vérifiée.

La conclusion de la partie a) du théorème est vérifiée.

Partie b). On choisit ici des polynômes trigonométriques

$$\varphi_\ell(t) = \sum_a^\ell \varepsilon_n \cos nt\,, \qquad \varepsilon_n = \pm 1\,,$$

avec la propriété de

$$\|\varphi_\ell\|_v \leq C\ell^{1/2} = C\|\varphi_\ell\|_2\,,$$

et on choisit

$$X_k(t) = a_k\,\varphi_{\ell_k}(\ell_k t)\,, \qquad a_k = \frac{1}{4C}\,2^{-\ell_k/2}\,.$$

Les X_k ne sont plus des fonctions indépendantes, mais, si la suite (ℓ_k) est assez rapidement croissante, elles sont orthogonales au sens fort qui a été décrit ci-dessus. On définit donc

$$\mu = \prod_1^\infty (1 + a_k\,\varphi_{\ell_k}(\ell_k\,t))$$

et on vérifie par les mêmes calculs que ci-dessus que μ est singulière, que ses sommes partielles sont positives, et que $\sum \varphi(|\widehat{\mu}(n)|) < \infty$.

C'est exactement la méthode de Katznelson. □

Références

[1] H. Helson, *Proof of a conjecture of Steinhaus.* Proc. Nat. Acad. Sci. USA **40** (1954), 205–206.

[2] Y. Katznelson, *Trigonometric series with positive partial sums.* Bull. Amer. Math. Soc. **71** (1965), 718–719.

[3] J. Peyrière, *Étude de quelques propriétés des produits de Riesz.* Ann. Inst. Fourier (Grenoble) **25** (1975), no. 2, 127–169.

Laboratoire de Mathématique, Université Paris–Sud à Orsay, Bât. 425 91405 Orsay Cedex, France

E-mail address: Jean-Pierre.Kahane@math.u-psud.fr

Amer. Math. Soc. Transl.
(2) Vol. **226**, 2009

Interpolation Sequences for the Bernstein Algebra

Xavier Massaneda and Joaquim Ortega-Cerdà

Dedicated to Victor Petrovich Havin on his 75th birthday

ABSTRACT. We give a description, in analytic and geometric terms, of the interpolation sequences for the algebra of entire functions of exponential type which are bounded on the real line.

1. Introduction

In this paper we describe interpolation sequences for the *Bernstein algebra* $\mathcal{B}$ of entire functions of exponential type which are bounded on the real line. The space $\mathcal{B}$ consists of the entire functions f such that for some $A, B > 0$,

$$\log |f(z)| \le A + B|\operatorname{Im} z|, \qquad z \in \mathbb{C}.$$

It can also be viewed as the union of the *Bernstein spaces* B_σ, $\sigma > 0$, of entire functions f such that

$$\sup_{z\in\mathbb{C}} |f(z)|e^{-\sigma|\operatorname{Im} z|} < \infty,$$

which are precisely the bounded functions with Fourier–Laplace transforms supported in $[-\sigma, \sigma]$.

We were motivated to study the interpolation in $\mathcal{B}$ by the recent characterisation of zero sets in the same class obtained by Sergei Yu. Favorov.

Given a discrete sequence $\Lambda \subset \mathbb{C}$ and a point $z \in \mathbb{C}$, consider the counting function $n_\Lambda(z,t) = \#(\Lambda \cap \overline{D(z,t)})$, $t > 0$, where $D(z,t)$ denotes the open disk with center z and radius t.

THEOREM ([**Fa08**, Theorem 2.2]). *A sequence Λ is the zero set of a function in $\mathcal{B}$ if and only if*:

(a) $\exists \lim_{R\to\infty} \sum_{\lambda\in\Lambda\cap D(0,R)\setminus\{0\}} \lambda^{-1}$;
(b) $n_\Lambda(0,t) = O(t)$, $t \to \infty$;

2000 *Mathematics Subject Classification.* Primary 30E05, 42A85.

Key words and phrases. Interpolation sequence, Bernstein spaces.

Both authors were supported by DGICYT grant MTM2008-05561-C02-01 and the CIRIT grant 2005-SGR 00611.

(c) $n_\Lambda(0,t+1) - n_\Lambda(0,t) = o(t)$, $t \to \infty$;
(d) $\sup_{x\in\mathbb{R}} \int_0^\infty [n_\Lambda(b,t) - n_\Lambda(x,t)]\frac{dt}{t} < \infty$ *for some* $b \in \mathbb{R}\setminus\Lambda$.

Notice that a subset of a zero set Λ of $\mathcal{B}$ is not necessarily a zero set of $\mathcal{B}$, since conditions (a) and (d) imply a certain regularity in the distribution of Λ, which can be destroyed when removing points.

In view of this result we ask ourselves whether interpolation sequences for $\mathcal{B}$ can be described in similar terms. Given the characteristic growth of functions in $\mathcal{B}$, the natural interpolation problem for this algebra is the following.

DEFINITION. A discrete sequence Λ in $\mathbb{C}$ is called an *interpolation sequence* for $\mathcal{B}$ if for every sequence of values $\{v_\lambda\}_{\lambda\in\Lambda}$ with

$$\sup_{\lambda\in\Lambda} |v_\lambda| e^{-C|\operatorname{Im}\lambda|} < \infty \tag{1}$$

for some $C > 0$, there exists $f \in \mathcal{B}$ with

$$f(\lambda) = v_\lambda, \quad \lambda \in \Lambda.$$

If $\mathcal{B}(\Lambda)$ denotes the space of sequences $\{v_\lambda\}_{\lambda\in\Lambda}$ satisfying (1) for some $C > 0$, we can equivalently define interpolating sequences as those for which the following restriction operator is onto:

$$\begin{aligned} \mathcal{R}_\Lambda : \mathcal{B} &\longrightarrow \mathcal{B}(\Lambda), \\ f &\mapsto \{f(\lambda)\}_{\lambda\in\Lambda}. \end{aligned}$$

Applying the open mapping theorem to $\mathcal{R}_\Lambda$ [**BrGa95**, Lemma 2.2.6] one sees that the interpolation is stable in the following sense: for each $C > 0$ there exist $M, \sigma > 0$ such that if $\{v_\lambda\}_{\lambda\in\Lambda}$ satisfies (1) then there exists $f \in \mathcal{B}$ with $f(\lambda) = v_\lambda$, $\lambda \in \Lambda$, and $\sup_{z\in\mathbb{C}} |f(z)|e^{-\sigma|\operatorname{Im} z|} \le M \sup_{\lambda\in\Lambda} |v_\lambda| e^{-C|\operatorname{Im}\lambda|}$.

There is an alternative (and equivalent) definition of $\mathcal{B}$-interpolation:

DEFINITION. A sequence Λ is a *free interpolation* (denoted $\Lambda \in Int_{\ell^\infty}\mathcal{B}$) if the trace $\mathcal{R}_\Lambda(\mathcal{B})$ is *ideal*, that is, if $\ell^\infty \cdot \mathcal{R}_\Lambda(\mathcal{B}) \subset \mathcal{R}_\Lambda(\mathcal{B})$.

Thus, $\Lambda \in Int_{\ell^\infty}\mathcal{B}$ if and only if whenever $f \in \mathcal{B}$ and $\{\alpha_\lambda\} \in \ell^\infty$ there exists $g \in \mathcal{B}$ such that $g(\lambda) = \alpha_\lambda f(\lambda)$, $\lambda \in \Lambda$. Since $\mathcal{B}$ is an algebra, this happens if and only if for any sequence of values $\{\alpha_\lambda\}_{\lambda\in\Lambda} \in \ell^\infty$ there exists $f \in \mathcal{B}$ with $f(\lambda) = \alpha_\lambda$, $\lambda \in \Lambda$.

It is worth mentioning that interpolation in $\mathcal{B}$ is of a"soft" nature, in the sense that the constant C in (1) and the type σ of the function f such that $f(\lambda) = v_\lambda$, $\lambda \in \Lambda$, need not be the same. Interpolation in a fixed B_σ was described in [**OS99**] in the following terms.

Consider the distance function $\rho(z,\zeta) = \frac{|z-\zeta|}{1+|z-\zeta|}$. A sequence Λ is called *ρ-separated* if

$$\inf_{\substack{\lambda,\lambda'\in\Lambda \\ \lambda\neq\lambda'}} \rho(\lambda,\lambda') > 0,$$

and it is said to satisfy the *two-sided Carleson condition* if for any disk D centered on the real axis and of radius $r(D)$, we have

$$\sum_{\lambda\in D\cap\Lambda} |\operatorname{Im}\lambda| \le r(D),$$

with C independent of D.

Given a ρ-separated sequence Λ and a constant $A>0$, denote by $n_A^+(r)$ the maximum number of points from Λ to be found in a rectangle of the form $\{z = x+iy : t<x<t+r\,,|y|<A\}$, where t is any real number.

The following result can be viewed as an extension of [**Beu**, pp. 351–365].

THEOREM ([**OS99**, Theorem 1]). *A sequence is interpolating for B_σ if and only if it is ρ-separated, it satisfies the two-sided Carleson condition and*

$$D^+(\Lambda) = \lim_{A\to\infty}\lim_{r\to+\infty}\frac{n_A^+(r)}{r} < \frac{\sigma}{\pi}.$$

There exists a general theory (and in particular an analytic description of interpolating sequences) for the algebras

$$A_p = \{f\in \mathrm{Hol}(\mathbb{C}) :\ \exists C>0 : \sup_{z\in\mathbb{C}}|f(z)|e^{-Cp(z)}<\infty\},$$

where $p:\mathbb{C}\longrightarrow\mathbb{R}_+$ is a subharmonic weight such that:

(a) $\log(1+|z|^2) = O(p(z))$;

(b) there are constants $C,D>0$ such that whenever $|\zeta-z|\le 1$, then $p(\zeta)\le Cp(z)+D$.

For this we refer to [**BrGa95**, Chapter 2].

Notice that $\mathcal{B}$ does not fall into this general theory, since $p(z) = |\operatorname{Im} z|+1$ does not satisfy (a) (in particular, $\mathcal{B}$ does not contain polynomials). However, the techniques used in the description of A_p-interpolation mentioned above (see [**BrLi95**, Corollary 3.5]) can be adapted to obtain an analogous result for $\mathcal{B}$.

THEOREM A. *The following statements are equivalent*:

(a) $\Lambda\in\operatorname{Int}\mathcal{B}$;

(b) $\Lambda\in\operatorname{Int}_{\ell^\infty}\mathcal{B}$;

(c) *there exist $A,B,C>0$ and peak-functions $f_\lambda\in\mathcal{B}$ such that $f_\lambda(\lambda')=\delta_{\lambda,\lambda'}$ and*

$$\sup_{z\in\mathbb{C}}|f_\lambda(z)|e^{-C|\operatorname{Im} z|}\le Ae^{B|\operatorname{Im}\lambda|};$$

(d) *there exist $F\in\mathcal{B}$ and ε, $C>0$ such that $F(\lambda)=0$ for all $\lambda\in\Lambda$ and*

$$|F'(\lambda)|\ge\varepsilon e^{-C|\operatorname{Im}\lambda|}.$$

As soon as the analytic conditions of Theorem A hold we can proceed as in [**MOO06**] and obtain geometric conditions. That paper deals with A_p-interpolation for weights of the form

$$p(z) = |\operatorname{Im} z| + \omega(|z|),$$

$\omega(t)$ being a subadditive increasing continuous function such that $\log(1+t)\lesssim\omega(t)$ and $\int_0^\infty\frac{\omega(t)}{1+t^2}dt<\infty$. Taking $p(z) = |\operatorname{Im} z|+1$ (i.e. $\omega\equiv 1$) and using Theorem A we can imitate the proof of [**MOO06**, Theorem 1] and obtain a geometric characterisation.

In order to state this result consider the counting function $n_\Lambda(z,r)$ defined previously and the integrated version

$$N_\Lambda(z,r) = \int_0^r\frac{n_\Lambda(z,t)-n_\Lambda(z,0)}{t}\,dt + n_\Lambda(z,0)\log r.$$

Hereinafter we assume that $\Lambda\cap\mathbb{R}=\emptyset$; otherwise replace $\mathbb{R}$ by any horizontal line that does not contain any of the points in Λ.

THEOREM B. *The sequence* Λ *is* $\mathcal{B}$*-interpolation if and only if:*

(i) *there is* $C > 0$ *such that*

$$N_\Lambda(\lambda, |\operatorname{Im}\lambda| + 1) \leq C(|\operatorname{Im}\lambda| + 1) \qquad \forall\lambda \in \Lambda;$$

(ii) *the following Carleson-type condition holds:*

$$\sup_{x\in\mathbb{R}} \sum_{\substack{\lambda\in\Lambda\\ |\operatorname{Im}\lambda|>1}} \frac{|\operatorname{Im}\lambda|}{|x-\lambda|^2} < \infty.$$

Since the Poisson kernel at λ in the corresponding half-plane (upper half-plane if $\operatorname{Im}\lambda > 1$ and lower half-plane when $\operatorname{Im}\lambda < -1$) is $P(\lambda, x) = \frac{|\operatorname{Im}\lambda|}{|x-\lambda|^2}$, a restatement of condition (ii) is that the measure $\sum \delta_\lambda$, where the sum runs over $\lambda \in \Lambda$ with $|\operatorname{Im}\lambda| > 1$, has bounded Poisson balayage on $\mathbb{R}$.

This result shows that, as expected, there exist interpolating sequences for $\mathcal{B}$ which are not interpolating for any B_σ, $\sigma > 0$. A simple example is given by $\Lambda = \{2^k(n+i)\}_{k\in\mathbb{N},n\in\mathbb{Z}}$. Straightforward computations show that conditions (i) and (ii) in Theorem B are satisfied, while the two-sided Carleson condition required by Theorem 1 in [**OS99**] fails, since for a disk D centered at a real point and with large radius r we have

$$\sum_{\lambda\in\Lambda\cap D} |\operatorname{Im}\lambda| \simeq r\log r.$$

The paper is structured as follows. In Section 2 we prove Theorem A and describe other analytic properties of $\mathcal{B}$-interpolation sequences which are relevant to the proof of Theorem B. For the sake of completeness, we discuss the proof of Theorem B in Section 3.

A final remark about notation: C will always denote a positive constant, and its actual value may change from one occurrence to the next; $A = O(B)$ and $A \lesssim B$ mean that $A \leq cB$ for some $c > 0$, and $A \simeq B$ is $A \lesssim B \lesssim A$.

Acknowledgment. We thank Anton Baranov for spotting a mistake in an earlier version of the manuscript and for many valuable comments.

2. Proof of the analytic conditions

Before the proof of Theorem A we need to spell out some consequences of its conditions.

LEMMA 1. *Let* Λ *be a discrete sequence in* $\mathbb{C}$ *for which* (c) *in Theorem A holds. Then*

(i) *There exist* α, $\varepsilon > 0$ *such that the disks* $D_\lambda := D(\lambda, \delta_\lambda)$, *with* $\delta_\lambda = \varepsilon e^{-\alpha|\operatorname{Im}\lambda|}$, *are pairwise disjoint.*

(ii) *There exist* $C > 0$ *such that for all* $\beta > 2$

$$\sup_{z\in\mathbb{C}} \sum_{\lambda\in\Lambda} \frac{e^{-C|\operatorname{Im}\lambda|}}{1+|z-\lambda|^\beta} < \infty.$$

When Λ satisfies (i) we say that Λ is *weakly separated.*

PROOF. (i) Fix $\lambda \in \Lambda$ and consider any λ' such that $|\lambda - \lambda'| \leq 1$. Then

$$1 = |f_\lambda(\lambda) - f_\lambda(\lambda')| \leq \sup_{\zeta\in D(\lambda,1)} |f'_\lambda(\zeta)|\,|\lambda - \lambda'|.$$

By the Cauchy estimates,

$$|f'_\lambda(\zeta)| \leq \sup_{\eta:|\eta-\zeta|=1} |f_\lambda(\eta)| \leq \sup_{z\in D(\lambda,2)} |f_\lambda(z)|$$
$$\leq \sup_{z\in D(\lambda,2)} Ae^{B|\operatorname{Im}\lambda|}e^{C\alpha|\operatorname{Im} z|} \lesssim e^{(B+C)|\operatorname{Im}\lambda|},$$

which yields the desired estimate.

(ii) This is an immediate consequence of (i): if $C \geq 2\alpha$, we have

$$\sum_{\lambda\in\Lambda} \frac{e^{-C|\operatorname{Im}\lambda|}}{1+|z-\lambda|^\beta} \lesssim \sum_{\lambda\in\Lambda} \int_{D_\lambda} \frac{e^{-(C-2\alpha)|\operatorname{Im}\zeta|}}{1+|z-\zeta|^\beta}\, dm(\zeta) \lesssim \int_{\mathbb{C}} \frac{dm(\zeta)}{1+|z-\zeta|^\beta}. \qquad \square$$

PROOF OF THEOREM A. (a) $\Longrightarrow$ (b). Obvious.

(b) $\Longrightarrow$ (c). Here we use the scheme of [**BrGa95**, Lemma 2.2.6]. Consider the complete metric space $S = \{v = \{v_\lambda\}_\lambda : \|v\|_\infty \leq 1\}$, and for $n \in \mathbb{N}$ let:

$$S_n = \{\{v_\lambda\}_\lambda \in S : \exists f \in \mathcal{B} : f(\lambda) = v_\lambda,\ \lambda \in \Lambda, \text{ and } \sup_{z\in\mathbb{C}} |f(z)|e^{-n|\operatorname{Im} z|} \leq n\}.$$

By the hypotheses, $S = \bigcup_n S_n$ and by Baire's Category Theorem there exists $n \in \mathbb{N}$ such that $\overset{\circ}{S_n} \neq \emptyset$. Since $0 \in S_n$ for all n, there exists $\varepsilon \in (0,1)$ such that

$$\|v\|_\infty \leq \varepsilon \Longrightarrow v \in S_n.$$

Let $v^\lambda = \{\varepsilon\delta_{\lambda,\lambda'}\}_{\lambda'}$. Then there exists $g_\lambda \in \mathcal{B}$ with

$$g_\lambda(\lambda') = \varepsilon\delta_{\lambda,\lambda'} \quad \text{and} \quad \sup_z |g_\lambda(z)|e^{-n|\operatorname{Im} z|} \leq n.$$

Taking $f_\lambda = g_\lambda/\varepsilon$ we get the desired properties.

(c) $\Longrightarrow$ (d). Define $F(z) = \sum_{\lambda\in\Lambda} w_\lambda(z)g_\lambda(z)$, where

$$w_\lambda(z) = \left(\frac{\sin(z-\lambda)}{z-\lambda}\right)^3 e^{i\sigma_\lambda M(z-\lambda)}, \qquad g_\lambda(z) = f_\lambda^2(z)\sin(z-\lambda),$$

$M > 0$ will be chosen later on, and

$$\sigma_\lambda = \begin{cases} 1 & \text{if } \operatorname{Im}\lambda < 0, \\ -1 & \text{if } \operatorname{Im}\lambda > 0. \end{cases}$$

Notice that w_λ is a holomorphic function in $\mathcal{B}$ with $w_\lambda(\lambda) = 1$, while g_λ is a function in $\mathcal{B}$ vanishing of order 1 at λ and at least at order 2 at all $\lambda' \neq \lambda$.

Let us see first that $F \in \mathcal{B}$. A direct estimate shows that

$$|w_\lambda(z)| \lesssim \frac{e^{(M+3)|\operatorname{Im} z|}e^{-(M-3)|\operatorname{Im}\lambda|}}{1+|z-\lambda|^3},$$
$$|g_\lambda(z)| \lesssim e^{(2C+1)|\operatorname{Im} z|}e^{(2B+1)|\operatorname{Im}\lambda|},$$

and therefore

$$|F(z)| \lesssim e^{(M+2C+4)|\operatorname{Im} z|} \sum_{\lambda\in\Lambda} \frac{e^{-(M-2B-4)|\operatorname{Im}\lambda|}}{1+|z-\lambda|^3}.$$

Taking M big enough and applying Lemma 1(ii) we see that $F \in \mathcal{B}$.

It is clear that $F(\lambda) = 0$ for all $\lambda \in \Lambda$. On the other hand,

$$F'(\lambda) = \sum_{\lambda'\in\Lambda} w_{\lambda'}(\lambda)g'_{\lambda'}(\lambda) = w_\lambda(\lambda)g'_\lambda(\lambda) = 1.$$

(d) $\Longrightarrow$ (a). Let $\{v_\lambda\}_{\lambda\in\Lambda}$ satisfy (1) and define

$$f(z) = \sum_{\lambda\in\Lambda} v_\lambda w_\lambda(z) \frac{F(z)}{F'(\lambda)(z-\lambda)},$$

where w_λ is defined as before.

It is clear that $F(\lambda) = v_\lambda$ for all $\lambda \in \Lambda$, and a computation similar to the one in the previous implication shows that $f \in \mathcal{B}$, since $|F(z)/(F'(\lambda)(z-\lambda))| \lesssim e^{C(|\operatorname{Im} z|+|\operatorname{Im}\lambda|)}$. □

We finish this section by showing that a weakly separated union of two $\mathcal{B}$-interpolation sequences is still a $\mathcal{B}$-interpolation.

THEOREM 2. *Let $\Lambda_1, \Lambda_2 \in \operatorname{Int}\mathcal{B}$ be such that $\Lambda_1 \cup \Lambda_2$ is weakly separated. Then $\Lambda_1 \cup \Lambda_2 \in \operatorname{Int}\mathcal{B}$.*

For the proof of this result we will need the following lemma.

LEMMA 3. *Let $\Lambda \in \operatorname{Int}\mathcal{B}$ and let $z_0 \notin \Lambda$. There exist $A, B, C > 0$ independent of z_0 and $f \in \mathcal{B}$ such that*

(i) *$f(z_0) = 1$ and $f(\lambda) = 0$ for all $\lambda \in \Lambda$;*
(ii) *$\sup_{z\in\mathbb{C}} |f(z)|e^{-C|\operatorname{Im} z|} \le Be^{-C|\operatorname{Im} z_0|}/d(z_0, \Lambda)$.*

PROOF. Assume first that $z_0 = 0$. Since Λ is weakly separated, there exists $\delta > 0$ such that $|\lambda - \lambda'| \ge 10\delta$ for all $\lambda, \lambda' \in \Lambda \cap \{|\operatorname{Im} z| < 1\}$.

LEMMA 4. *Let $\Lambda \in \operatorname{Int}\mathcal{B}$. There exists $F \in \mathcal{B}$ with $F(0) = 0$ and such that*

$$\begin{aligned} |F(\lambda)| &\ge \varepsilon && \text{if } |\lambda| \ge \delta, \\ |F(\lambda)| &\ge \varepsilon|\lambda| && \text{if } |\lambda| < \delta. \end{aligned}$$

We defer the proof of this technical result to the end of the section.

Consider the values $v_\lambda = 1/F(\lambda)$. Since $\sup_{\lambda\in\Lambda} |v_\lambda| \le \frac{1}{\min(\varepsilon, \varepsilon d(0,\Lambda))}$, there exist $A, B > 0$ and $G \in \mathcal{B}$ such that $G(\lambda) = v_\lambda = 1/F(\lambda)$ and $\sup_{z\in\mathbb{C}} |G(z)|e^{-A|\operatorname{Im} z|} \le B/(\varepsilon d(0, \Lambda))$.

The function $f(z) = 1 - F(z)G(z)$ then satisfies the required properties.

For $z_0 \notin \Lambda$ arbitrary, apply the previous case to the sequence $\Lambda - z_0$. Notice that $\Lambda - z_0 \in \operatorname{Int}\mathcal{B}$, since there exist the peak-functions required by Theorem A(c): just take $f_{\lambda - z_0}(z) = f_\lambda(z + z_0)$, where f_λ are the peak-functions associated to Λ. □

PROOF OF THEOREM 2. Let us show that there exist peak-functions as in Theorem A(c).

Fix $\lambda_1 \in \Lambda_1$. Applying the previous lemma to Λ_2 and $z_0 = \lambda_1$, we see that there exist $A, B > 0$ and $g_{\lambda_1} \in \mathcal{B}$ such that

$$g_{\lambda_1}(\lambda_1) = 1, \quad \text{and} \quad g_{\lambda_1}(\lambda_2) = 0 \ \forall \lambda_2 \in \Lambda_2,$$
$$\sup_{z\in\mathbb{C}} |g_{\lambda_1}(z)|e^{-A|\operatorname{Im} z|} \le B\frac{e^{C|\operatorname{Im}\lambda_1|}}{d(\lambda_1, \Lambda_2)} \le Be^{C'|\operatorname{Im}\lambda_1|}.$$

Since $\Lambda_1 \in \operatorname{Int}\mathcal{B}$, there exists a function $f_{\lambda_1} \in \mathcal{B}$, as given by Theorem A(c). Then $h_{\lambda_1} = f_{\lambda_1} g_{\lambda_1}$ is a peak-function for $\Lambda_1 \cup \Lambda_2$. □

PROOF OF LEMMA 4. Consider

$$\sin(\pi z) = \pi z \prod_{k\in\mathbb{Z}\setminus\{0\}} \left(1 - \frac{z}{k}\right).$$

If $|\lambda - k| \geq \delta$ for all $\lambda \in \Lambda$ and $k \in \mathbb{Z}$, we can just take $F(z) = \sin(\pi z)$.

For each $k \in \mathbb{Z}$ such that there is $\lambda \in \Lambda$ with $|\lambda - k| < \delta$, we replace in $\sin(\pi z)$ the factor $1 - z/k$ by a small perturbation $1 - z/p_k$, where $p_k = k + i\epsilon_k$ and $\epsilon_k > 0$ is chosen so that

(a) $|\epsilon_k| \leq 5\delta$;
(b) $|p_k - \lambda| \geq \delta$ for all $\lambda \in \Lambda$.

Denoting by $\mathcal{I}$ the indices of $\mathbb{Z} \setminus \{0\}$ affected by this modification, we define

$$F(z) = \sin(\pi z) \prod_{k\in\mathcal{I}} \frac{1 - z/p_k}{1 - z/k}.$$

Notice that the zeros of this new sine-type function are at a distance of at least δ from Λ.

In order to see that $F \in \mathcal{B}$, it is enough to show that the latter product converges and defines a function which has exponential growth away from the points k and p_k and is bounded above and below on $\mathbb{R}$.

Since the product converges to 1 for $z = 0$, it is enough to prove that the modulus of the product converges. We have

$$\begin{aligned} \log\left|\prod_{k\in\mathcal{I}} \frac{1 - z/p_k}{1 - z/k}\right| &= \sum_{k\in\mathcal{I}} \log\left|\frac{1 - z/p_k}{1 - z/k}\right| + \log\left|\frac{k}{p_k}\right| \\ &= \sum_{k\in\mathcal{I}} \log\left|1 - \frac{i\epsilon_k}{z-k}\right| + \sum_{k\in\mathcal{I}} \log\left|1 + \frac{i\epsilon_k}{k}\right|. \end{aligned}$$

The second sum is clearly bounded:

$$\sum_{k\in\mathcal{I}} \log\left|1 + \frac{i\epsilon_k}{k}\right| = \frac{1}{2}\sum_{k\in\mathcal{I}} \log\left(1 + \frac{\epsilon_k^2}{k^2}\right) \lesssim \sum_{k\in\mathbb{N}} \frac{1}{k^2}.$$

For the first sum we have

$$\begin{aligned} \sum_{k\in\mathcal{I}} \log\left|1 - \frac{i\epsilon_k}{z-k}\right| &= \frac{1}{2}\sum_{k\in\mathcal{I}} \log\left(1 + \frac{\epsilon_k^2}{|z-k|^2} - 2\operatorname{Re}\left(\frac{i\epsilon_k}{z-k}\right)\right) \\ &\simeq \sum_{k\in\mathcal{I}} \frac{\epsilon_k^2}{|z-k|^2} + 2\epsilon_k \operatorname{Im}\left(\frac{1}{z-k}\right) \\ &\lesssim \sum_{k\in\mathcal{I}} \frac{1}{|z-k|^2} + \frac{|\operatorname{Im} z|}{|z-k|^2}. \end{aligned}$$

This shows that, away from k and p_k, the product has exponential growth and is bounded on $\mathbb{R}$, as required. □

3. Proof of the geometric conditions

In this section we briefly recall the proof of Theorem B.

3.1. Necessity. The necessity of (i) is a standard (and straightforward) application of Jensen's formula to the functions f_λ provided by Theorem A(c) on the disks $D(\lambda, |\operatorname{Im}\lambda| + 1)$.

To see that (ii) is also necessary consider the functions $g_\lambda(z) = f_\lambda(z)e^{iCz}$, which according to Theorem A(c) are bounded in the upper half-plane $\mathbb{H}$ and satisfy $|g_\lambda(\lambda)| = e^{-C\operatorname{Im}\lambda}$.

Apply now Jensen's formula in the half-plane to the function g_λ:

$$\log|g_\lambda(\lambda)| = \int_{\mathbb{R}} P(\lambda, x)\log|g_\lambda(x)|\,dx - \int_{\mathbb{H}} G(\lambda,\zeta)\Delta\log|g_\lambda(\zeta)|,$$

where $P(\lambda, x)$ denotes the Poisson kernel and $G(\lambda,\zeta) = \log\left|\frac{\lambda-\zeta}{\lambda-\bar\zeta}\right|^{-1}$ is the Green function in $\mathbb{H}$ with pole in λ.

Since g_λ vanishes on $\Lambda\setminus\{\lambda\}$, the equality above yields

$$\sum_{\substack{\lambda'\in\Lambda\cap\mathbb{H}\\ \lambda'\neq\lambda}} \log\left|\frac{\lambda-\lambda'}{\lambda-\bar\lambda'}\right|^{-1} \leq \sup_{\mathbb{R}}\log|g_\lambda| - \log|g_\lambda(\lambda)| \lesssim \operatorname{Im}\lambda + 1.$$

Notice that this implies that $\Lambda\cap\mathbb{H}$ is a Blaschke sequence in the upper half-plane.

The estimate $1 - t \leq \log t^{-1}$ for $t \in (0,1)$ then yields

$$\sup_{\lambda\in\Lambda\cap\mathbb{H}} \sum_{\substack{\lambda'\neq\lambda\\ \operatorname{Im}\lambda'>1}} \frac{\operatorname{Im}\lambda'}{|\lambda-\bar\lambda'|^2} < +\infty.$$

Given $x\in\mathbb{R}$, consider the $\lambda\in\Lambda\cap\mathbb{H}$ closest to x. Then $\frac{\operatorname{Im}\lambda'}{|x-\bar\lambda'|^2} \leq 2\frac{\operatorname{Im}\lambda'}{|\lambda-\bar\lambda'|^2}$. This and the corresponding computation for the lower half-plane imply (ii).

3.2. Sufficiency. Let us see first that condition (ii) implies the weak separation of Λ. Let $\delta_\lambda = \inf_{\lambda'\in\Lambda\setminus\{\lambda\}}|\lambda-\lambda'|$ and assume that $\delta_\lambda \leq 1$. Then

$$N_\Lambda(\lambda, |\operatorname{Im}\lambda| + 1) \geq \int_{\delta_\lambda}^1 \frac{dt}{t} = \log\frac{1}{\delta_\lambda},$$

whence follows the existence of α, $\varepsilon > 0$ as in Lemma 1(i).

Then, according to Theorem 2, it will be enough to show that

$$\Lambda_+ = \Lambda\cap\{\operatorname{Im} z > 1\},\quad \Lambda_0 = \Lambda\cap\{|\operatorname{Im} z| < 1\},\quad\text{and}\quad \Lambda_- = \Lambda\cap\{\operatorname{Im} z < -1\}$$

are $\mathcal{B}$-interpolations.

We start with the sequence Λ_0. The points in Λ_0 are uniformly separated by condition (i), i.e., there is a $\delta > 0$ such that $|\lambda - \lambda'| > \delta$ for all λ, $\lambda' \in \Lambda_0$, $\lambda \neq \lambda'$. Therefore the Beurling–Landau upper density $D^+(\Lambda_0)$ is finite. By the theorem in [**OS99**] stated in the Introduction, the sequence Λ_0 is interpolating for the Bernstein space B_σ if σ is greater than the density of Λ_0. Thus condition (c) of Theorem A is satisfied and the sequence Λ_0 is interpolating for $\mathcal{B}$.

We now deal with the sequence Λ_+. By Theorem A(d), in order to prove that Λ_+ is a $\mathcal{B}$-interpolation, it is enough to construct a function $F \in \mathcal{B}$ such that $\Lambda_+ \subset \mathcal{Z}(F)$ and

$$|F'(\lambda)| \geq \varepsilon e^{-K\operatorname{Im}\lambda},\qquad \lambda\in\Lambda_+,$$

for some constants ε, $k > 0$.

We start with any entire function G with zero set $\mathcal{Z}(G) = \Lambda_+$. Condition (ii) implies that Λ_+ is a Blaschke sequence in $\mathbb{H}$; i.e., the Blaschke product

$$B(z) = \prod_{\lambda \in \Lambda_+} \frac{z-\lambda}{z-\bar{\lambda}}, \quad z \in \mathbb{H},$$

converges. Define

$$\Psi(z) = N|\operatorname{Im} z| - \phi(z),$$

where

$$\phi(z) = \begin{cases} \log\left|\frac{G(z)}{B(z)}\right|, & \operatorname{Im} z > 0, \\ \log|G(z)|, & \operatorname{Im} z \leq 0. \end{cases}$$

A computation shows that ϕ is harmonic outside the real axis, subharmonic on $\mathbb{C}$, and $\Delta\phi(x) = \sum_{\lambda\in\Lambda_+} \frac{\operatorname{Im}\lambda}{|x-\lambda|^2}\, dx$ (see [**MOO06**, Lemma 10]). Thus, by condition (ii), $\Delta\Psi \simeq dx$ when $N \in \mathbb{N}$ is big enough. In this situation, according to [**OS99**, Lemma 3], there exists a multiplier associated with Ψ, i.e., an entire function h such that:

(a) $\mathcal{Z}(h)$ is a separated sequence contained in $\mathbb{R}$ and separated from Λ (i.e., $\inf_{z\in\mathcal{Z}(h)} d(z,\Lambda) > 0$).

(b) Given any $\varepsilon > 0$, $|h(z)| \simeq \exp(\Psi(z))$ for all points z such that $d(z, \mathcal{Z}(h)) > \varepsilon$.

Define now $F = hG$. It is clear that $F \in \mathcal{B}$:

$$|F(z)| \lesssim e^{\Psi(z)+\log|G(z)|} \leq e^{\Psi(z)+\phi(z)} \leq e^{N|\operatorname{Im} z|}, \quad z \in \mathbb{C}.$$

It is also clear that $\Lambda_+ \subset \mathcal{Z}(F)$, since $\Lambda_+ \subset \mathcal{Z}(G)$.

In order to prove that there exist $\varepsilon, C > 0$ such that

$$|F'(\lambda)| \geq \varepsilon e^{-C\operatorname{Im}\lambda}, \tag{2}$$

consider the disjoint disks $D_\lambda = D(\lambda, \delta_\lambda)$, $\delta_\lambda = \varepsilon e^{-\alpha\operatorname{Im}\lambda}$ provided by Lemma 1(i). Since Λ_+ is far from $\mathcal{Z}(h)$, the estimate

$$|F(z)| = |h(z)|e^{\phi(z)}|B(z)| \simeq e^{N|\operatorname{Im} z|}|B(z)|, \quad z \in \partial D_\lambda,$$

holds.

From here we finish as in [**MOO06**]: since the hypotheses imply that there exists $C > 0$ such that $|B(z)| \geq \epsilon e^{-C|\operatorname{Im} z|}$, $z \in \partial D_\lambda$, we have $|F(z)| \gtrsim e^{-C|\operatorname{Im} z|}$ for all $z \in \partial D_\lambda$. The function $g(z) = F(z)/(z-\lambda)$ is then holomorphic, non-vanishing in D_λ, and with $|g(z)| \gtrsim e^{-c\operatorname{Im}\lambda}$ for $z \in \partial D_\lambda$. By the minimum principle

$$|F'(\lambda)| = |g(0)| \gtrsim e^{-c\operatorname{Im}\lambda},$$

as desired.

The sequence Λ_- is treated exactly as Λ_+.

References

[Beu] A. Beurling, *The collected works of Arne Beurling*, Vol. 2. Birkhäuser, Boston, 1989.

[BrGa95] C. A. Berenstein and R. Gay, *Complex analysis and special topics in harmonic analysis*. Springer-Verlag, 1995.

[BrLi95] C. A. Berenstein and B. Q. Li, *Interpolating varieties for spaces of meromorphic functions*, J. Geom. Anal. **5** (1995), 1–48.

[Fa08] S. Yu. Favorov, *Zero sets of entire functions of exponential type with some conditions on the real axis*, Algebra i Analiz **20** (2008), no. 1, 138–145; English transl., St. Petersburg Math. J. **20** (2009), 95–100.

[MOO06] X. Massaneda, J. Ortega-Cerdà, and M. Ounaïes, *A geometric characterization of interpolation in $\hat{\mathcal{E}}'(\mathbb{R})$*, Trans. Amer. Math. Soc. **358** (2006), 3459–3472.

[OS99] J. Ortega-Cerdà and K. Seip, *Multipliers for entire functions and an interpolation problem of Beurling*, J. Funct. Anal. **162** (1999), 400–415.

Departament de Matemàtica Aplicada i Anàlisi, Universitat de Barcelona, Gran Via 585, 08071-Barcelona, Spain

E-mail address: xavier.massaneda@ub.edu

Departament de Matemàtica Aplicada i Anàlisi, Universitat de Barcelona, Gran Via 585, 08071-Barcelona, Spain

E-mail address: jortega@ub.edu

Amer. Math. Soc. Transl.
(2) Vol. **226**, 2009

Integral and Isocapacitary Inequalities

Vladimir Maz'ya

To Victor Havin on his 75th birthday with 45-year affection and admiration

ABSTRACT. It is shown by a counterexample that isocapacitary and isoperimetric constants of a multidimensional Euclidean domain starshaped with respect to a ball are not equivalent. Sharp integral inequalities involving the harmonic capacity which imply the Faber–Krahn property of the fundamental Dirichlet–Laplace eigenvalue are obtained. Necessary and sufficient conditions ensuring integral inequalities between a difference seminorm and the L_p-norm of the gradient are found.

1. Introduction

Isocapacitary inequalities are intimately connected with various properties of Sobolev spaces, especially with norms of embedding operators [**8**], [**12**]–[**16**], [**18**], [**19**], [**21**], [**26**]. For instance, the best constants in some of these inequalities give two-sided estimates for the eigenvalues of boundary value problems [**15**], [**16**], [**18**], [**19**]. Recently, in [**7**] and [**3**], isocapacitary inequalities were applied to the theory of quasi-linear second order elliptic equations.

The present paper deals with three topics related to isocapacitary inequalities. First we show by a counterexample in Section 2 that the fundamental eigenvalue of the Dirichlet Laplacian is not equivalent to an isoperimetric constant, usually called Cheeger's constant [**6**], in contrast with an isocapacitary constant introduced in [**15**] (see also [**18**]).[1] This equivalence, even uniform with respect to the dimension, holds for convex domains, as proved recently by B. Klartag and E. Milman (oral communication), but, as I show, it fails even in the class of domains starshaped with respect to a ball.

2000 *Mathematics Subject Classification.* Primary 54C40, 14E20; Secondary 46E25, 20C20.

Key words and phrases. Dirichlet problem, Laplace operator, eigenvalue, isoperimetric inequalities, isocapacitary inequalities, area minimizing function, Besov spaces, difference seminorms.

The author was partially supported by the UK Engineering and Physical Sciences Research Council grant EP/F005563/1.

[1]By the equivalence of set functions a and b defined on subsets of $\mathbb{R}^n$, I mean the existence of positive constants c_1 and c_2 depending only on n and such that $c_1 a \le b \le c_2 a$.

Section 3 is devoted to certain integral capacitary inequalities which are stronger than the classical Faber–Krahn property of the fundamental Dirichlet–Laplace eigenvalue (see [**28**]).

In Sections 4 and 5 one can find necessary and sufficient conditions for the inequality

$$\left(\int_\Omega\int_\Omega |u(x)-u(y)|^q \mu(dx,dy)\right)^{1/q} \le C\,\|\nabla u\|_{L_p(\Omega)} \tag{1}$$

formulated in terms of the isoperimetric ($q \ge p = 1$) and isocapacitary ($q > p > 1$) inequalities of a new type. Here Ω is a subdomain of a Riemannian manifold, μ is a given measure of two subsets of Ω and u is an arbitrary smooth function.

No characterization of (1) was known previously even for functions on the real line $\mathbb{R}$ (see Problem 3 in [**11**]). A special case of a result obtained at the end of Section 5 is the criterion of the validity of (1) for all $u \in C_0^\infty(\mathbb{R})$:

$$\mu\big([\alpha,\beta], \mathbb{R}\setminus(\alpha - r, \beta + r)\big) \le \mathrm{const}\, r^{-q(1-p)/p},$$

where $r > 0$, $\alpha < \beta$, and the constant factor does not depend on α, β, and r.

The marginal value $q = p > 1$ in (1) has special features and a sufficient condition for (1) is given in Section 6. The article ends with a short discussion of the inequality

$$\left(\int_\Omega |u|^q\,\nu(dx)\right)^{1/q} \le \left(\int_\Omega\int_\Omega |u(x)-u(y)|^p \mu(dx,dy)\right)^{1/p}$$

with a nonnegative measure ν in Ω, μ as above, and $q \ge p \ge 1$.

It is worth mentioning that a Riemannian structure of Ω is not very important for most of the results presented in Sections 3–5. It can be replaced by some natural requirements on the p-energy integral on a metric space (see [**24**], [**10**]).

In this article, I use a number of assertions from the book [**22**], which are not formulated in detail but supplied with references. It is therefore helpful to read the paper with [**22**] close at hand.

2. The first Dirichlet–Laplace eigenvalue and an isoperimetric constant

Let Ω be a subdomain of an n-dimensional Riemannian manifold $\mathfrak{R}_n$, and let $\Lambda(\Omega)$ be the first eigenvalue of the Dirichlet problem for the Laplace operator $-\Delta$ in Ω or, more generally, the greatest lower bound of the spectrum of this operator:

$$\Lambda(\Omega) = \inf_{u\in C_0^\infty(\Omega)} \frac{\|\nabla u\|^2_{L_2(\Omega)}}{\|u\|^2_{L_2(\Omega)}}. \tag{2}$$

By [**15**] (see also [**18**] and [**22**, Corollary 2.3.3]), $\Lambda(\Omega)$ admits the two-sided estimate

$$\frac{1}{4}\Gamma(\Omega) \le \Lambda(\Omega) \le \Gamma(\Omega) \tag{3}$$

with

$$\Gamma(\Omega) := \inf_{\{F\}} \frac{\mathrm{cap}(F;\Omega)}{m_n(F)}.$$

Here m_n denotes n-dimensional Lebesgue measure on $\mathfrak{R}_n$, the infimum is taken over all compact subsets of Ω, and $\mathrm{cap}(F;\Omega)$ stands for the relative harmonic capacity of F with respect to Ω:

$$\operatorname{cap}(F;\Omega) = \inf\Big\{\int_\Omega |\nabla u|^2 dx : u \in C_0^\infty(\Omega),\ \ u \geq 1 \text{ on } F\Big\}.$$

We write $\operatorname{cap}(F)$ instead of $\operatorname{cap}(F;\mathbb{R}^n)$.

By [**22**, Theorem 2.2.1], the set function

(4) $$\gamma(\Omega) = \inf_{u\in C_0^\infty(\Omega)} \frac{\|\nabla u\|_{L_1(\Omega)}}{\|u\|_{L_1(\Omega)}}$$

admits the geometric representation

(5) $$\gamma(\Omega) = \inf_{\{g\}} \frac{H_{n-1}(\partial g)}{m_n(g)},$$

where g is an arbitrary open subset of $\mathfrak{R}_n$ with compact closure $\overline{g}$ in Ω and smooth boundary ∂g, and H_{n-1} is $(n-1)$-dimensional Hausdorff measure. Obviously, for all $u \in C_0^\infty(\Omega)$,

$$\gamma(\Omega) \leq \frac{\displaystyle\int_\Omega |\nabla(u^2)|\, dx}{\displaystyle\int_\Omega u^2 dx} \leq 2\frac{\|\nabla u\|_{L_2(\Omega)}}{\|u\|_{L_2(\Omega)}}.$$

Hence

(6) $$\gamma(\Omega)^2 \leq 4\Lambda(\Omega),$$

which shows, together with (3) and (4), that

(7) $$\gamma(\Omega)^2 \leq 4\Gamma(\Omega)$$

(the square of the isoperimetric constant is dominated by the isocapacitary constant).

One can ask whether an upper bound for $\Gamma(\Omega)$ formulated in terms of $\gamma(\Omega)$ exists. A negative answer is obtained easily if the class of sets Ω is not restricted. In fact, let F be a compact subset of the open n-dimensional unit cube Q in the Euclidean space $\mathbb{R}^n$, and let

$$H_{n-1}(F) = 0 \qquad \text{and} \qquad \operatorname{cap}(F) > 0.$$

By Ω we mean the complement of the union of all integral shifts of F. Now, by [**22**, Theorem 11.2], $\gamma(\Omega) = 0$ and $\Gamma(\Omega) \geq \Lambda(\Omega) > 0$, which gives a negative answer to the question formulated above.

Let us put the same question for domains in $\mathbb{R}^n$ starshaped with respect to balls. We show that the answer stays negative in a certain sense.

EXAMPLE. Let Ω be a subdomain of the n-dimensional unit ball B, starshaped with respect to a concentric ball $B(0;\rho) = \{x : |x| < \rho\}$. Here we show that the inequality opposite to (7):

(8) $$\gamma(\Omega)^2 \geq C\Gamma(\Omega)$$

is impossible with C independent of ρ. Moreover, we shall construct a sequence of domains $\{\Omega_N\}_{n\geq 1}$ situated in B and such that

(i) Ω_N is starshaped with respect to a ball $B(0,\rho_N)$, where $\rho_N \to 0$;

(ii) $\Gamma(\Omega_N) \to \infty$;

(iii) $\gamma(\Omega_N) \leq c$, where c depends only on n.

Let N stand for a sufficiently large integer. By $\{\omega_j\}_{j=1}^{N^{n-1}}$ we denote a collection of points on the unit sphere S^{n-1} placed uniformly in the sense that the distance from every point ω_j to the set of other points of the collection lies between $c_1 N^{-1}$ and $c_2 N^{-1}$, where c_1 and c_2 are positive constants, depending only on n. Consider a closed rotational cone C_j with the axis $O\omega_j$ and the vertex at the distance $c_0 N^{-1}$ from O, where c_0 is an absolute constant large enough. The opening of C_j will be independent of j and denoted by ε_N. Let $\varepsilon_N = o(N^{\frac{1-n}{n-2}})$. Clearly, the complement of C_j is visible from a sufficiently small ball $B(0;\rho_N)$. Therefore, the domain

$$\Omega_N := B \setminus \bigcup_j C_j$$

is starshaped with respect to $B(0,\rho_N)$ (see Figure 1).

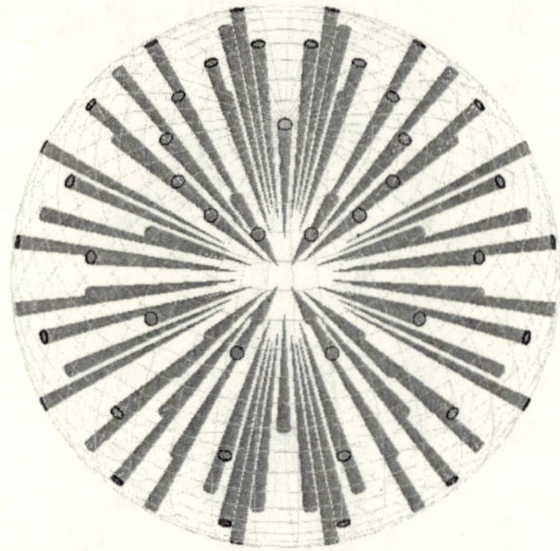

FIGURE 1. The boundary of the subdomain Ω_N of the unit ball is the union of many thin conic surfaces and a spherical part.

We shall find the limit of $\gamma(\Omega_N)$ as $N \to \infty$ as well as a lower estimate for $\Gamma(\Omega_N)$. Clearly, $\gamma(\Omega_N) \geq \gamma(B) = n$. Furthermore, by (5),

$$\gamma(\Omega_N) \leq \frac{H_{n-1}(\partial\Omega_N)}{m_n(\Omega_N)} = \frac{H_{n-1}(\partial B) + H_{n-1}\big(\bigcup_j (B \operatorname{cap} \partial C_j)\big)}{m_n(B) - m_n\big(\bigcup_j (B \operatorname{cap} C_j)\big)}$$
$$\leq \frac{|S^{n-1}| + c_1 \varepsilon_N^{n-1} N^{n-1}}{|S^{n-1}|/n - c_2 \varepsilon_N^{n-2} N^{n-1}},$$

and therefore,

$$\lim_{N\to\infty} \gamma(\Omega_N) = n.$$

In order to estimate $\Gamma(\Omega_N)$ from below, we construct a covering of B by the balls $\mathcal{B}_k := B(x_k, 4c_0 N^{-1})$, whose multiplicity does not exceed a constant depending only on n. Suppose $|x_k| \geq c_0 N^{-1}$. Theorem 10.1.2 in [**22**] implies

$$cN^n \operatorname{cap}(\mathcal{B}_k \setminus \Omega_N) \int_{\mathcal{B}_k} u^2\,dx \leq \int_{\mathcal{B}_k} |\nabla u|^2\,dx \tag{9}$$

for all $u \in C_0^\infty(\Omega_N)$, and the result will stem from a proper lower bound for $\operatorname{cap}(\mathcal{B}_k \setminus \Omega_N)$.

First, consider $n = 3$. Clearly, $\mathcal{B}_k \setminus \Omega_N$ contains a right rotational cylinder T_k with height $c_0 N^{-1}$ and diameter of the base $\varepsilon_N N^{-1}$. Now, by [**22**, Proposition 9.1.3/1],

$$\operatorname{cap}(T_k) \geq cN^{-1}|\log \varepsilon_N|^{-1}.$$

This estimate in combination with (9) gives

$$cN^2|\log \varepsilon_N|^{-1} \int_{\mathcal{B}_k} u^2\,dx \leq \int_{\mathcal{B}_k} |\nabla u|^2\,dx. \tag{10}$$

Choosing $\varepsilon_N = \exp(-N)$ and summing (10) over all balls $\mathcal{B}_k$, we obtain $\lambda(\Omega_N) \geq cN$. Hence $\lambda(\Omega_N) \to \infty$ whereas $\gamma(\Omega_N) \leq c$. Thus, in particular, there is no inequality

$$\Bigl(\inf_{\{g\}} \frac{s(\partial g)}{m_3(g)}\Bigr)^2 \geq C \inf_{\{F\}} \frac{\operatorname{cap}(F;\Omega)}{m_3(F)}$$

and, equivalently,

$$\Bigl(\inf_{\{g\}} \frac{s(\partial g)}{m_3(g)}\Bigr)^2 \geq C\Lambda(\Omega)$$

with constant factors C independent of the radius ρ.

For dimensions greater than 3, the very end of the argument remains intact, but the estimation of $\operatorname{cap}(\mathcal{B}_k \setminus \Omega_N)$ becomes a bit more complicated and the choice of ε_N will be different.

Let α, $\mathcal{B}_k$ stand for the ball concentric with $\mathcal{B}_k$ and dilated with coefficient α. We introduce the set $s_k = \{j : C_j \operatorname{cap} \frac{1}{2}\mathcal{B}_k \neq \emptyset\}$. With every j in s_k we associate a right rotational cylinder T_j coaxial with the cone C_j and situated in $C_j \operatorname{cap} \frac{1}{2}\mathcal{B}_k$. The height of T_j will be equal to $c_0 N^{-1}$ and the diameter of the base equal to $\varepsilon_N |x_k|$. We define a cut-off function η_j, equal to 1 on the $\varepsilon_N|x_k|$-neighbourhood of T_j, zero outside the $2\varepsilon_N|x_k|$-neighbourhood of T_j and satisfying the estimate

$$|\nabla \eta_j(x)| \leq c\delta(x)^{-1},$$

where $\delta(x)$ is the distance from x to the intersection of T_j with the axis of C_j.

By $\mathcal{P}_k$ we denote the equilibrium potential of $\mathcal{B}_k \setminus \Omega_N$. We have

$$\begin{aligned}\sum_{j\in s_k} \operatorname{cap}(C_j \operatorname{cap} \mathcal{B}_k) &\leq \sum_{j \in s_k} \int_{\mathbb{R}^n} |\nabla(\mathcal{P}_k \eta_j)|^2\,dx \\ &\leq c\Bigl(\int_{\mathbb{R}^n} |\nabla \mathcal{P}_k|^2\,dx + \int_{\mathbb{R}^n} \mathcal{P}_k^2 \delta^{-2}\,dx\Bigr).\end{aligned}$$

Changing the constant c, one can majorize the last integral by the previous one due to Hardy's inequality. Hence,

$$\operatorname{cap}(\mathcal{B}_k \setminus \Omega_N) \geq c \sum_{j\in s_k} \operatorname{cap}(T_j). \tag{11}$$

By [**22**, Proposition 9.1.3/1],

$$\operatorname{cap}(T_j) \geq c\bigl(\varepsilon_N|x_k|\bigr)^{n-3} N^{-1}.$$

Furthermore, it is seen that the number of integers in s_k is between two multiples of $|x_k|^{1-n}$. Now, by (11),

$$\operatorname{cap}(\mathcal{B}_k \setminus \Omega_N) \geq c|x_k|^{1-n}\bigl(\varepsilon_N|x_k|\bigr)^{n-3} N^{-1},$$

and by (9),

$$cN^{n-1}|x_k|^{-2}\varepsilon_N^{n-3}\int_{\mathcal{B}_k} u^2\,dx \le \int_{\mathcal{B}_k} |\nabla u|^2\,dx. \tag{12}$$

Since $|x_k| \le 1$, it follows by summation of (12) over k that

$$\lambda(\Omega_N) \ge c\varepsilon_N^{n-3}N^{n-1}.$$

Putting, for instance,

$$\varepsilon_N = N^{(1-n)/(n-5/2)},$$

we see that $\Gamma(\Omega_N) \to \infty$, and the desired counterexample is constructed for $n > 3$.

3. Capacitary improvement of the Faber–Krahn inequality

We state and prove the main result of this section. Here Ω is an open subset of an arbitrary n-dimensional Riemannian manifold.

THEOREM 1. *Suppose $\mathcal{R} > 0$, $u \in C_0^\infty(\Omega)$, and $N_t = \{x \in \Omega : |u(x)| \ge t\}$. If $n > 2$, then*

$$\Big(\frac{j_{(n-2)/2}}{\mathcal{R}}\Big)^2 m_n(B_{\mathcal{R}})\int_0^\infty \Big(\frac{\operatorname{cap}(N_t;\Omega)}{\operatorname{cap}(B_{\mathcal{R}}) + \operatorname{cap}(N_t;\Omega)}\Big)^{\frac{n}{n-2}} d(t^2) \le \|\nabla u\|^2_{L_2(\Omega)}, \tag{13}$$

where j_ν is the first positive root of the Bessel function J_ν. If $n = 2$, then

$$\pi j_0^2 \int_0^\infty \exp\Big(\frac{-4\pi}{\operatorname{cap}(N_t;\Omega)}\Big)\, d(t^2) \le \|\nabla u\|^2_{L_2(\Omega)}. \tag{14}$$

PROOF. Let w be an arbitrary absolutely continuous function on $(0, \mathcal{R}]$ such that $w(\mathcal{R}) = 0$. The inequality

$$\Big(\frac{j_{(n-2)/2}}{\mathcal{R}}\Big)^2 \int_0^{\mathcal{R}} w(\rho)^2 \rho^{n-1} d\rho \le \int_0^{\mathcal{R}} w'(\rho)^2\rho^{n-1} d\rho, \tag{15}$$

where $n > 2$, is equivalent to the fact that the first eigenvalue of the Dirichlet–Laplace operator in the unit ball B equals $j^2_{(n-2)/2}$. Similarly, with $n = 2$ the inequality

$$\Big(\frac{j_0}{\mathcal{R}}\Big)^2 \int_0^{\mathcal{R}} w(\rho)^2 \rho\, d\rho \le \int_0^{\mathcal{R}} w'(\rho)^2 \rho\, d\rho \tag{16}$$

is associated.

In the case of $n > 2$, we introduce the new variables

$$\psi = \frac{\rho^{2-n} - \mathcal{R}^{2-n}}{(n-2)|S^{n-1}|}, \qquad t(\psi) = w(\rho(\psi)),$$

and write (15) in the form

$$\Big(\frac{|S^{n-1}| j_{(n-2)/2}}{\mathcal{R}}\Big)^2 \int_0^\infty \frac{t(\psi)^2 d\psi}{\big((n-2)|S^{n-1}|\psi + \mathcal{R}^{2-n}\big)^{2(n-1)/(n-2)}} \le \int_0^\infty t'(\psi)^2 d\psi. \tag{17}$$

Similarly, for $n = 2$, putting

$$\psi = (2\pi)^{-1}\log\frac{\mathcal{R}}{\rho}, \qquad t(\psi) = w(\rho(\psi)),$$

we write (16) as

$$(2\pi j_0)^2 \int_0^\infty t(\psi)^2 \exp(-4\pi\psi)\, d\psi \le \int_0^\infty t'(\psi)^2 d\psi. \tag{18}$$

Note that the function t in (17) and (18) is subject to the boundary condition $t(0) = 0$. We write (17) and (18) as

(19)

$$n^{-1}|S^{n-1}| \left(\frac{j_{(n-2)/2}}{\mathcal{R}}\right)^2 \int_0^\infty \frac{dt(\psi)^2}{\big((n-2)|S^{n-1}|\psi + \mathcal{R}^{2-n}\big)^{n/(n-2)}} \le \int_0^\infty t'(\psi)^2\, d\psi$$

and

$$\pi j_0^2 \int_0^\infty \exp(-4\pi\psi)\, dt(\psi)^2 \le \int_0^\infty t'(\psi)^2 d\psi. \tag{20}$$

Now, as in [**22**, Section 2.2.1], we introduce the function

$$\psi(t) = \int_0^t \frac{d\tau}{\int_{|u|=\tau} |\nabla u|\, dH_{n-1}}, \tag{21}$$

as well as its inverse $\psi \to t(\psi)$, and replace the integral on the right-hand side of (19) and (20) by $\|\nabla u\|^2_{L_2(\Omega)}$. It remains to note that

$$\psi \le \big(\text{cap}(N_{t(\psi)}; \Omega)\big)^{-1} \tag{22}$$

by [**22**, Lemma 2.2.2/1]. □

Let us use the area minimizing function of Ω:

$$\lambda(v) = \inf H_{n-1}(\partial g), \tag{23}$$

where the infimum is extended over all sets g with smooth boundaries and compact closures $\overline{g} \subset \Omega$, subject to the inequality $m_n(g) \ge v$. This and related geometrical characterizations of Ω proved to be useful in the theory of Sobolev spaces and elliptic equations; see [**5**], [**12**], [**14**], and [**20**]. The function λ appears in the lower estimate of the capacity

$$\text{cap}(F; \Omega) \ge \Big(\int_{m_n(F)}^{m_n(\Omega)} \frac{dv}{\lambda(v)^2}\Big)^{-1}$$

(see [**22**, Corollary 2.2.3/2]). Therefore, (13), (14), and the identity

$$\text{cap}(B_{\mathcal{R}}) = (n-2)|S^{n-1}|\mathcal{R}^{n-2}$$

lead to the following Lorentz-type estimates.

COROLLARY 1. *If $n > 2$ and $\mathcal{R} > 0$, then, for all $u \in C_0^\infty(\Omega)$,*

(24)

$$\Big(\frac{j_{(n-2)/2}}{\mathcal{R}}\Big)^2 m_n(B_{\mathcal{R}}) \int_0^\infty \Big(\text{cap}(B_{\mathcal{R}}) \int_{m_n(N_t)}^{m_n(\Omega)} \frac{dv}{\lambda(v)^2} + 1\Big)^{\frac{n}{2-n}} d(t^2) \le \|\nabla u\|^2_{L_2(\Omega)}.$$

If $n = 2$, then, for all $u \in C_0^\infty(\Omega)$,

$$\pi j_0^2 \int_0^\infty \exp\Big(-4\pi \int_{m_n(N_t)}^{m_n(\Omega)} \frac{dv}{\lambda(v)^2}\Big) d(t^2) \le \|\nabla u\|^2_{L_2(\Omega)}. \tag{25}$$

REMARK 1. Since

$$\lambda(v) \geq n^{\frac{n-1}{n}} |S^{n-1}|^{\frac{1}{n}} v^{\frac{n-1}{n}} \tag{26}$$

by the classical isoperimetric inequality for $\mathbb{R}^n$, the estimates (24) and (25) imply the Faber–Krahn property

$$\Lambda(\Omega) \geq \left(\frac{j_{(n-2)/2}}{\mathcal{R}}\right)^2$$

for any n-dimensional Euclidean domain Ω with $m_n(\Omega) = n^{-1}|S^{n-1}|\mathcal{R}^n$.

Theorem 1 is a very special case of the following general assertion.

THEOREM 2. *Let* $\mathbf{M}$ *be a decreasing nonnegative function on* $[0, \infty)$ *and let* $q > 0$ *and* $p \geq 1$. *Suppose that for all absolutely continuous functions* $\psi \to t(\psi)$ *on* $[0, \infty)$, *the inequality*

$$\left(-\int_0^\infty |t(\psi)|^q d\mathbf{M}(\psi)\right)^{1/q} \leq \left(\int_0^\infty |t'(\psi)|^p d\psi\right)^{1/p} \tag{27}$$

holds. Then, for all $u \in C_0^\infty(\Omega)$,

$$\left(\int_0^\infty \mathbf{M}\Big(\big(\mathrm{cap}_p(N_t;\Omega)\big)^{1/(1-p)}\Big) d(t^q)\right)^{1/q} \leq \|\nabla u\|_{L_p(\Omega)}, \tag{28}$$

where cap_p *is the p-capacity defined by*

$$\mathrm{cap}_p(F;\Omega) = \inf\Big\{\int_\Omega |\nabla u|^p dx : u \in C_0^\infty(\Omega),\ \ u \geq 1 \ \ on \ \ F\Big\}. \tag{29}$$

PROOF. The role of the function ψ given by (21) is played in the present proof by

$$\psi(t) = \int_0^t \frac{d\tau}{\left(\int_{|u|=\tau} |\nabla u|^{p-1}\, dH_{n-1}\right)^{1/(p-1)}}. \tag{30}$$

We write the left-hand side of (27) in the form

$$\left(\int_0^\infty \mathbf{M}(\psi)\, d(t(\psi))^q\right)^{1/q}$$

and use the monotonicity of $\mathbf{M}$ and the inequality

$$\psi \leq \big(\mathrm{cap}_p(N_{t(\psi)};\Omega)\big)^{1/(1-p)} \tag{31}$$

proved in [**22**, Lemma 2.2.2/1]. It remains to apply (27) and the identity

$$\int_0^\infty |f'(\psi)|^p\, d\psi = \int_\Omega |\nabla u|^p\, dx \tag{32}$$

found in [**22**, Lemma 2.3.1]. □

Using the area minimizing function λ defined by (23) and the estimate

$$\mathrm{cap}_p(F;\Omega) \geq \left(\int_{m_n(F)}^{m_n(\Omega)} \frac{dv}{\lambda(v)^{p/(p-1)}}\right)^{1-p} \tag{33}$$

(see [**22**, Corollary 2.2.3/2]), from Theorem 2 we deduce the following.

COROLLARY 2. *Let μ, p, and q be the same as in Theorem 2 and let (27) hold. Then*

$$\Big(\int_0^\infty \mathbf{M}\Big(\int_{m_n(N_t)}^{m_n(\Omega)} \frac{dv}{\lambda(v)^{p/(p-1)}}\Big) d(t^q)\Big)^{1/q} \le \|\nabla u\|_{L_p(\Omega)} \tag{34}$$

for all $u \in C_0^\infty(\Omega)$.

Clearly, (34) is a generalization of the estimates (24) and (25), which were obtained for $p = 2$ with a particular choice of μ. Another obvious remark is that (27), where $\mathbf{M}$ is defined on the interval $0 < t < m_n(\Omega)$ by

$$\mathbf{M}\Big(\int_t^{m_n(\Omega)} \frac{dv}{\lambda(v)^{p/(p-1)}}\Big) = \Lambda_{p,q} t$$

with a constant $\Lambda_{p,q}$ depending on $m_n(\Omega)$, implies the inequality

$$\Lambda_{p,q}^{1/q} \|u\|_{L_q(\Omega)} \le \|\nabla u\|_{L_p(\Omega)} \tag{35}$$

for all $u \in C_0^\infty(\Omega)$.

4. Criterion for an upper estimate of a difference seminorm (the case $p = 1$)

Consider the seminorm

$$\langle u\rangle_{q,\mu} = \Big(\int_\Omega\int_\Omega |u(x) - u(y)|^q \mu(dx, dy)\Big)^{1/q}, \tag{36}$$

where Ω is an open subset of a Riemannian manifold and μ is a nonnegative measure on $\Omega \times \Omega$, locally finite outside the diagonal $\{(x, y) : x = y\}$. By definition, the product $0 \cdot \infty$ equals zero.

In this section, we first characterize both μ and Ω subject to the inequality

$$\langle u\rangle_{q,\mu} \le C\, \|\nabla u\|_{L_1(\Omega)}, \tag{37}$$

where $q \ge 1$ and u is an arbitrary function in $C^\infty(\Omega)$. We show that (37) is equivalent to a somewhat unusual relative isoperimetric inequality.

THEOREM 3. *Inequality (37) holds for all $u \in C^\infty(\Omega)$ with $q \ge 1$ if and only if for any open subset g of Ω such that Ω cap ∂g is smooth, the inequality*

$$\big(\mu(g, \Omega \setminus \overline{g}) + \mu(\Omega \setminus \overline{g}, g)\big)^{1/q} \le C H_{n-1}(\Omega \cap \partial g) \tag{38}$$

holds with the same value of C as in (37). In particular, a constant C in (37) exists if and only if

$$\sup_{\{g\}} \frac{\mu(g, \Omega \setminus \overline{g})^{1/q}}{H_{n-1}(\Omega \cap \partial g)} < \infty. \tag{39}$$

PROOF. *Sufficiency.* Denote by u_+ and u_- the positive and negative parts of u, so that $u = u_+ - u_-$. We notice that

$$\langle u\rangle_{q,\mu} \le \langle u_+\rangle_{q,\mu} + \langle u_-\rangle_{q,\mu} \tag{40}$$

and

$$\int_\Omega |\nabla u|\, dx = \int_\Omega |\nabla u_+|\, dx + \int_\Omega |\nabla u_-|\, dx. \tag{41}$$

First, we obtain (37) separately for $u = u_+$ and $u = u_-$. Let $a > b$ and let $\chi_t(a,b) = 1$ if $a > t > b$ and $\chi_t(a,b) = 0$ otherwise.

Clearly,

$$\begin{aligned}\langle u\rangle_{q,\mu} &= \left(\int_\Omega\int_\Omega \left|\int_{u(x)}^{u(y)} dt\right|^q \mu(dx,dy)\right)^{1/q}\\ &= \left(\int_\Omega\int_\Omega \left|\int_0^\infty (\chi_t(u(x),u(y)) + \chi_t(u(y),u(x)))\, dt\right|^q \mu(dx,dy)\right)^{1/q}.\end{aligned}$$

By Minkowski's inequality,

$$\begin{aligned}\langle u\rangle_{q,\mu} &\le \int_0^\infty \left(\int_\Omega\int_\Omega (\chi_t(u(x),u(y)) + \chi_t(u(y),u(x)))^q \mu(dx,dy)\right)^{1/q} dt\\ &= \int_0^\infty \left(\int_\Omega\int_\Omega (\chi_t(u(x),u(y)) + \chi_t(u(y),u(x)))\mu(dx,dy)\right)^{1/q} dt\\ &= \int_0^\infty \left(\mu(M_t, \Omega\setminus N_t) + \mu(\Omega\setminus N_t, M_t)\right)^{1/q} dt,\end{aligned}$$

where $M_t = \{x \in \Omega : u(x) > t\}$ and $N_t = \{x \in \Omega : u(x) \ge t\}$.

By (38) and the coarea formula, the last integral does not exceed

$$C\int_0^\infty H_{n-1}\big(\{x\in\Omega : u(x) > t\}\big)\, dt = C\int_\Omega |\nabla u(x)|\, dx.$$

Therefore,

$$\langle u_\pm\rangle_{q,\mu} \le C\int_\Omega |\nabla u_\pm(x)|\, dx$$

and the reference to (40) and (41) completes the proof of sufficiency.

Necessity. Let $\{w_m\}$ be the sequence of locally Lipschitz functions in Ω constructed in [**22**, Lemma 3.2.2] with the following properties:

(1) $w_m(x) = 0$ in $\Omega\setminus g$;
(2) $w_m(x) \in [0,1]$ in Ω;
(3) for any compact set $K \subset g$ there exists an integer $N(e)$ such that $w_m(x) = 1$ for $x \in K$ and $m \ge N(e)$;
(4) the limit relation holds,

$$\limsup_{m\to\infty} \int_\Omega |\nabla w_m(x)|\, dx = H_{n-1}(\Omega\cap\partial g).$$

By [**22**, Theorem 1.1.5/1], the inequality (37) holds for all locally Lipschitz functions. Therefore,

$$\langle w_m\rangle_{q,\mu} \le C\,\|\nabla w_m\|_{L_1(\Omega)} \tag{42}$$

and due to property (4),

$$\limsup_{m\to\infty} \langle w_m\rangle_{q,\mu} \le CH_{n-1}(\Omega\cap\partial g). \tag{43}$$

On the other hand,

$$\langle w_m\rangle_{q,\mu}^q = \int_{x\in g}\int_{y\in\Omega\setminus g} w_m(x)^q\mu(dx,dy) + \int_{x\in\Omega\setminus g}\int_{y\in g} w_m(y)^q\mu(dx,dy)$$
$$+\int_g\int_g |w_m(x)-w_m(y)|^q\mu(dx,dy),$$

which implies

$$\langle w_m\rangle_{q,\mu}^q \geq \int_g w_m(x)^q\mu(dx,\Omega\setminus\overline{g}) + \int_g w_m(y)^q\mu(\Omega\setminus\overline{g},dy).$$

This, along with property (3), leads to

$$\liminf_{m\to\infty}\langle w_m\rangle_{q,\mu}^q \geq \mu(g,\Omega\setminus\overline{g}) + \mu(\Omega\setminus g,\overline{g}).$$

Combining this relation with (42) and (43), we arrive at (38). □

COROLLARY 3 (One-dimensional case). *Let*

$$\Omega = (\alpha,\beta), \quad \textit{where} \quad -\infty\leq\alpha<\beta\leq\infty.$$

The inequality

$$\left(\int_\Omega\int_\Omega |u(x)-u(y)|^q\mu(dx,dy)\right)^{1/q} \leq C\int_\Omega |u'(x)|\,dx \tag{44}$$

with $q\geq 1$ *holds for all* $u\in C^\infty(\Omega)$ *if and only if*

$$\big(\mu(I,\Omega\setminus\overline{I}) + \mu(\Omega\setminus\overline{I},I)\big)^{1/q} \leq 2C \tag{45}$$

for all open intervals I *such that* $\overline{I}\subset\Omega$, *and*

$$\big(\mu(I,\Omega\setminus\overline{I}) + \mu(\Omega\setminus\overline{I},I)\big)^{1/q} \leq C \tag{46}$$

for all intervals $I\subset\Omega$ *such that* $\overline{I}$ *contains one of the endpoints of* Ω.

In particular, a constant in (44) *exists if and only if*

$$\sup_{\{I\}}\mu(I,\Omega\setminus\overline{I}) < \infty.$$

PROOF. Necessity follows directly from (38) by setting $g=I$. Let us check the sufficiency of (45). We represent an arbitrary open set g of Ω as the union of nonoverlapping open intervals I_k. Then by (45) and (46),

$$\big(\mu(g,\Omega\setminus\overline{g}) + \mu(\Omega\setminus\overline{g},g)\big)^{1/q} = \left(\sum_k\big(\mu(I_k,\Omega\setminus\overline{g}) + \mu(\Omega\setminus\overline{g},I_k)\big)\right)^{1/q}$$
$$\leq \sum_k\big(\mu(I_k,\Omega\setminus\overline{g}) + \mu(\Omega\setminus\overline{g},I_k)\big)^{1/q} \leq C\sum_k H_0(\Omega\cap\partial I_k),$$

which is the same as (38). The result follows from Theorem 3. □

REMARK 2. Suppose that the class of admissible functions in Theorem 3 is restricted by the requirement that $u=0$ *in a neighbourhood of a closed subset* F *of* $\overline{\Omega}$. Then the same proof leads to the same criterion (38) with the only difference that the admissible sets g should be at a positive distance from F. For the example

$F = \partial\Omega$, i.e., for the inequality (37) with any $u \in C_0^\infty(\Omega)$, the necessary and sufficient condition (38) becomes the isoperimetric inequality

$$(47) \qquad \big(\mu(g, \Omega \setminus \overline{g}) + \mu(\Omega \setminus \overline{g}, g)\big)^{1/q} \le C H_{n-1}(\partial g)$$

for all open sets g with smooth boundary and compact closure $\overline{g} \subset \Omega$. Thus, in particular, in Corollary 3, the criterion of the validity of (44) for all $u \in C_0^\infty(\Omega)$ is the inequality (45) for every interval I, $\overline{I} \subset \Omega$. In the case of $u = 0$ near one of the endpoints $\Omega = (\alpha, \beta)$, we must require both (45) and (46) but the intervals I should be at a positive distance from that endpoint.

Needless to say, the condition (38) is simplified as follows for a symmetric measure μ, i.e., under the assumption $\mu(\mathcal{E}, \mathcal{F}) = \mu(\mathcal{F}, \mathcal{E})$:

$$\mu(g, \Omega \setminus \overline{g})^{1/q} \le 2^{-1/q} C H_{n-1}(\Omega \cap \partial g)$$

for the same open sets g as in Theorem 3.

REMARK 3. The integration domain $\Omega \times \Omega$ in (36) excludes inequalities for integrals taken over $\partial\Omega$. This can be easily avoided assuming additionally that μ is defined on compact subsets of $\overline{\Omega} \times \overline{\Omega}$ and that $u \in C(\overline{\Omega}) \cap C^\infty(\Omega)$. Then, with the same proof, one obtains the corresponding criterion, similar to (38):

$$\big(\mu(\overline{g}, \overline{\Omega} \setminus \overline{g}) + \mu(\overline{\Omega} \setminus \overline{g}, \overline{g})\big)^{1/q} \le C H_{n-1}(\Omega \cap \partial g).$$

As an application, consider the inequality

$$(48) \qquad \int_{\partial\Omega} \int_{\partial\Omega} |u(x) - u(y)|\, H_{n-1}(dx)\, H_{n-1}(dy) \le C \int_\Omega |\nabla u|\, dx,$$

which holds if and only if

$$(49) \qquad H_{n-1}(\partial\Omega \cap \partial g)\, H_{n-1}(\partial\Omega \setminus \partial g) \le 2^{-1} C H_{n-1}(\Omega \cap \partial g)$$

for the same sets g as in Theorem 3.

By [**22**, Corollary 6.4.4/3], which appeared first in [**5**],

(i) if Ω is the unit ball in $\mathbb{R}^3$, then

$$4\pi H_2(\Omega \cap \partial g) \ge H_2(\partial\Omega \cap \partial g)\, H_2(\partial\Omega \setminus \partial g),$$

and

(ii) if Ω is the unit disk on the plane, then

$$H_1(\Omega \cap \partial g) \ge 2 \sin\Big(\frac{1}{2} H_1(\partial\Omega \cap \partial g)\Big).$$

Moreover, the last two inequalities are sharp. Hence, the inequality (48) holds with the best constant $C = 8\pi$ if $\Omega = B$. In the case (ii),

$$H_1(\Omega \cap \partial g) \ge 2^{-1} \min_{0 \le \varphi \le \pi} \frac{\sin\varphi}{\varphi(\pi - \varphi)} H_1(\partial\Omega \cap \partial g) H_1(\partial\Omega \setminus \partial g).$$

Since the last minimum equals π^{-1}, it follows that the best value of C in the inequality (48) for the unit disk is 4π.

We can simplify the criterion (38) for $\Omega = \mathbb{R}^n$, replacing arbitrary sets g by arbitrary balls $B(x,\rho)$ as in [**22**, Theorem 1.4.2/2], where the norm

$$\|u\|_{L_q(\mu)} = \Big(\int_{\mathbb{R}^n} |u|^q \, d\mu\Big)^{1/q}$$

is treated in place of $\langle u\rangle_{q,\mu}$. Unfortunately, the best constant in the sufficiency part will be lost.

COROLLARY 4. (i) *If* $q \geq 1$ *and*

$$\sup_{x\in\mathbb{R}^n,\rho>0} \rho^{(1-n)q}\big(\mu(B(x,\rho),\mathbb{R}^n\setminus B(x,\rho)) + \mu(\mathbb{R}^n\setminus B(x,\rho),B(x,\rho))\big) < \infty, \tag{50}$$

then the inequality

$$\Big(\int_{\mathbb{R}^n}\int_{\mathbb{R}^n} |u(x)-u(y)|^q \mu(dx,dy)\Big)^{1/q} \leq C\,\|\nabla u\|_{L_1(\mathbb{R}^n)} \tag{51}$$

holds for all $u \in C^\infty(\mathbb{R}^n)$ *and*

$$C^q \leq c^q \sup_{x\in\mathbb{R}^n,\rho>0} \rho^{(1-n)q}\big(\mu(B(x,\rho),\mathbb{R}^n\setminus B(x,\rho)) + \mu(\mathbb{R}^n\setminus B(x,\rho),B(x,\rho))\big), \tag{52}$$

where c *depends only on* n.

(ii) *If* (51) *holds for all* $u \in C^\infty(\mathbb{R}^n)$, *then*

$$C^q \geq |S^{n-1}|^{-q} \sup_{x\in\mathbb{R}^n,\rho>0} \rho^{(1-n)q}\big(\mu(B(x,\rho),\mathbb{R}^n\setminus B(x,\rho)) + \mu(\mathbb{R}^n\setminus B(x,\rho),B(x,\rho))\big).$$

PROOF. Let g be an arbitrary open set in $\mathbb{R}^n$ with smooth boundary and let $\{B(x_j,\rho_j)\}$ be the Gustin covering of g subject to

$$\sum_j \rho_j^{n-1} \leq c\,H_{n-1}(\partial g), \tag{53}$$

where c depends only on n (see [**22**, Theorem 1.2.2/2]). Then

$$\begin{aligned}
\mu(g,\mathbb{R}^n\setminus g) &\leq \sum_j \mu(B(x_j,\rho_j),\mathbb{R}^n\setminus g)\\
&\leq \Big(\sum_j \mu(B(x_j,\rho_j),\mathbb{R}^n\setminus g)^{1/q}\Big)^q\\
&\leq \Big(\sum_j \mu(B(x_j,\rho_j),\mathbb{R}^n\setminus B(x_j,\rho_j))^{1/q}\Big)^q\\
&\leq (c\,B)^q\big(\sum_j \rho_j^{n-1}\big)^q,
\end{aligned}$$

where B is the value of the supremum in (50). This and (53) imply

$$\mu(g,B(x_j,\rho_j)) \leq (cBH_{n-1}(\partial g))^q.$$

Similarly,

$$\mu(\mathbb{R}^n\setminus g,g) \leq (cBH_{n-1}(\partial g))^q$$

and the result follows from Theorem 3.

The assertion (ii) stems from (38) by setting $g = B(x,\rho)$. □

5. Criterion for an upper estimate of a difference norm (the case $p > 1$)

Now we deal with the inequality

$$\langle u \rangle_{q,\mu} \le C \, \|\nabla u\|_{L_p(\Omega)}, \tag{54}$$

where $q > p > 1$, and show that it is equivalent to a certain isocapacitary inequality.

The capacity to appear in the present context is defined as follows. Let F_1 and F_2 be nonoverlapping subsets of Ω, closed in Ω. The p-capacity of the pair (F_1, F_2) with respect to Ω is given by

$$\operatorname{cap}_p(F_1, F_2; \Omega) = \inf_{\{u\}} \int_\Omega |\nabla u(x)|^p \, dx,$$

where $\{u\}$ is the set of all $u \in C^\infty(\Omega)$ such that $u \ge 1$ on F_1 and $u \le 0$ on F_2.

Obviously, this capacity does not change if F_1 and F_2 change places. Furthermore, if F is a closed set in $\mathbb{R}^n$ and $F \subset G$, where G is an open set such that $\overline{G} \subset \Omega$, then $\operatorname{cap}_p(F, \Omega \setminus G; \Omega)$ coincides with the p-capacity $\operatorname{cap}_p(F; G)$ defined in (29).

THEOREM 4. *Inequality* (54) *with* $p \in (1, q)$ *holds for all* $u \in C^\infty(\Omega)$ *if and only if for any pair* (F_1, F_2) *of nonoverlapping sets closed in* Ω *we have*

$$\mu(F_1, F_2)^{p/q} \le B \operatorname{cap}_p(F_1, F_2; \Omega), \tag{55}$$

where B *depends only on* p *and* q. *In the sufficiency part we may assume that* F_1 *and* F_2 *are sets with smooth* $\Omega \cap \partial F_i$.

In the proof of this theorem, we use the inequality

$$\left(\int_{\mathbb{R}_+} |f(\psi)|^q \psi^{-1-q/p'} \, d\psi \right)^{1/q} \le c \, \|f'\|_{L_p(\mathbb{R}_+)} \tag{56}$$

due to Bliss [**4**] and the inequality

$$\left(\int_{\mathbb{R}_+} \int_{\mathbb{R}_+} \frac{|f(\psi) - f(\phi)|^q}{|\psi - \phi|^{2+q/p'}} \, d\phi \, d\psi \right)^{1/q} \le c \, \|f'\|_{L_p(\mathbb{R}_+)}, \tag{57}$$

where $q > p > 1$, $p' = p/(p-1)$ and f is an arbitrary absolutely continuous function on $\overline{\mathbb{R}}_+$.

A short argument leading to (57) is as follows. Clearly, (57) results from the same inequality with $\mathbb{R}$ in place of $\mathbb{R}_+$, which in turn follows from the estimate

$$\|f\|_{B_q^{1-(q-p)/pq}(\mathbb{R})} \le c \, \|f\|_{W_p^1(\mathbb{R})} \tag{58}$$

by dilation with a coefficient λ and the limit passage as $\lambda \to 0_+$. (The standard notation B and W for Besov and Sobolev spaces with nonhomogeneous norms is used in (57).) In order to obtain (58), we recall the well-known Sobolev type inequality

$$\|h\|_{L_{p'}(\mathbb{R}_+)} \le c \, \|h\|_{B_{q'}^{(q-p)/pq}(\mathbb{R})}$$

(see [**29**, Sec. 5.1, Theorem 4′]) and put $h = (-\Delta + 1)^{-1/2} f$, which shows that

$$\|f\|_{W_{p'}^{-1}(\mathbb{R})} \le c \, \|f\|_{B_{q'}^{-1+(q-p)/pq}(\mathbb{R})}. \tag{59}$$

By duality, (59) is equivalent to (58).

With (57) at hand, we return to Theorem 4.

PROOF. *Sufficiency.* Arguing as at the beginning of the the proof of Theorem 2, we see that it suffices to prove (54) for a nonnegative u. By the definition of the Lebesgue integral

$$\int_\Omega u\,d\nu = \int_{\mathbb{R}_+} \nu(N_\tau)\,d\tau = \int_{\mathbb{R}_+} \nu(M_\tau)\,d\tau,$$

where ν is a measure, and therefore

$$\int_\Omega P(u)\,d\nu = \int_{\mathbb{R}_+} \nu(N_\tau)\,dP(\tau), \tag{60}$$

where P is a nondecreasing function on $\mathbb{R}_+$. Putting here $u = 1/v$ and $Q(\tau) = P(\tau^{-1})$, we deduce that

$$\int_\Omega Q(u)\,d\nu = -\int_{\mathbb{R}_+} \nu(\Omega\setminus M_\tau)\,dQ(\tau), \tag{61}$$

where Q is nonincreasing. We obtain

$$\begin{aligned}&\int_\Omega\int_\Omega |u(x)-u(y)|^q\mu(dx,dy)\\ &\quad= \int_\Omega\int_\Omega (u(x)-u(y))_+^q\mu(dx,dy) + \int_\Omega\int_\Omega (u(y)-u(x))_+^q\mu(dx,dy)\\ &\quad= \int_\Omega\int_\Omega (u(x)-u(y))_+^q(\mu(dx,dy)+\mu(dy,dx)).\end{aligned}$$

By (60) and (61), the last double integral is equal to

$$\begin{aligned}&q\int_{\mathbb{R}_+}\int_\Omega (t-u(y))_+^{q-1}(\mu(N_\tau,dy)+\mu(dy,N_\tau))\,d\tau\\ &\quad= q(q-1)\int_{\mathbb{R}_+}\int_{\mathbb{R}_+} (\tau-\sigma)_+^{q-2}\big(\mu(N_\tau,\Omega\setminus M_\sigma)+\mu(\Omega\setminus M_\sigma,N_\tau)\big)\,d\tau\,d\sigma.\end{aligned}$$

Now, (55) implies

$$\langle u\rangle_{q,\mu}^q \le 2q(q-1)B\int_{\mathbb{R}_+}\int_{\mathbb{R}_+}(\tau-\sigma)_+^{q-2}\operatorname{cap}_p(N_\tau,\Omega\setminus M_\sigma;\Omega)\,d\tau\,d\sigma$$

and using the inverse function $\psi\mapsto t(\psi)$ to (30), we arrive at the inequality

$$\begin{aligned}&\langle u\rangle_{q,\mu}^q \le 2q(q-1)B^{q/p}\\ &\qquad\times\int_{\mathbb{R}_+}\int_0^\psi (t(\psi)-t(\phi))^{q-2}\big(\operatorname{cap}(N_{t(\psi)},\Omega\setminus M_{t(\phi)};\Omega)\big)^{q/p}t'(\phi)t'(\psi)\,d\phi\,d\psi.\end{aligned}$$

By [**22**, Lemma 2.2.2/1],

$$\operatorname{cap}(N_{t(\psi)},\Omega\setminus M_{t(\phi)};\Omega) \le (\psi-\phi)^{1-p}$$

for $\psi>\phi$, and therefore

$$\langle u\rangle_{q,\mu}^q \le 2q(q-1)B^{q/p}\int_{\mathbb{R}_+}\int_0^\psi (\psi-\phi)^{-q/p'}(t(\psi)-t(\phi))^{q-2}t'(\phi)t'(\psi)\,d\phi\,d\psi. \tag{62}$$

Integrating by parts twice on the right-hand side of (62), we obtain

$$\begin{aligned}\langle u\rangle_{q,\mu}^q &\le 2B^{q/p}\frac{q}{p'}\Big(\Big(\frac{q}{p'}+1\Big)\int_{\mathbb{R}_+}\int_0^{\psi}\frac{(t(\psi)-t(\phi))^q}{(\psi-\phi)^{2+q/p'}}\,d\phi\,d\psi+\int_{\mathbb{R}_+}\psi^{-q/p'}t(\psi)^{q-1}t'(\psi)\,d\psi\Big)\\ &= B^{q/p}\frac{q}{p'}\Big(\Big(\frac{q}{p'}+1\Big)\int_{\mathbb{R}_+}\int_{\mathbb{R}_+}\frac{|t(\psi)-t(\phi)|^q}{|\psi-\phi|^{2+q/p'}}\,d\phi\,d\psi+\frac{1}{p'}\int_{\mathbb{R}_+}t(\psi)^q\psi^{-1-q/p'}\,d\psi\Big).\end{aligned}$$

Hence, we deduce from (56) and (57) that

$$\langle u\rangle_{q,\mu}\le cB^{1/p}\|t'\|_{L_p(\mathbb{R}_+)},\tag{63}$$

where c depends only on p and q. It remains to refer to (32).

Necessity. Let F_1 and F_2 be subsets of Ω, closed in Ω. We take an arbitrary function $u\in C^\infty(\Omega)$ such that $u\ge 1$ on F_1 and $u\le 0$ on F_2, and put it into (54):

$$\mu(F_1,F_2;\Omega)^{p/q}\le\Big(\int_{F_1}\int_{F_2}|u(x)-u(y)|^q\mu(dx,dy)\Big)^{1/q}\le C\int_\Omega|\nabla u|^p\,dx.$$

It remains to minimize the right-hand side, in order to obtain

$$\mu(F_1,F_2;\Omega)^{p/q}\le C\,\mathrm{cap}_p(F_1,F_2;\Omega).$$

The result follows. □

A direct consequence of Theorem 4 and the isocapacitary inequality for $\mathrm{cap}_p(F;G)$ (see [**22**, Sec. 2.2.3, (5) and (6)]) is the following sufficient condition for (54) formulated in terms of n-dimensional Lebesgue measure:

$$\mu(F,\Omega\setminus G)\le c\Big(\log\frac{m_n(G)}{m_n(F)}\Big)^{q(1-n)/n}\quad\text{if } p=n,\tag{64}$$

and

$$\mu(F,\Omega\setminus G)\le c\,\big|m_n(G)^{(p-n)/n(p-1)}-m_n(F)^{(p-n)/n(p-1)}\big|^{1-p}\quad\text{if } p\ne n.\tag{65}$$

Choosing two concentric balls situated in Ω for the sets F_1 and $\Omega\setminus F_2$ in (55) and using the explicit formulae for the p-capacity of spherical condensers (see [**22**, Sec. 2.2.4, (1) and (2)]) we see that the inequalities (64) and (65), with concentric balls F and G placed in Ω, are a necessary condition for (54).

In the one-dimensional case Theorem 4 can be written in a much simpler form.

COROLLARY 5. *Let $\Omega=(\alpha,\beta)$, $-\infty\le\alpha<\beta\le\infty$. The inequality*

$$\Big(\int_\Omega\int_\Omega|u(x)-u(y)|^q\mu(dx,dy)\Big)^{1/q}\le C\Big(\int_\Omega|u'(x)|^p\,dx\Big)^{1/p}\tag{66}$$

holds for every $u\in C^\infty(\Omega)$ if and only if for all pairs of intervals I and J of the three types:

$$I=[x-d,x+d]\quad\text{and}\quad J=(x-d-r,x+d+r),$$

$$I=(\alpha,x]\quad\text{and}\quad J=(\alpha,x+r),\tag{67}$$

$$I=[x,\beta)\quad\text{and}\quad J=(x-r,\beta),\tag{68}$$

where d and r are positive and $J\subset\Omega$, we have

$$r^{(p-1)/p}\big(\mu(I,\Omega\setminus J)\big)^{1/q}\le B,\tag{69}$$

where B does not depend on I and J.

PROOF. The necessity of (69) follows directly from the necessity in Theorem 4 and the inequality

$$\operatorname{cap}_p(I, \Omega \setminus J; \Omega) \le 2r^{1-p}$$

(see [**22**, Lemma 2.2.2/2]).

Let us prove sufficiency. By G_1 we mean an open subset of Ω such that $F_1 \subset G_1$ and $\overline{G}_1 \subset \Omega \setminus F_2$. Connected components of $\Omega \setminus F_2$ will be denoted by J_k. Let J_k contain the closed convex hull $\tilde{J}_k$ of those connected components of G_1 which are situated in J_k. Then

$$\mu(F_1, F_2)^{p/q} \le \mu(G_1, F_2)^{p/q} \le \Big(\sum_k \mu(\tilde{J}_k, \Omega \setminus J_k)\Big)^{p/q} \le \sum_k \mu(\tilde{J}_k, \Omega \setminus J_k)^{p/q},$$

and since by (69)

$$\mu\left(\tilde{J}_k, \Omega \setminus J_k\right)^{p/q} \le B^p \big(\operatorname{dist}\{I_k, \mathbb{R} \setminus J_k\}\big)^{1-p},$$

we obtain

$$\mu(F_1, F_2)^{p/q} \le B^p \sum_k \big(\operatorname{dist}\{I_k, \mathbb{R} \setminus J_k\}\big)^{1-p}. \tag{70}$$

Consider an arbitrary function $u \in C^\infty(\Omega)$ such that $u = 1$ on G_1 and $u = 0$ on F_2. Clearly, $u = 0$ on ∂J_k. We have

$$\int_\Omega |u'|^p dx \ge \sum_k \int_{J_k} |u'|^p dx \ge \sum_k \int_{J_k} |\tilde{u}_k'|^p\, dx, \tag{71}$$

where $\tilde{u} = u$ on $J_k \setminus \tilde{I}_k$, $\tilde{u}_k = 1$ on $\tilde{I}_k$, and $\tilde{u}_k = 0$ on ∂J_k. Hence

$$\int_\Omega |u'|^p dx \ge \sum_k \big(\operatorname{dist}\{\tilde{I}_k, \mathbb{R} \setminus J_k\}\big)^{1-p}.$$

Comparing this estimate with (70), we arrive at

$$\int_\Omega |u'|^p dx \ge \mu(F_1, F_2)^{p/q},$$

and minimizing the integral on the left-hand side over all functions u, we obtain (69). $\square$

REMARK 4. It is straightforward but somewhat cumbersome to obtain a more general criterion by replacing the seminorm on the right-hand side of (66) with

$$\Big(\int_\Omega |u'(x)|^p \sigma(dx)\Big)^{1/p}, \tag{72}$$

where σ is a measure in Ω. In fact, one can replace σ by its absolutely continuous part $(d\sigma^*/dx)dx$ and further, roughly speaking, the criterion will follow from Corollary 5 by the change of variables $x \to \xi$, where

$$d\xi = (d\sigma^*/dx)^{1/(1-p)} dx.$$

Restricting myself to this hint, I leave details to the interested reader.

6. Capacitary sufficient condition in the case $q = p$

In the marginal case $q = p$ the condition (55) in Theorem 4 is necessary but is not sufficient in general. In fact, let $n = 1$, $\Omega = \mathbb{R}$, and let

$$\mu(dx, dy) = \frac{dx\,dy}{|x-y|^{p+1}}.$$

Then as shown in the proof of Corollary 4, (55) is equivalent to (69), and (69) holds, since

$$\begin{aligned}\mu(I, \mathbb{R}\setminus J) &= \int_{|t-x|<d} dt \int_{|\tau-x|>d+r} \frac{d\tau}{|t-\tau|^{p+1}} \\ &= \int_{|t|<d} dt \int_{|\tau|>d+r} \frac{d\tau}{|t-\tau|^{p+1}} \le cr^{1-p}\end{aligned}$$

and the same estimate holds for I and J defined by (67) and (68).

On the other hand, (47) fails, because

$$\int_{\mathbb{R}}\int_{\mathbb{R}} \frac{|u(x)-u(y)|^p}{|x-y|^{p+1}}\,dx\,dy = \infty$$

for every nonconstant function u.

In the next theorem we give a sufficient condition for (54) with $q = p > 1$ formulated in terms of an isocapacitary inequality.

THEOREM 5. *Given $p \in (1, \infty)$ and a positive, vanishing at infinity, nonincreasing absolutely continuous function ν on $\mathbb{R}_+$ such that*

$$S := \sup_{\tau>0}\Big(\int_0^\tau |\nu'(\sigma)|^{1/(1-p)}\frac{d\sigma}{\sigma}\Big)^{p-1}\int_\tau^\infty |\nu'(\sigma)|\frac{d\sigma}{\sigma} < \infty,$$

suppose that

$$\mu(F_1, F_2) \le \nu\big((\mathrm{cap}_p(F_1, F_2; \Omega))^{1-p}\big) \tag{73}$$

for all nonoverlapping sets F_1 and F_2 closed in Ω. Assume also that

$$\mathcal{K} := \int_0^\infty |\nu'(\sigma)|\,\sigma^{p-1} d\sigma < \infty. \tag{74}$$

Then

$$\|u\|_{p,\mu} \le 2^{1/p} p\,\Big(\frac{S}{(p-1)^{p-1}}\Big)^{1/pp'} \mathcal{K}^{1/p}\|\nabla u\|_{L_p(\Omega)} \tag{75}$$

for all $u \in C^\infty(\Omega)$.

PROOF. We assume that $\nabla u \in L_p(\Omega)$ and the integral in (75) involving derivatives of ν is convergent. Arguing as in the proof of Theorem 4 and using (73) instead of (55), we obtain

$$\langle u\rangle_{p,\mu}^p \le 2p(p-1)\int_0^\infty\int_\phi^\infty \nu(\psi-\phi)\big(t(\psi)-t(\phi)\big)^{p-2} t'(\psi)\,d\psi\, t'(\phi)\,d\phi. \tag{76}$$

In view of (74), we can integrate by parts the inner integral in (76) and obtain

$$\begin{aligned}\langle u\rangle_{p,\mu}^p &\le 2p\int_0^\infty\int_\phi^\infty |\nu'(\psi-\phi)|\big(t(\psi)-t(\phi)\big)^{p-1}\,d\psi\, t'(\phi)\,d\phi \\ &= 2p\int_0^\infty\int_0^\psi |\nu'(\psi-\phi)|\big(t(\psi)-t(\phi)\big)^{p-1}\, t'(\phi)\,d\phi\,d\psi.\end{aligned}$$

By Hölder's inequality,

$$\langle u\rangle_{p,\mu}^p \le 2p\int_0^\infty \mathcal{A}(\phi)^{1/p'}\mathcal{B}(\phi)^{1/p}\,d\phi, \tag{77}$$

where

$$\mathcal{A} = \int_0^\psi \frac{|\nu'(\psi-\phi)|}{\psi-\phi}(t(\psi)-t(\phi))^p\,d\phi \quad \text{and}$$

$$\mathcal{B} = \int_0^\psi |\nu'(\psi-\phi)|\,(\psi-\phi)^{p-1}\,|t'(\psi)|^p\,d\phi.$$

Using [**22**, Theorem 1.3.1/1] concerning a two-weight Hardy inequality, we obtain

$$\mathcal{A} \le \frac{p^p}{(p-1)^{p-1}}\,S\,\mathcal{B},$$

which together with (77) gives

$$\langle u\rangle_{p,\mu}^p \le 2p^p(p-1)^{(1-p)/p'}S^{1/p'}\int_0^\infty\int_0^\psi |\nu'(\psi-\phi)|\,(\psi-\phi)^{p-1}\,|t'(\psi)|^p\,d\phi\,d\psi.$$

Changing the order of integration, we arrive at

$$\langle u\rangle_{p,\mu} \le 2^{1/p}p\left((p-1)^{1-p}S\right)^{1/pp'}\mathcal{K}^{1/p}\|\,t'\,\|_{L_p(\mathbb{R}_+)}.$$

It remains to apply (32). □

REMARK 5. If the requirement

$$u = 0 \quad \text{on a neighbourhood of a closed subset } E \text{ of } \overline{\Omega}$$

is added in the formulation of Theorems 4 and 5, the same proofs give conditions for the validity of (54), similar to (55) and (69). The only new feature is the following restriction:

$$\Omega\,\mathrm{cap}\,\partial(\Omega\setminus F_2) \quad \text{is at a positive distance from } E.$$

In the important special case where $E = \partial\Omega$, which corresponds to zero Dirichlet data on $\partial\Omega$, the conditions (55) and (73) become

$$\mu(F,\Omega\setminus G)^{p/q} \le B\ \mathrm{cap}_p(F;G) \quad \text{and} \tag{78}$$

$$\mu(F,\Omega\setminus G) \le \nu\big((\mathrm{cap}_p(F;G))^{1-p}\big), \tag{79}$$

respectively, where F is closed and G is open, $G \supset F$, and the closure of G is compact and lies in Ω. The capacity $\mathrm{cap}_p(F;G)$ is defined by (29) with $\Omega = G$.

Using lower estimates for the p-capacity in terms of area minimizing functions, one obtains sufficient conditions from (55), (69), (78), and (79), formulated in geometrical terms in the spirit of Corollary 2. For example, by (78) and (79), inequalities (55) and (73) hold for all $u \in C_0^\infty(\Omega)$ if, respectively,

$$\mu(F,\Omega\setminus G)^{p/q} \le B\Big(\int_{m_n(F)}^{m_n(\Omega\setminus G)}\frac{dv}{\lambda(v)^{p/(p-1)}}\Big) \quad \text{and}$$

$$\mu(F,\Omega\setminus G) \le \nu\Big(\int_{m_n(F)}^{m_n(\Omega\setminus G)}\frac{dv}{\lambda(v)^{p/(p-1)}}\Big),$$

where F and G are the same as in (78) and (79).

By obvious modifications of the proof of sufficiency in Corollary 4 one deduces the following assertion from Theorem 5.

COROLLARY 6 (One-dimensional case). *With the notation used in Corollary 5, suppose that*

$$\mu(I, \Omega \setminus J) \le \nu(r).$$

Then there exists a positive constant c depending only on p and such that

$$\langle u \rangle_{p,\mu} \le c\, S^{1/pp'} K^{1/p}\, \|u'\|_{L_p(\Omega)}$$

for all $u \in C^\infty(\Omega)$.

REMARK 6. Let us show that the condition $K < \infty$, which appeared in Theorem 5, is sharp. Suppose that there exists a positive constant C independent of u and such that

$$\int_{\mathbb{R}} \int_{\mathbb{R}} |u(t) - u(\tau)|^p\, \nu''(t-\tau)\, dt\, d\tau \le C \int_{\mathbb{R}} |u'(t)|^p\, dt, \tag{80}$$

where ν is a convex function in $C^2(\mathbb{R})$. We take an arbitrary $N > 0$ and put $u(t) = \min\{|t|, N\}$ into (80). Then

$$\int_0^{N/2} \int_\tau^N (t-\tau)^p\, \nu''(t-\tau)\, dt\, d\tau \le 2CN$$

and setting $t = \tau + s$ here, we obtain

$$\frac{1}{2} pN \int_0^{N/2} s^{p-1}\, |\nu'(s)|\, ds \le p \int_0^{N/2} \int_0^{N-\tau} s^{p-1} |\nu'(s)|\, ds\, d\tau \le 2CN.$$

Hence $K \le 4p^{-1}C$.

REMARK 7. It seems appropriate, in conclusion, to say a few words about the lower estimate for the difference seminorm $\langle u \rangle_{p,\mu}$, similar to the classical Sobolev inequality:

$$\Big(\int_\Omega |u|^q\, \nu(dx)\Big)^{1/q} \le C\, \langle u \rangle_{p,\mu}, \tag{81}$$

where Ω is a subdomain of a Riemannian manifold, μ and ν are measures in $\Omega \times \Omega$ and Ω, respectively, and u is an arbitrary function in $C_0^\infty(\Omega)$. Suppose $q \ge p \ge 1$. Then a necessary and sufficient condition for (81) is the isocapacitary inequality

$$\sup_{\{F\}} \frac{\nu(F)^{p/q}}{\operatorname{cap}_{p,\mu}(F;\Omega)} < \infty, \tag{82}$$

where F is an arbitrary compact set in Ω and the capacity is defined by

$$\operatorname{cap}_{p,\mu}(F;\Omega) = \inf\big\{\langle u \rangle_{p,\mu}^p : u \in C_0^\infty(\Omega),\, u \ge 1 \text{ on } F\big\}.$$

The necessity of (82) is obvious and sufficiency results directly from the inequality

$$\int_0^\infty \operatorname{cap}_{p,\mu}(N_t;\Omega)\, d(t^p) \le c(p)\, \langle u \rangle_{p,\mu}^p \tag{83}$$

(see [**24**] for the proof and history of (83)).

Although providing a universal characterization of (81), the condition (82) does not seem satisfactory when dealing with concrete measures and domains. This is related even to the one-dimensional case (cf. [**11**, Problem 2]). As an example of a more visible criterion, consider the measure μ on $\mathbb{R}^n \times \mathbb{R}^n$ given by

$$\mu(dx, dy) = |x-y|^{-n-p\alpha}\, dx\, dy \tag{84}$$

with $0 < \alpha < 1$ and $\alpha p < n$. This measure generates a seminorm in the homogeneous Besov space $b_p^\alpha(\mathbb{R}^n)$. With this particular choice of μ, we see from [**22**, Theorem 8.7.1 and Remark 8.6/3] that (81) holds with $q > p > 1$ and $q \geq p = 1$ if and only if

$$\sup_{x\in\mathbb{R}^n,\, \rho>0} \frac{\nu(B(x,\rho))^{p/q}}{\rho^{n-p\alpha}} < \infty. \tag{85}$$

The inequality (85) is the same as (82) with $F = B(x, \rho)$. It is unknown whether the replacement of arbitrary sets by balls in (82) is possible for a general μ and $\Omega = \mathbb{R}^n$ in (82). If not, what are the exact requirements allowing this replacement?

Let $q = p > 1$, $\Omega = \mathbb{R}^n$, and let μ be given by (84). Then (81) holds simultaneously with the inequality

$$\int_{\mathbb{R}^n} |u|^p\, \nu(dx) \leq c\, \|(-\Delta)^{\alpha/2} u\|^p_{L_p(\mathbb{R}^n)} \tag{86}$$

because both (81) and (86) are equivalent to isocapacitary inequalities of the type (82) with equivalent capacities on the right-hand side (see [**2**, Sec. 4.4]).

Note that (86) is the so-called trace inequality for the Riesz potential operator $I_\alpha := (-\Delta)^{-\alpha/2}$. This inequality has been studied intensively (see [**30**] for a survey of this area). First of all, the simplest estimate

$$\nu(B) \leq c\, m_n(B)^{1-p\alpha/n} \quad \text{for all balls } B,$$

that is necessary for (86), is not sufficient for it (see [**1**] and [**2**]). However, there exist other conditions involving no capacity, which are necessary and sufficient for (86). They are as follows:

(i) For every ball B,

$$\int_B (I_\alpha \nu_B)^p dx \leq c\, \nu(B),$$

where ν_B is the restriction of ν to B; see [**9**].

(ii) Almost everywhere in $\mathbb{R}^n$,

$$I_\alpha (I_\alpha \nu)^{p'} \leq c\, I_\alpha \nu;$$

see [**27**].

(iii) For every dyadic cube P of side length $\ell(P)$,

$$\sum_{Q\subset P} \big(\nu(Q)\, \ell(Q)^{\alpha-n/p}\big)^{p'} \leq c\nu(P),$$

where the sum is taken over all dyadic cubes Q contained in P; see [**30**, Sec. 3].

In accordance with the equivalence of (81) and (86) mentioned previously, the criteria (i)–(iii) characterize not only (86) but also (81) with $q = p > 1$ and μ defined by (84). It is unclear how these criteria could be modified to characterize (81) with an arbitrary μ.

References

[1] D. R. Adams, *My love affair with the Sobolev inequality*, Sobolev Spaces in Mathematics I, Sobolev Type Inequalities, Springer, 2008, pp. 1–24.

[2] D. R. Adams and L.-I. Hedberg, *Function Spaces and Potential Theory*, Springer, 1996.

[3] A. Alvino, A. Cianchi, V. Maz′ya, and A. Mercaldo, *Well-posed elliptic Neumann problems involving irregular data and domains*, to appear.

[4] G. A. Bliss, *An integral inequality*, J. London Math. Soc. **5** (1930), 40–46.

[5] Yu. Burago and V. Maz′ya, *Certain Questions of Potential Theory and Function Theory for Regions with Irregular Boundaries*, Zap. Naučn. Sem. Leningrad. Otdel. Mat. Inst. Steklov. (LOMI) **3**, 1967; English translation: Potential Theory and Function Theory for Irregular Regions. Seminars in Mathematics, V. A. Steklov Mathematical Institute, Leningrad, Vol. 3, Consultants Bureau, New York, 1969.

[6] J. Cheeger, *A lower bound for the smallest eigenvalue of the Laplacian*, Problems in Analysis, R. Gunning ed., Princeton Univ. Press, 1970, pp. 195–199.

[7] A. Cianchi and V. Maz′ya, *Neumann problems and isocapacitary inequalities*, J. Math. Pures Appl. **(9) 89** (2008), no. 1, 71–105.

[8] A. Grigor′yan, *Isoperimetric inequalities and capacities on Riemannian manifolds*, Operator Theory, Advances and Applications, Vol. 110, The Maz′ya Anniversary Collection, vol. 1, Birkhäuser, 1999, pp. 139–153.

[9] R. Kerman and E. Sawyer, *The trace inequality and eigenvalue estimates for Schrödinger operators*, Ann. Inst. Fourier (Grenoble) **36** (1986), 207–228.

[10] J. Kinnunen and R. Korte, *Characterization of Sobolev inequalities on metric spaces*, `arXiv: 0709.2013v1`.

[11] A. Kufner, L. Maligranda, and L.-E. Persson, *The Hardy inequality, about its history and some related results*, Pilsen, 2007.

[12] V. Maz′ya, *Classes of domains and imbedding theorems for function spaces*, Sov. Math. Dokl. **1** (1960), 882–885.

[13] ———, *On p-conductivity and theorems on embedding certain functional spaces into a C-space*, Sov. Math. Dokl. **2** (1961), 1200–1203.

[14] ———, *Some estimates of solutions of second-order elliptic equations*, Sov. Math. Dokl. **2** (1961), 413–415.

[15] ———, *The negative spectrum of the higher-dimensional Schrödinger operator*, Sov. Math. Dokl. **3** (1962), 808–810.

[16] ———, *On the solvability of the Neumann problem*, Sov. Math. Dokl. **3** (1962), 1595–1598.

[17] ———, *The Dirichlet problem for elliptic equations of arbitrary order in unbounded domains*, Sov. Math. Dokl. **4** (1963), 860–863.

[18] ———, *On the theory of the higher-dimensional Schrödinger operator*, Izv. Akad. Nauk SSSR Ser. Mat. **28** (1964), no. 4, 1145–1172.

[19] ———, *The Neumann problem in regions with nonregular boundaries*, Sibirsk. Mat. Ž. **9** (1968), 1322–1350.

[20] ———, *Weak solutions of the Dirichlet and Neumann problems*, Trudy Moskov. Mat. Obšč. **20** (1969), 137–172.

[21] ———, *Certain integral inequalities for functions of several variables*, Problems of mathematical analysis, no. 3: Integral and differential operators, Differential equations, pp. 33–68, Izdat. Leningrad. Univ., Leningrad, 1972; English translation: J. Soviet Math. **1** (1973), 205–234.

[22] ———, *Sobolev Spaces*, Springer, 1985.

[23] ———, *Lectures on isoperimetric and isocapacitary inequalities in the theory of Sobolev spaces*, Contemporary Mathematics, **338**, Heat Kernels and Analysis on Manifolds, Graphs, and Metric Spaces, American Math. Society, Providence, RI, 2003, pp. 307–340.

[24] ———, *Conductor and capacitary inequalities for functions on topological spaces and their applications to Sobolev type imbeddings*, J. Funct. Analysis **224** (2005), no. 2, 408–430.

[25] ———, *Conductor inequalities and criteria for Sobolev type two-weight imbeddings*, J. Comput. Appl. Math. **194** (2006), no. 1, 94–114.

[26] ———, *Analytic criteria in the qualitative spectral analysis of the Schrödinger operator*, Spectral Theory and Mathematical Physics: a Festschrift in honor of Barry Simon's 60th birthday, Proc. Sympos. Pure Math., Part 1, Amer. Math. Soc., Providence, RI **76** (2007), pp. 257–288.

[27] V. Maz′ya and I. Verbitsky, *Capacitary estimates for fractional integrals, with applications to partial differential equations and Sobolev multipliers*, Arkiv för Matem. **33** (1995), 81–115.

[28] G. Pólya and G. Szegö, *Isoperimetric Inequalities in Mathematical Physics*, Ann. Math. Stud. 27, Princeton Univ. Press, Princeton, 1951.

[29] E. M. Stein, *Singular Integrals and Differentiability Properties of Functions*, Princeton Univ. Press, Princeton, NJ, 1970.

[30] I. Verbitsky, *Nonlinear potentials and trace inequalities*, Operator Theory, Advances and Applications, Vol. 110, The Maz′ya Anniversary Collection, vol. 2, Birkhäuser, 1999, pp. 323–343.

Department of Mathematical Sciences, University of Liverpool, Liverpool L69 7ZL, United Kingdom, and Department of Mathematics, Linköping University, Linköping SE-581 83, Sweden

E-mail address: vlmaz@mai.liu.se

Amer. Math. Soc. Transl.
(2) Vol. **226**, 2009

Differentiability of Functions of Contractions

V. V. Peller

ABSTRACT. In this paper we study differentiability properties of the map $T \mapsto \varphi(T)$, where φ is a given function in the disk algebra and T ranges over the set of contractions on Hilbert space. We obtain sharp conditions (in terms of Besov spaces) for differentiability and existence of higher derivatives. We also find explicit formulae for directional derivatives (and higher derivatives) in terms of double (and multiple) operator integrals with respect to semispectral measures.

1. Introduction

The purpose of this paper is to study differentiability properties of functions

$$T \mapsto \varphi(T),$$

for a given function φ analytic in the open unit disk $\mathbb{D}$ and continuous in the closed disk (in other words, φ belongs to the disk algebra C_A), where T ranges over the set of contractions (i.e., operators of norm at most 1) on Hilbert space.

Recall that by von Neumann's inequality,

$$\|\varphi(T)\| \le \max_{|\zeta|\le 1} |\varphi(\zeta)| \tag{1.1}$$

for an arbitrary contraction T on Hilbert space and an arbitrary polynomial φ. This allows one to define a functional calculus

$$\varphi \mapsto \varphi(T), \quad \varphi \in C_A,$$

for an arbitrary contraction T. Moreover, for this functional calculus, von Neumann's inequality (1.1) holds.

For contractions T and R on Hilbert space, we consider the one-parameter family of contractions

$$T_t = (1-t)T + tR, \quad 0 \le t \le 1,$$

and we study differentiability properties of the map

$$t \mapsto \varphi(T_t) \tag{1.2}$$

for a given function φ in C_A.

2000 *Mathematics Subject Classification.* Primary 47A55, 47B49; Secondary 47A60.
The author is partially supported by NSF grant DMS 0700995.

The study of the problem of differentiability for functions of self-adjoint operators on Hilbert space was initiated by Daletskii and S. G. Krein in [**DK**]. They showed that for a function f of class C^2 on the real line $\mathbb{R}$ and for bounded self-adjoint operators A and B, the function

$$t \mapsto f(A+tB) \tag{1.3}$$

is differentiable in the operator norm and the derivative can be computed in terms of double operator integrals:

$$\frac{d}{dt}f(A+tB)\Big|_{t=0} = \iint_{\mathbb{R}\times\mathbb{R}} \frac{f(x)-f(y)}{x-y}\,dE_A(x)\,B\,dE_A(y), \tag{1.4}$$

where E_A is the spectral measure of A. The expression on the right is a double operator integral. The beautiful theory of double operator integrals due to Birman and Solomyak was created later in [**BS1, BS2, BS3**] (see also the survey article [**BS4**]).

The condition $f \in C^2$ was relaxed by Birman and Solomyak in [**BS3**]: they proved that the function (1.3) is differentiable and the Daletskii–Krein formula (1.4) holds under the condition that f is differentiable and the derivative f' satisfies a Hölder condition of order α for some $\alpha > 0$. The approach of Birman and Solomyak is based on their formula

$$f(A+B) - f(A) = \iint_{\mathbb{R}\times\mathbb{R}} \frac{f(x)-f(y)}{x-y}\,dE_{A+B}(x)\,B\,dE_A(y). \tag{1.5}$$

Actually, Birman and Solomyak showed in [**BS3**] that formula (1.5) is valid under the condition that the divided difference $\mathfrak{D}f$ is a Schur multiplier of the space of all bounded linear operators (see § 2.2 for the definitions).

However, it follows from the results of Farforovskaya in [**Fa**] that the condition $f \in C^1$ is not sufficient for the differentiability of the map (1.3).

A further improvement was obtained in [**Pe2**]: it was shown that the function (1.3) is differentiable and (1.4) holds under the assumption that f belongs to the Besov space $B^1_{\infty 1}(\mathbb{R})$ (see § 2.5). Moreover, in [**Pe2**], a necessary condition was also found: f must locally belong to the Besov space $B^1_1(\mathbb{R}) = B^1_{11}(\mathbb{R})$. This necessary condition also implies that the condition $f \in C^1$ is not sufficient. Actually, in [**Pe2**] a stronger necessary condition was also obtained; see § 2.3 for a further discussion. Finally, we mention another sufficient condition obtained in [**ABF**], which is slightly better than the condition $f \in B^1_{\infty 1}(\mathbb{R})$, though I believe it is more convenient to work with Besov spaces.

Note that similar results were obtained in [**Pe5**] in the case where A is an unbounded self-adjoint operator and B is a bounded self-adjoint operator.

The problem of the existence of higher derivatives of the function (1.3) was studied in [**St**], where it was shown that under certain assumptions on f, the function (1.3) has second derivative that can be expressed in terms of the following triple operator integral:

$$\frac{d^2}{dt^2}f(A+tB)\Big|_{t=0} = 2\iiint_{\mathbb{R}\times\mathbb{R}\times\mathbb{R}} \left(\mathfrak{D}^2 f\right)(x,y,z)\,dE_A(x)\,B\,dE_A(y)\,B\,dE_A(z),$$

where $\mathfrak{D}^2 f$ stands for the divided difference of order 2 (see § 2 for the definition). To interpret triple operator integrals, repeated integration was used in [**St**]. However, the class of integrable functions in [**St**] was rather narrow, and the assumption on f imposed in [**St**] for the existence of the second operator derivative was too restrictive. Similar results were also obtained in [**St**] for the nth derivative and multiple operator integrals.

In [**Pe9**] a new approach to multiple operator integrals was given. It is based on integral projective tensor products of L^∞ spaces and gives a much larger class of integrable functions than under the approach of [**St**]. It was shown in [**Pe9**] that under the assumption that f belongs to the Besov space $B^n_{\infty 1}(\mathbb{R})$ the function (1.3) has n derivatives and the nth derivative can be expressed in terms of a multiple operator integral. Similar results were also obtained in [**Pe9**] in the case of an unbounded self-adjoint operator A.

Note that Besov spaces $B^n_{\infty 1}(\mathbb{R})$ arise in operator theory on many different occasions; see [**Pe1, Pe4, Pe6, Pe8**].

Let us mention here another formula by Birman and Solomyak (see [**BS4**]) for commutators. Suppose that A is a self-adjoint operator and Q is a bounded linear operator. Then

$$f(A)Q - Qf(A) = \iint_{\mathbb{R}\times\mathbb{R}} \frac{f(x)-f(y)}{x-y}\, dE_A(x)\,(AQ-QA)\,dE_A(y), \tag{1.6}$$

which is also valid under the assumption that $\mathfrak{D}f$ is a Schur multiplier of the space of all bounded linear operators.

To study the problem of differentiability of functions of unitary operators, we should consider a Borel function f on the unit circle $\mathbb{T}$ and the map

$$U \mapsto f(U),$$

where U is a unitary operator on Hilbert space. If U and V are unitary operators and $V = e^{\mathrm{i}A}U$, where A is a self-adjoint operator, we can consider the one-parameter family of unitary operators

$$e^{\mathrm{i}tA}U, \quad 0 \le t \le 1,$$

and study the question of the differentiability of the function

$$t \mapsto \left(e^{\mathrm{i}tA}U\right)$$

and the question of the existence of its higher derivatives. The results in the case of unitary operators are similar to the results for self-adjoint operators; see [**BS3, Pe2, ABF, Pe9**].

In this paper we study the case of functions of contractions. This study was initiated in [**Pe3**], where the Lipschitz property was studied. Recently, in [**KS**], new results on operator Lipschitz functions of contractions were obtained; see § 2.3 for more detailed information.

It turns out that the right tool to study differentiability properties of functions of contractions is double (and multiple) operator integrals with respect to semispectral measures. Note that even if both contractions T and R are unitary operators, the differentiability problem for this pair is different from the differentiability problem for unitary operators.

In § 3 we define double and multiple operator integrals with respect to semispectral measures. In § 4 we obtain an analog of the Birman–Solomyak formulae (1.5) and (1.6) for semispectral measures. Then, in § 5, we obtain conditions on a function $\varphi \in C_A$ for the differentiability of the map (1.2) in the operator norm, as well as conditions for the existence of higher operator derivatives. We also obtain in § 5 formulae for the derivatives of (1.2) in terms of multiple operator integrals with respect to semispectral measures. Finally, in § 6 we study the problem of differentiability of the function (1.2) in the Hilbert–Schmidt norm.

In § 2 we give necessary information on Besov spaces, double and multiple operator integrals, and semispectral measures.

I would like to express my gratitude to Victor Shulman for stimulating discussions.

2. Preliminaries

In this section, we are going to collect necessary information on Besov spaces, double operator integrals, multiple operator integrals, and semispectral measures.

2.1. Besov spaces. Let $1 \le p,\, q \le \infty$ and $s \in \mathbb{R}$. The Besov class B^s_{pq} of functions (or distributions) on $\mathbb{T}$ can be defined in the following way. Let w be a continuous piecewise linear function on $\mathbb{R}$ such that

$$w \ge 0, \quad \operatorname{supp} w \subset \left[\frac{1}{2}, 2\right], \quad w(1) = 1,$$

and w is a linear function on the intervals $[1/2, 1]$ and $[1, 2]$.

Consider the trigonometric polynomials W_n and $W_n^\#$ defined by

$$W_n(z) = \sum_{k\in\mathbb{Z}} w\Big(\frac{k}{2^n}\Big) z^k, \quad n \ge 1, \quad W_0(z) = \bar{z} + 1 + z, \quad \text{and}$$
$$W_n^\#(z) = \overline{W_n(z)}, \quad n \ge 0.$$

Then for each distribution φ on $\mathbb{T}$,

$$\varphi = \sum_{n\ge 0} \varphi * W_n + \sum_{n \ge 1} \varphi * W_n^\#.$$

The Besov class B^s_{pq} consists of functions (in the case of $s > 0$) or distributions φ on $\mathbb{T}$ such that

$$\big\{\|2^{ns}\varphi * W_n\|_{L^p}\big\}_{n\ge 0} \in \ell^q \quad \text{and} \quad \big\{\|2^{ns}\varphi * W_n^\#\|_{L^p}\big\}_{n\ge 1} \in \ell^q.$$

Besov classes admit many other descriptions. In particular, for $s > 0$, the space B^s_{pq} admits the following characterization. A function f belongs to B^s_{pq}, $s > 0$, if and only if

$$\int_{\mathbb{T}} \frac{\|\Delta_\tau^n f\|_{L^p}^q}{|1-\tau|^{1+sq}}\, d\boldsymbol{m}(\tau) < \infty \quad \text{for} \quad q < \infty$$

and

$$\sup_{\tau \ne 1} \frac{\|\Delta_\tau^n f\|_{L^p}}{|1-\tau|^s} < \infty \quad \text{for} \quad q = \infty,$$

where $\boldsymbol{m}$ is normalized Lebesgue measure on $\mathbb{T}$, n is an integer greater than s, and Δ_τ is the difference operator: $(\Delta_\tau f)(\zeta) = f(\tau\zeta) - f(\zeta)$, $\zeta \in \mathbb{T}$.

We are going to use the notation B^s_p for B^s_{pp}.

It is easy to see from the definition of Besov classes that the Riesz projection $\mathbb{P}_+$,

$$\mathbb{P}_+ f = \sum_{n \ge 0} \hat{f}(n) z^n,$$

is bounded on B^s_{pq} and the functions in $\big(B^s_{pq}\big)_+ \stackrel{\text{def}}{=} \mathbb{P}_+ B^s_{pq}$ admit a natural extension to analytic functions in the unit disk $\mathbb{D}$. It is well known that the functions in $\big(B^s_{pq}\big)_+$ admit the following description:

$$f \in \big(B^s_{pq}\big)_+ \Leftrightarrow \int_0^1 (1-r)^{q(n-s)-1} \|f_r^{(n)}\|_p^q \, dr < \infty, \quad q < \infty,$$

and

$$f \in \big(B^s_{p\infty}\big)_+ \Leftrightarrow \sup_{0<r<1} (1-r)^{n-s} \|f_r^{(n)}\|_p < \infty,$$

where $f_r(\zeta) \stackrel{\text{def}}{=} f(r\zeta)$ and n is a nonnegative integer greater than s.

In a similar way one can define the (homogeneous) Besov space B^s_{pq} of functions (distributions) on $\mathbb{R}$.

We refer the reader to [**Pee**] and [**Pe7**] for more detailed information on Besov spaces.

2.2. Double operator integrals. In this subsection we give a brief introduction to the theory of double operator integrals developed by Birman and Solomyak in [**BS1, BS2, BS3**]; see also their survey [**BS5**].

Let $(\mathcal{X}_1, E_1)$ and $(\mathcal{X}_2, E_2)$ be spaces with spectral measures E_1 and E_2 on Hilbert spaces $\mathcal{H}_1$ and $\mathcal{H}_2$. Let us first define double operator integrals

$$\int_{\mathcal{X}_1}\int_{\mathcal{X}_2} \Phi(\lambda,\mu)\, dE_1(\lambda)\, Q\, dE_2(\mu) \tag{2.1}$$

for bounded measurable functions Φ and operators $Q : \mathcal{H}_2 \to \mathcal{H}_1$ of Hilbert–Schmidt class $\boldsymbol{S}_2$. Consider the set function F whose values are orthogonal projections on the Hilbert space $\boldsymbol{S}_2(\mathcal{H}_2, \mathcal{H}_1)$ of Hilbert–Schmidt operators from $\mathcal{H}_2$ to $\mathcal{H}_1$ and which is defined on measurable rectangles by

$$F(\Delta_1 \times \Delta_2)T = E_1(\Delta_1) Q E_2(\Delta_2), \quad Q \in \boldsymbol{S}_2(\mathcal{H}_2, \mathcal{H}_1),$$

Δ_1 and Δ_2 being measurable subsets of $\mathcal{X}_1$ and $\mathcal{X}_2$. It was shown in [**BS4**] that F extends to a spectral measure on $\mathcal{X}_1 \times \mathcal{X}_2$, and if Φ is a bounded measurable function on $\mathcal{X}_1 \times \mathcal{X}_2$, we define

$$\int_{\mathcal{X}_1}\int_{\mathcal{X}_2} \Phi(\lambda,\mu)\, dE_1(\lambda)\, Q\, dE_2(\mu) = \left(\int_{\mathcal{X}_1\times\mathcal{X}_2} \Phi\, dF\right) Q.$$

Clearly,

$$\left\|\int_{\mathcal{X}_1}\int_{\mathcal{X}_2} \Phi(\lambda,\mu)\, dE_1(\lambda)\, Q\, dE_2(\mu)\right\|_{\boldsymbol{S}_2} \le \|\Phi\|_{L^\infty} \|Q\|_{\boldsymbol{S}_2}.$$

It is easy to see from the definition of double operator integrals in the case $Q \in \boldsymbol{S}_2$ that if $\{\Phi_n\}_{n\ge1}$ is a sequence of measurable functions such that

$$\lim_{n\to\infty} \Phi_n(\lambda,\mu) = \Phi(\lambda,\mu), \quad \lambda \in \mathcal{X}_1,\ \mu \in \mathcal{X}_2,$$

and

$$\sup_n \sup_{\lambda,\,\mu\in\mathcal{X}_1\times\mathcal{X}_2} |\Phi_n(\lambda,\mu)| < \infty,$$

then

(2.2)

$$\lim_{n\to\infty}\left\|\int_{\mathcal{X}_1}\int_{\mathcal{X}_2} \Phi_n(\lambda,\mu)\, dE_1(\lambda)\, Q\, dE_2(\mu) - \int_{\mathcal{X}_1}\int_{\mathcal{X}_2} \Phi(\lambda,\mu)\, dE_1(\lambda)\, Q\, dE_2(\mu)\right\|_{\boldsymbol{S}_2} = 0.$$

If the transformer

$$Q \mapsto \int_{\mathcal{X}_1}\int_{\mathcal{X}_2} \Phi(\lambda,\mu)\,dE_1(\lambda)\,Q\,dE_2(\mu)$$

maps the trace class $\boldsymbol{S}_1$ into itself, we say that Φ is a *Schur multiplier of* $\boldsymbol{S}_1$ *associated with the spectral measures* E_1 *and* E_2. In this case the transformer

$$Q \mapsto \int_{\mathcal{X}_2}\int_{\mathcal{X}_1} \Phi(\lambda,\mu)\,dE_2(\mu)\,Q\,dE_1(\lambda), \quad Q \in \boldsymbol{S}_2(\mathcal{H}_1,\mathcal{H}_2), \tag{2.3}$$

extends by duality to a bounded linear transformer on the space of bounded linear operators from $\mathcal{H}_1$ to $\mathcal{H}_2$, and we say that the function Ψ on $\mathcal{X}_2 \times \mathcal{X}_1$ defined by

$$\Psi(\mu,\lambda) = \Phi(\lambda,\mu)$$

is *a Schur multiplier of the space of bounded linear operators associated with* E_2 *and* E_1. We denote the space of such Schur multipliers by $\mathfrak{M}(E_2,E_1)$.

In [**BS3**] it was shown that if A is a self-adjoint operator (not necessarily bounded), K is a bounded self-adjoint operator and φ is a continuously differentiable function on $\mathbb{R}$ such that the divided difference $\mathfrak{D}\varphi$ defined by

$$(\mathfrak{D}\varphi)(\lambda,\mu) = \frac{\varphi(\lambda)-\varphi(\mu)}{\lambda-\mu}$$

is a Schur multiplier of the space of bounded linear operators with respect to the spectral measures of $A+K$ and A, then

$$\varphi(A+K)-\varphi(A) = \iint_{\mathbb{R}\times\mathbb{R}} \frac{\varphi(\lambda)-\varphi(\mu)}{\lambda-\mu}\,dE_{A+K}(\lambda)K\,dE_A(\mu) \tag{2.4}$$

and

$$\|\varphi(A+K)-\varphi(A)\| \le \text{const}\,\|\varphi\|_{\mathfrak{M}}\|K\|,$$

where $\|\varphi\|_{\mathfrak{M}}$ is the norm of φ in $\mathfrak{M}(E_{A+K},E_A)$.

The same formula (2.4) holds in the case where K is a Hilbert–Schmidt operator and φ is a Lipschitz function (in this case $\mathfrak{D}\varphi$ is not necessarily defined on the diagonal of $\mathbb{R}\times\mathbb{R}$, and one can define $\mathfrak{D}\varphi$ to be zero on the diagonal).

It is easy to see that if a function Φ on $\mathcal{X}\times\mathcal{Y}$ belongs to the *projective tensor product* $L^\infty(E)\hat{\otimes}L^\infty(F)$ of $L^\infty(E)$ and $L^\infty(F)$ (i.e., Φ admits a representation

$$\Phi(\lambda,\mu) = \sum_{n\ge0} f_n(\lambda)g_n(\mu),$$

where $f_n \in L^\infty(E)$, $g_n \in L^\infty(F)$, and $\sum_{n\ge0}\|f_n\|_{L^\infty}\|g_n\|_{L^\infty} < \infty$), then $\Phi \in \mathfrak{M}(E,F)$, i.e., Φ is a Schur multiplier of the space of bounded linear operators. For such functions Φ we have

$$\int_{\mathcal{X}}\int_{\mathcal{Y}} \Phi(\lambda,\mu)\,dE(\lambda)Q\,dF(\mu) = \sum_{n\ge0}\left(\int_{\mathcal{X}} f_n\,dE\right)Q\left(\int_{\mathcal{Y}} g_n\,dF\right).$$

More generally, Φ is a Schur multiplier if Φ belongs to the *integral projective tensor product* $L^\infty(E)\hat{\otimes}_{\rm i}L^\infty(F)$ of $L^\infty(E)$ and $L^\infty(F)$, i.e., Φ admits a representation

$$\Phi(\lambda,\mu) = \int_\Omega f(\lambda,\omega)g(\mu,\omega)\,d\sigma(\omega), \tag{2.5}$$

where (Ω, σ) is a measure space, f is a measurable function on $\mathcal{X} \times \Omega$, g is a measurable function on $\mathcal{Y} \times \Omega$, and

$$\int_\Omega \|f(\cdot,\omega)\|_{L^\infty(E)} \|g(\cdot,\omega)\|_{L^\infty(F)}\, d\sigma(\omega) < \infty. \tag{2.6}$$

If $\Phi \in L^\infty(E)\hat{\otimes}_{\mathrm{i}} L^\infty(F)$, then

$$\int_{\mathcal{X}}\int_{\mathcal{Y}} \Phi(\lambda,\mu)\, dE(\lambda)\, Q\, dF(\mu) = \int_\Omega \left(\int_{\mathcal{X}} f(\lambda,\omega)\, dE(\lambda)\right) Q \left(\int_{\mathcal{Y}} g(\mu,\omega)\, dF(\mu)\right) d\sigma(\omega).$$

Clearly, the function

$$\omega \mapsto \left(\int_{\mathcal{X}} f(\lambda,\omega)\, dE(\lambda)\right) Q \left(\int_{\mathcal{Y}} g(\mu,\omega)\, dF(\mu)\right)$$

is weakly measurable and

$$\int_\Omega \left\| \left(\int_{\mathcal{X}} f(\lambda,\omega)\, dE(\lambda)\right) T \left(\int_{\mathcal{Y}} g(\mu,\omega)\, dF(\mu)\right)\right\| d\sigma(\omega) < \infty.$$

Moreover, it can easily be seen that such functions Φ are Schur multipliers of any symmetrically normed ideal of operators.

It turns out that all Schur multipliers of the space of bounded linear operators can be obtained in this way. More precisely, the following result holds (see [**Pe2**]):

THEOREM ON SCHUR MULTIPLIERS. *Let Φ be a measurable function on $\mathcal{X} \times \mathcal{Y}$. The following are equivalent:*

(i) *$\Phi \in \mathfrak{M}(E, F)$;*

(ii) *$\Phi \in L^\infty(E)\hat{\otimes}_{\mathrm{i}} L^\infty(F)$;*

(iii) *there exist measurable functions f on $\mathcal{X} \times \Omega$ and g on $\mathcal{Y} \times \Omega$ such that* (2.5) *holds and*

$$\left\| \int_\Omega |f(\cdot,\omega)|^2\, d\sigma(\omega)\right\|_{L^\infty(E)} \left\| \int_\Omega |g(\cdot,\omega)|^2\, d\sigma(\omega)\right\|_{L^\infty(F)} < \infty. \tag{2.7}$$

Note that the implication (iii)$\Rightarrow$(ii) was established in [**BS3**]. Note also that in the case of matrix Schur multipliers (this corresponds to discrete spectral measures of multiplicity 1) the equivalence of (i) and (ii) was proved in [**Be**].

It is interesting to observe that if f and g satisfy (2.6), then they also satisfy (2.7), but the converse is false. However, if Φ admits a representation of the form (2.5) with f and g satisfying (2.7), then it also admits a (possibly different) representation of the form (2.5) with f and g satisfying (2.6).

Note that in a similar way we can define the *projective tensor product* $A\hat{\otimes}B$ and the *integral projective tensor product* $A\hat{\otimes}_{\mathrm{i}}B$ of arbitrary Banach function spaces A and B.

Similar results also hold in the case of unitary operators.

2.3. Sufficient conditions and necessary conditions. We state here results in the case of unitary operators. Birman and Solomyak proved in [**BS3**] that if φ is a function on the unit circle such that the derivative of φ satisfies a Hölder condition of order $\alpha > 0$, then the divided difference $\mathfrak{D}\varphi$ belongs to $\mathfrak{M}(E, F)$, which implies that if U and V are unitary operators, then

$$\varphi(U) - \varphi(V) = \iint_{\mathbb{T}\times\mathbb{T}} \frac{\varphi(\zeta) - \varphi(\tau)}{\zeta - \tau}\, dE_U(\zeta)\,(U - V)\, dE_V(\tau),$$

and so

$$\|\varphi(U) - \varphi(V)\| \le \text{const}\,\|U - V\|,$$

i.e., φ is *an operator Lipschitz function.* Moreover, it was shown in [**BS3**] that under the same assumptions the function φ is *operator differentiable*, i.e., if $V = e^{\mathrm{i}A}U$, then the function $s \to \varphi\big(e^{\mathrm{i}sA}U\big)$ is differentiable and

$$\frac{d}{ds}\Big(\varphi(e^{\mathrm{i}sA}U)\Big)\bigg|_{s=o} = \mathrm{i}\left(\iint \frac{\varphi(\zeta) - \varphi(\tau)}{\zeta - \tau}\, dE_U(\zeta) A\, dE_U(\tau)\right) U.$$

Later, a much stronger result was obtained in [**Pe2**]. It was shown in [**Pe2**] that the same conclusions can be made under the assumption that $\varphi \in B^1_{\infty 1}$. Moreover, it was shown in [**Pe2**] that if $\varphi \in B^1_{\infty 1}$, then φ belongs to the projective tensor product $C(\mathbb{T})\hat{\otimes}C(\mathbb{T})$ of the space of continuous functions on $\mathbb{T}$ with itself, i.e., there exist functions f_n and g_n, $n \ge 1$, in $C(\mathbb{T})$ such that

$$\sum_{n\ge 1} \|f_n\|_\infty \|g_n\|_\infty < \infty$$

and

$$\frac{\varphi(\zeta) - \varphi(\tau)}{\zeta - \tau} = \sum_{n \ge 1} f_n(\zeta) g_n(\tau).$$

This implies that the functions in $B^1_{\infty 1}$ are operator Lipschitz and operator differentiable. Moreover, this also implies that the functions in $B^1_{\infty 1}$ satisfy the inequality

$$\|\varphi(U) - \varphi(V)\|_{\boldsymbol{S}} \le \text{const}\,\|U - V\|_{\boldsymbol{S}}$$

for arbitrary unitary operators U and V and for an arbitrary symmetrically normed ideal $\boldsymbol{S}$.

Similar results hold for (not necessarily bounded) self-adjoint operators; see [**Pe5**] and [**Pe9**].

It was proved in [**Pe2**] that if φ is an operator Lipschitz function on $\mathbb{T}$ (or if φ is operator differentiable), then $\varphi \in B^1_1$. This implies that

$$\sum_{n \ge 0} 2^n |\hat{\varphi}(2^n)| < \infty,$$

and so the condition $\varphi \in C^1(\mathbb{T})$ is not sufficient for φ to be operator Lipschitz or operator differentiable. An even stronger necessary condition found in [**Pe2**] (see also [**Pe6**], where that necessary condition was reformulated with the help of a remark by S. Semmes) says that if φ is operator Lipschitz (or operator differentiable), then

$$\varphi \in \mathcal{L} \stackrel{\text{def}}{=} \{\varphi :\ |\nabla^2 \varphi|\, dx\, dy \ \text{ is a Carleson measure in } \mathbb{D}\},$$

where $\nabla^2\varphi$ is the second gradient of the harmonic extension of φ to the unit disk.

M. Frazier observed that actually $\mathcal{L}$ is the Triebel–Lizorkin space $\dot{F}^1_{\infty 1}$. Note that the definition of the Triebel–Lizorkin spaces $\dot{F}^s_{pq}$ on $\mathbb{R}^n$ for $p = \infty$ and $q > 1$ can be found in [**T**], § 5.1. A definition for all $q > 0$, which is equivalent to Triebel's definition if $q > 1$, was given by Frazier and Jawerth in [**FrJ**]. Their approach does not use harmonic extensions, but a straightforward exercise in comparing kernels shows that Frazier and Jawerth's definition of $\dot{F}^1_{\infty 1}$ is equivalent to the definition requiring $|\nabla^2 u|\, dx\, dy$ to be a Carleson measure on the upper half-space. Our space $\mathcal{L}$ is the analogue for the unit disk.

It was observed in [**Pe3**] that if φ is an analytic function in $\left(B_{\infty 1}^{1}\right)_{+}$, then the divided difference $\mathfrak{D}\varphi$ belongs to the projective tensor product $C_A\hat{\otimes}C_A$ of the disk algebra C_A with itself, and so if $\varphi \in \left(B_{\infty 1}^{1}\right)_{+}$, then for arbitrary contractions T and R and an arbitrary symmetrically normed ideal $\boldsymbol{S}$, the following inequality holds:

$$\|\varphi(T) - \varphi(R)\|_{\boldsymbol{S}} \le \text{const}\,\|T - R\|_{\boldsymbol{S}}.$$

Recently it was proved in [**KS**] that if $\varphi \in C_A$, then the following conditions are equivalent:

(i) $\|\varphi(U) - \varphi(V)\| \le \text{const}\,\|U - V\|$ for any unitary operators U and V;
(ii) $\|\varphi(T) - \varphi(R)\| \le \text{const}\,\|T - R\|$ for any contractions T and R.

2.4. Multiple operator integrals. The equivalence of (i) and (ii) in the theorem on Schur multipliers suggests the idea, explored in [**Pe9**], to define multiple operator integrals.

To simplify the notation, we consider here the case of triple operator integrals; the case of arbitrary multiple operator integrals can be treated in the same way.

Let $(\mathcal{X}, E)$, $(\mathcal{Y}, F)$, and $(\mathcal{Z}, G)$ be spaces with spectral measures E, F, and G on Hilbert spaces $\mathcal{H}_1$, $\mathcal{H}_2$, and $\mathcal{H}_3$. Suppose that Φ belongs to the integral projective tensor product $L^\infty(E)\hat{\otimes}_{\rm i}L^\infty(F)\hat{\otimes}_{\rm i}L^\infty(G)$, i.e., Φ admits a representation

$$\Phi(\lambda, \mu, \nu) = \int_\Omega f(\lambda, \omega)g(\mu, \omega)h(\nu, \omega)\,d\sigma(\omega), \tag{2.8}$$

where (Ω, σ) is a measure space, f is a measurable function on $\mathcal{X} \times \Omega$, g is a measurable function on $\mathcal{Y} \times \Omega$, h is a measurable function on $\mathcal{Z} \times \Omega$, and

$$\int_Q \|f(\cdot, x)\|_{L^\infty(E)}\|g(\cdot, x)\|_{L^\infty(F)}\|h(\cdot, x)\|_{L^\infty(G)}\,d\sigma(x) < \infty.$$

Suppose that T_1 is a bounded linear operator from $\mathcal{H}_2$ to $\mathcal{H}_1$, and T_2 is a bounded linear operator from $\mathcal{H}_3$ to $\mathcal{H}_2$. For a function Φ of the form (2.8) in $L^\infty(E)\hat{\otimes}_{\rm i}L^\infty(F)\hat{\otimes}_{\rm i}L^\infty(G)$, we put

(2.9)
$$\begin{aligned}&\int_{\mathcal{X}}\int_{\mathcal{Y}}\int_{\mathcal{Z}} \Phi(\lambda, \mu, \nu)\,dE(\lambda)T_1\,dF(\mu)T_2\,dG(\nu)\\ &\stackrel{\text{def}}{=} \int_\Omega\left(\int_{\mathcal{X}} f(\lambda, \omega)\,dE(\lambda)\right)T_1\left(\int_{\mathcal{Y}} g(\mu, \omega)\,dF(\mu)\right)T_2\left(\int_{\mathcal{Z}} h(\nu, \omega)\,dG(\nu)\right)d\sigma(\omega).\end{aligned}$$

The following lemma from [**Pe9**] (see also [**ACDS**] for a different proof) shows that this definition does not depend on the choice of a representation (2.8).

LEMMA 2.1. *Suppose that* $\Phi \in L^\infty(E)\hat{\otimes}_{\rm i}L^\infty(F)\hat{\otimes}_{\rm i}L^\infty(G)$. *Then the right-hand side of* (2.9) *does not depend on the choice of a representation* (2.8).

It is easy to see that the following inequality holds:

$$\left\|\int_{\mathcal{X}}\int_{\mathcal{Y}}\int_{\mathcal{Z}} \Phi(\lambda, \mu, \nu)\,dE(\lambda)T_1\,dF(\mu)T_2\,dG(\nu)\right\| \le \|\Phi\|_{L^\infty\hat{\otimes}_{\rm i}L^\infty\hat{\otimes}_{\rm i}L^\infty}\cdot\|T_1\|\cdot\|T_2\|.$$

In particular, the triple operator integral on the left-hand side of (2.9) can be defined if Φ belongs to the projective tensor product $L^\infty(E)\hat{\otimes}L^\infty(F)\hat{\otimes}L^\infty(G)$, i.e., Φ admits a representation

$$\Phi(\lambda, \mu, \nu) = \sum_{n\ge 1} f_n(\lambda)g_n(\mu)h_n(\nu),$$

where $f_n \in L^\infty(E)$, $g_n \in L^\infty(F)$, $h_n \in L^\infty(G)$, and

$$\sum_{n\ge1} \|f_n\|_{L^\infty(E)} \|g_n\|_{L^\infty(F)} \||h_n\|_{L^\infty(G)} < \infty.$$

In a similar way one can define multiple operator integrals; see [**Pe9**].

For a function φ on the circle, the *divided differences* $\mathfrak{D}^k\varphi$ *of order* k are defined inductively as follows:

$$\mathfrak{D}^0\varphi \stackrel{\text{def}}{=} \varphi;$$

if $k \ge 1$, then in the case where $\lambda_1, \dots, \lambda_{k+1}$ are distinct points in $\mathbb{T}$,

$$(\mathfrak{D}^k\varphi)(\lambda_1, \dots, \lambda_{k+1}) \stackrel{\text{def}}{=} \frac{(\mathfrak{D}^{k-1}\varphi)(\lambda_1, \dots, \lambda_{k-1}, \lambda_k) - (\mathfrak{D}^{k-1}\varphi)(\lambda_1, \dots, \lambda_{k-1}, \lambda_{k+1})}{\lambda_k - \lambda_{k+1}}$$

(the definition does not depend on the order of the variables). Clearly,

$$\mathfrak{D}\varphi = \mathfrak{D}^1\varphi.$$

If $\varphi \in C^k(\mathbb{T})$, then $\mathfrak{D}^k\varphi$ extends by continuity to a function defined for all points $\lambda_1, \dots, \lambda_{k+1}$.

It was shown in [**Pe9**] that if $\varphi \in B^n_{\infty 1}$, then $\mathfrak{D}^n\varphi$ belongs to the projective tensor product $C(\mathbb{T})\hat{\otimes}\cdots\hat{\otimes}C(\mathbb{T})_{n+1}$ ($n+1$ factors). Moreover, it was shown in [**Pe9**] that if U and V are unitary operators, $V = e^{iA}U$, then the function

$$t \mapsto e^{itA}U$$

has n derivatives in the norm and

$$\frac{d^n}{dt^n}\Big(\big(\varphi(e^{itA}U)\big)\Big|_{s=0}$$

$$= i^n n! \left(\underbrace{\int \cdots \int}_{n+1} (\mathfrak{D}^n\varphi)(\lambda_1, \dots, \lambda_{n+1})\, dE_U(\lambda_1) A \cdots A\, dE_U(\lambda_{n+1}) \right) U^n.$$

The reasoning given in [**Pe9**] shows that

$$\varphi \in \left(B^n_{\infty 1}\right)_+ \quad \Rightarrow \quad \mathfrak{D}^n\varphi \in \underbrace{C_A\hat{\otimes}\cdots\hat{\otimes}C_A}_{n+1}. \tag{2.10}$$

Note that recently in [**JTT**] Haagerup tensor products were used to define multiple operator integrals. However, it is not clear whether this can lead to a larger class of functions φ for which $\mathfrak{D}^n\varphi$ can be integrated.

2.5. Semispectral measures. Let $\mathcal{H}$ be a Hilbert space and let $(\mathcal{X}, \mathcal{B})$ be a measurable space. A map $\mathcal{E}$ from $\mathcal{B}$ to the algebra $B(\mathcal{H})$ of all bounded operators on $\mathcal{H}$ is called a *semispectral measure* if

$$\mathcal{E}(\Delta) \ge \mathbf{0}, \quad \Delta \in \mathcal{B},$$

$$\mathcal{E}(\varnothing) = \mathbf{0} \quad \text{and} \quad \mathcal{E}(\mathcal{X}) = I,$$

and for a sequence $\{\Delta_j\}_{j\ge1}$ of disjoint sets in $\mathcal{B}$,

$$\mathcal{E}\left(\bigcup_{j=1}^\infty \Delta_j\right) = \lim_{N\to\infty} \sum_{j=1}^N \mathcal{E}(\Delta_j) \quad \text{in the weak operator topology.}$$

If $\mathcal{K}$ is a Hilbert space, $(\mathcal{X},\mathcal{B})$ is a measurable space, $E:\mathcal{B}\to B(\mathcal{K})$ is a spectral measure, and $\mathcal{H}$ is a subspace of $\mathcal{K}$, then it is easy to see that the map $\mathcal{E}:\mathcal{B}\to B(\mathcal{H})$ defined by

$$\mathcal{E}(\Delta)=P_{\mathcal{H}}E(\Delta)\big|\mathcal{H},\quad \Delta\in\mathcal{B}, \tag{2.11}$$

is a semispectral measure. Here $P_{\mathcal{H}}$ stands for the orthogonal projection onto $\mathcal{H}$.

Naimark proved in [N] (see also [**SNF**]) that all semispectral measures can be obtained in this way, i.e., a semispectral measure is always a *compression* of a spectral measure. A spectral measure E satisfying (2.11) is called a *spectral dilation of the semispectral measure* $\mathcal{E}$.

A spectral dilation E of a semispectral measure $\mathcal{E}$ is called *minimal* if

$$\mathcal{K}=\operatorname{clos\,span}\{E(\Delta)\mathcal{H}:\ \Delta\in\mathcal{B}\}.$$

It was shown in [**MM**] that if E is a minimal spectral dilation of a semispectral measure $\mathcal{E}$, then E and $\mathcal{E}$ are mutually absolutely continuous and all minimal spectral dilations of a semispectral measure are isomorphic in the natural sense.

If φ is a bounded complex-valued measurable function on $\mathcal{X}$ and $\mathcal{E}:\mathcal{B}\to B(\mathcal{H})$ is a semispectral measure, then the integral

$$\int_{\mathcal{X}} f(x)\,d\mathcal{E}(x) \tag{2.12}$$

can be defined as

$$\int_{\mathcal{X}} f(x)\,d\mathcal{E}(x)=P_{\mathcal{H}}\left(\int_{\mathcal{X}} f(x)\,dE(x)\right)\Big|\mathcal{H}, \tag{2.13}$$

where E is a spectral dilation of $\mathcal{E}$. It is easy to see that the right-hand side of (2.13) does not depend on the choice of a spectral dilation. The integral (2.12) can also be computed as the limit of sums

$$\sum f(x_\alpha)\mathcal{E}(\Delta_\alpha),\quad x_\alpha\in\Delta_\alpha,$$

over all finite measurable partitions $\{\Delta_\alpha\}_\alpha$ of $\mathcal{X}$.

If T is a contraction on a Hilbert space $\mathcal{H}$, then by the Sz.-Nagy dilation theorem (see [**SNF**]), T has a unitary dilation, i.e., there exists a Hilbert space $\mathcal{K}$ such that $\mathcal{H}\subset\mathcal{K}$ and a unitary operator U on $\mathcal{K}$ such that

$$T^n=P_{\mathcal{H}}U^n\big|\mathcal{H},\quad n\ge 0, \tag{2.14}$$

where $P_{\mathcal{H}}$ is the orthogonal projection onto $\mathcal{H}$. Let E_U be the spectral measure of U. Consider the operator set function $\mathcal{E}$ defined on the Borel subsets of the unit circle $\mathbb{T}$ by

$$\mathcal{E}(\Delta)=P_{\mathcal{H}}E_U(\Delta)\big|\mathcal{H},\quad \Delta\subset\mathbb{T}.$$

Then $\mathcal{E}$ is a semispectral measure. It follows immediately from (2.14) that

$$T^n=\int_{\mathbb{T}}\zeta^n\,d\mathcal{E}(\zeta)=P_{\mathcal{H}}\int_{\mathbb{T}}\zeta^n\,dE_U(\zeta)\Big|\mathcal{H},\quad n\ge 0. \tag{2.15}$$

Such a semispectral measure $\mathcal{E}$ is called a *semispectral measure* of $\mathbb{T}$. Note that it is not unique. To have uniqueness, we can consider a minimal unitary dilation U of T, which is unique up to an isomorphism (see [**SNF**]).

It follows easily from (2.15) that for an arbitrary function φ in the disk algebra C_A,

$$\varphi(T)=P_{\mathcal{H}}\int_{\mathbb{T}}\varphi(\zeta)\,dE_U(\zeta)\Big|\mathcal{H}.$$

3. Double operator integrals with respect to semispectral measures

In this section we extend the Birman–Solomyak theory of double operator integrals to the case of semispectral measures.

Suppose that $(\mathcal{X}_1,\mathcal{B}_1)$ and $(\mathcal{X}_2,\mathcal{B}_2)$ are measurable spaces, $\mathcal{H}_1$ and $\mathcal{H}_2$ are Hilbert spaces, and $\mathcal{E}_1:\mathcal{B}_1\to B(\mathcal{H}_1)$ and $\mathcal{E}_2:\mathcal{B}_2\to B(\mathcal{H}_2)$ are semispectral measures. For a bounded measurable function Φ on $\mathcal{X}_1\times\mathcal{X}_2$ and an operator $Q:\mathcal{H}_2\to\mathcal{H}_1$, we consider double operator integrals

$$\iint_{\mathcal{X}_1\times\mathcal{X}_2}\Phi(x_1,x_2)\,d\mathcal{E}_1(x_1)Q\,d\mathcal{E}_2(X_2). \tag{3.1}$$

In the case where Q is a Hilbert–Schmidt operator, integrals of the form (3.1) can be interpreted as in the case of double operator integrals with respect to spectral measures (see § 2.2). Indeed, we define the map $\mathcal{F}$ on the set of all measurable rectangles $\Delta_1\times\Delta_2$ by

$$\mathcal{F}(\Delta_1\times\Delta_2)Q=\mathcal{E}_1(\Delta_1)Q\mathcal{E}_2(\Delta_2),\quad Q\in\boldsymbol{S}_2(\mathcal{H}_2,\mathcal{H}_1).$$

Clearly, $\mathcal{F}(\Delta_1\times\Delta_2)$ is a bounded linear operator on $\boldsymbol{S}_2(\mathcal{H}_2,\mathcal{H}_1)$ that satisfies the inequalities $\mathbf{0}\le\mathcal{F}(\Delta_1\times\Delta_2)\le I$.

LEMMA 3.1. *$\mathcal{F}$ extends to a semispectral measure on $\mathcal{B}_1\times\mathcal{B}_2$.*

PROOF. Let E_1 and E_2 be spectral dilations of $\mathcal{E}_1$ and $\mathcal{E}_2$ on Hilbert spaces $\mathcal{K}_1$ and $\mathcal{K}_2$. Define the map F on measurable rectangles $\Delta_1\times\Delta_2$ by

$$F(\Delta_1\times\Delta_2)Q=E_1(\Delta_1)QE_2(\Delta_2),\quad Q\in\boldsymbol{S}_2(\mathcal{K}_2,\mathcal{K}_1).$$

By the theorem of Birman and Solomyak (see [**BS4**]), F extends to a spectral measure (which will also be denoted by F) defined on $\mathcal{X}_1\times\mathcal{X}_2$. Clearly,

$$\mathcal{F}(\Delta_1\times\Delta_2)Q=P_{\mathcal{H}_1}F(\Delta_1\times\Delta_2)Q,\quad Q\in\boldsymbol{S}_2(\mathcal{H}_2,\mathcal{H}_1).$$

We can now define the operator $\mathcal{F}(\Omega)$ for an arbitrary measurable $\Omega\subset\mathcal{X}_1\times\mathcal{X}_2$ by

$$\mathcal{F}(\Omega)Q=P_{\mathcal{H}_1}F(\Omega)Q,\quad Q\in\boldsymbol{S}_2(\mathcal{H}_2,\mathcal{H}_1).$$

As is easily seen, $\mathcal{F}$ is a semispectral measure on the Hilbert space $\boldsymbol{S}_2(\mathcal{H}_2,\mathcal{H}_1)$. $\square$

Given a bounded measurable function Φ on $\mathcal{X}_1\times\mathcal{X}_2$ and a Hilbert–Schmidt operator $Q:\mathcal{H}_2\to\mathcal{H}_1$, we can define the double operator integral

$$\iint_{\mathcal{X}_1\times\mathcal{X}_2}\Phi(x_1,x_2)\,d\mathcal{E}_1(x_1)Q\,d\mathcal{E}_2(x_2)$$

as in the case of integration with respect to spectral measures:

$$\iint_{\mathcal{X}_1\times\mathcal{X}_2}\Phi(x_1,x_2)\,d\mathcal{E}_1(x_1)Q\,d\mathcal{E}_2(x_2)\stackrel{\mathrm{def}}{=}\left(\int_{\mathcal{X}_1\times\mathcal{X}_2}\Phi\,d\mathcal{F}\right)Q.$$

It is easy to see that

$$\begin{aligned}&\iint_{\mathcal{X}_1\times\mathcal{X}_2}\Phi(x_1,x_2)\,d\mathcal{E}_1(x_1)Q\,d\mathcal{E}_2(x_2)\\&\qquad=P_{\mathcal{H}_1}\iint_{\mathcal{X}_1\times\mathcal{X}_2}\Phi(x_1,x_2)\,dE_1(x_1)(QP_{H_2})\,dE_2(x_2)\Big|\mathcal{H}_2.\end{aligned}$$

Clearly,

$$\left\|\iint_{\mathcal{X}_1\times\mathcal{X}_2}\Phi(x_1,x_2)\,d\mathcal{E}_1(x_1)Q\,d\mathcal{E}_2(x_2)\right\|_{\boldsymbol{S}_2}\le\sup_{x_1\in\mathcal{X}_1,x_2\in\mathcal{X}_2}|\Phi(x_1,x_2)|\cdot\|Q\|_{\boldsymbol{S}_2}. \tag{3.2}$$

We need the following fact.

LEMMA 3.2. *Suppose that $\mathcal{E}_1$ and $\mathcal{E}_2$ are semispectral measures as above and Q is a Hilbert–Schmidt operator. If $\{\Phi_n\}_{n\ge1}$ is a sequence of measurable functions such that*

$$\lim_{n\to\infty}\Phi_n(\lambda,\mu)=\Phi(\lambda,\mu),\qquad \lambda\in\mathcal{X}_1,\ \mu\in\mathcal{X}_2,$$

and

$$\sup_n\sup_{\lambda,\,\mu\in\mathcal{X}_1\times\mathcal{X}_2}|\Phi_n(\lambda,\mu)|<\infty,$$

then

$$\lim_{n\to\infty}\left\|\int_{\mathcal{X}_1}\int_{\mathcal{X}_2}\Phi_n(\lambda,\mu)\,d\mathcal{E}_1(\lambda)\,Q\,d\mathcal{E}_2(\mu)-\int_{\mathcal{X}_1}\int_{\mathcal{X}_2}\Phi(\lambda,\mu)\,d\mathcal{E}_1(\lambda)\,Q\,d\mathcal{E}_2(\mu)\right\|_{\boldsymbol{S}_2}=0.$$

PROOF. The result follows immediately from the same fact in the case of spectral measures; see (2.2). $\square$

Let us now proceed to double operator integrals (3.1) with bounded operators Q.

Suppose that $\mathcal{E}_1$ and $\mathcal{E}_2$ are semispectral measures and E_1 and E_2 are their minimal spectral dilations. If Φ is a Schur multiplier of the space of bounded linear operators, then the double operator integral

$$\iint_{\mathcal{X}_1\times\mathcal{X}_2}\Phi(x_1,x_2)\,d\mathcal{E}_1(x_1)Q\,d\mathcal{E}_2(x_2)\tag{3.3}$$

is defined as

$$P_{\mathcal{H}_1}\iint_{\mathcal{X}_1\times\mathcal{X}_2}\Phi(x_1,x_2)\,dE_1(x_1)\big(QP_{H_2}\big)\,dE_2(x_2)\Big|\mathcal{H}_2.$$

It is easy to see that if $\Phi\in L^\infty(\mathcal{E}_1)\hat{\otimes}_{\rm i}L^\infty(\mathcal{E}_2)$ and

$$\Phi(x_1,x_2)=\int_\Omega f(x_1,\omega)g(x_2,\omega)\,d\sigma(\omega)$$

with

$$\int_\Omega\|f(\cdot,\omega)\|_{L^\infty(\mathcal{E}_1)}\|g(\cdot,\omega)\|_{L^\infty(\mathcal{E}_2)}\,d\sigma(\omega)<\infty,$$

then Φ is a Schur multiplier of the space of bounded linear operators, the integral (3.3) is equal to

$$\int_\Omega\left(\int_{\mathcal{X}_1}f(x_1,\omega)\,d\mathcal{E}_1(x_1)\right)Q\left(\int_{\mathcal{X}_2}g(x_2,\omega)\,d\mathcal{E}_2(x_2)\right)\,d\sigma(\omega),$$

and its norm is less than or equal to

$$\|Q\|\int_\Omega\|f(\cdot,\omega)\|_{L^\infty(\mathcal{E}_1)}\|g(\cdot,\omega)\|_{L^\infty(\mathcal{E}_2)}\,d\sigma(\omega).$$

We can now define multiple operator integrals in the same way as in the case of spectral measures. Suppose that $\mathcal{H}_1,\dots,\mathcal{H}_n$ are Hilbert spaces and for $j=1,\dots,n-1$, Q_j is a bounded linear operator from $\mathcal{H}_{j+1}$ to $\mathcal{H}_j$. Suppose also that $\mathcal{E}_1,\dots,\mathcal{E}_n$ are semispectral measures defined on σ-algebras of $\mathcal{X}_1,\dots,\mathcal{X}_n$, $\mathcal{E}_j$ takes values in the space $\mathcal{B}(\mathcal{H}_j)$, and Φ is a function on $\mathcal{X}_1\times\cdots\times\mathcal{X}_n$ of class $L^\infty(\mathcal{E}_1)\hat{\otimes}_{\rm i}\cdots\hat{\otimes}_{\rm i}L^\infty(\mathcal{E}_n)$, i.e., Φ admits a representation

$$\Phi(x_1,\dots,x_n)=\int_\Omega f_1(x_1,\omega)\cdots f_n(x_n,\omega)\,d\sigma(\omega),\tag{3.4}$$

in which

$$\int_\Omega \|f_1(,\cdot,\omega)\|_{L^\infty(\mathcal{E}_1)}\cdots\|f_n(,\cdot,\omega)\|_{L^\infty(\mathcal{E}_n)}\,d\sigma(\omega)<\infty. \tag{3.5}$$

We define the multiple operator integral

$$\underbrace{\int\cdots\int}_{n}\Phi(x_1,\dots,x_n)\,d\mathcal{E}_1(x_1)Q_1\cdots Q_{n-1}\,d\mathcal{E}_n(x_n)$$

as

$$\int_\Omega\left(\int_{\mathcal{X}_1}f_1(x_1,\omega)\,d\mathcal{E}_1(x_1)\right)Q_1\cdots Q_{n-1}\left(\int_{\mathcal{X}_n}f_n(x_n,\omega)\,d\mathcal{E}_n(x_n)\right)d\sigma(\omega). \tag{3.6}$$

Certainly, we must prove that the multiple operator integral is well defined. In other words, we must show that the value of (3.6) does not depend on the choice of a representation (3.4), which is a consequence of the following lemma.

LEMMA 3.3. *Suppose that*

$$\int_\Omega f_1(x_1,\omega)\cdots f_n(x_n,\omega)\,d\sigma(\omega)=0,\quad x_1\in\mathcal{X}_1,\dots,x_n\in\mathcal{X}_n,$$

and (3.5) *holds. Then*

$$\int_\Omega\left(\int_{\mathcal{X}_1}f_1(x_1,\omega)\,d\mathcal{E}_1(x_1)\right)Q_1\cdots Q_{n-1}\left(\int_{\mathcal{X}_n}f_n(x_n,\omega)\,d\mathcal{E}_n(x_n)\right)d\sigma(\omega) \tag{3.7}$$

is the zero operator.

PROOF. We deduce the lemma from the corresponding fact for multiple operator integrals with respect to spectral measures. Suppose that E_j is a minimal spectral dilation of $\mathcal{E}_j$ and E_j takes values in $\mathcal{B}(\mathcal{K})$. Then the integral in (3.7) is equal to

(3.8)

$$P_\mathcal{H}\left(\int_\Omega\left(\int_{\mathcal{X}_1}f_1(x_1,\omega)\,dE_1(x_1)\right)B_1\cdots B_{n-1}\left(\int_{\mathcal{X}_n}f_n(x_n,\omega)\,dE_n(x_n)\right)d\sigma(\omega)\right)\Big|\mathcal{H},$$

where $B_j\stackrel{\text{def}}{=}P_{\mathcal{H}_j}Q_jP_{\mathcal{H}_{j+1}}$, $1\le j\le n-1$. It follows now from Lemma 2.1 that the operator in (3.8) is the zero operator. □

4. Analogs of the Birman–Solomyak formulae

In this section we obtain analogs of the Birman–Solomyak formulae (1.5) and (1.6) for contractions.

If T is a contraction on Hilbert space, then $\mathcal{E}_T$ denotes a semispectral measure of $\mathbb{T}$.

Recall that if $\varphi'\in C_A$, the function $\mathfrak{D}\varphi$ extends to the diagonal

$$\boldsymbol{\Delta}\stackrel{\text{def}}{=}\{(\zeta,\zeta):\zeta\in\mathbb{T}\}$$

by continuity: $(\mathfrak{D}\varphi)(\zeta,\zeta)=\varphi'(\zeta)$, $\zeta\in\mathbb{T}$.

THEOREM 4.1. *Let $\varphi\in\left(B^1_{\infty1}\right)_+$. Then for contractions T and R on Hilbert space the following formula holds:*

$$\varphi(T)-\varphi(R)=\iint_{\mathbb{T}\times\mathbb{T}}\frac{\varphi(\zeta)-\varphi(\tau)}{\zeta-\tau}\,d\mathcal{E}_T(\zeta)\,(T-R)\,d\mathcal{E}_R(\tau). \tag{4.1}$$

Recall that it follows from the results of **[Pe2]** that the function $\mathfrak{D}\varphi$ belongs to the projective tensor product $C_A \hat{\otimes} C_A$ (see § 2.2), and so the right-hand side of (4.1) is well defined.

PROOF OF THEOREM 4.1. Suppose that

$$(\mathfrak{D}\varphi)(\zeta, \tau) = \frac{\varphi(\zeta) - \varphi(\tau)}{\zeta - \tau} = \sum_{n \geq 0} f_n(\zeta) g_n(\tau),$$

where $f_n \in C_A$, $g_n \in C_A$, and

$$\sum_{n \geq 0} \|f_n\|_\infty \|g_n\|_\infty < \infty.$$

We have

$$\begin{aligned}
&\iint_{\mathbb{T}\times\mathbb{T}} \frac{\varphi(\zeta) - \varphi(\tau)}{\zeta - \tau} \, d\mathcal{E}_T(\zeta)(T - R) \, d\mathcal{E}_R(\tau) \\
&\qquad = \sum_{n \geq 0} f_n(T)(T - R) g_n(R) = \sum_{n \geq 0} T f_n(T) g_n(R) - \sum_{n \geq 0} f_n(T) g_n(R) R.
\end{aligned}$$

Clearly,

$$\begin{aligned}
\sum_{n \geq 0} T f_n(T) g_n(R) &= \sum_{n \geq 0} \iint_{\mathbb{T}\times\mathbb{T}} \zeta f_n(\zeta) g_n(\tau) \, d\mathcal{E}_T(\zeta) \, d\mathcal{E}_R(\tau) \\
&= \iint_{\mathbb{T}\times\mathbb{T}} \zeta (\mathfrak{D}\varphi)(\zeta, \tau) \, d\mathcal{E}_T(\zeta) \, d\mathcal{E}_R(\tau)
\end{aligned}$$

and

$$\begin{aligned}
\sum_{n \geq 0} f_n(T) g_n(R) R &= \sum_{n \geq 0} \iint_{\mathbb{T}\times\mathbb{T}} f_n(\zeta) g_n(\tau) \tau \, d\mathcal{E}_T(\zeta) \, d\mathcal{E}_R(\tau) \\
&= \iint_{\mathbb{T}\times\mathbb{T}} \tau (\mathfrak{D}\varphi)(\zeta, \tau) \, d\mathcal{E}_T(\zeta) \, d\mathcal{E}_R(\tau).
\end{aligned}$$

Thus

$$\begin{aligned}
&\iint_{\mathbb{T}\times\mathbb{T}} (\mathfrak{D}\varphi)(\zeta, \tau) \, d\mathcal{E}_T(\zeta)(T - R) \, d\mathcal{E}_R(\tau) \\
&\quad = \iint_{\mathbb{T}\times\mathbb{T}} \zeta (\mathfrak{D}\varphi)(\zeta, \tau) \, d\mathcal{E}_T(\zeta) \, d\mathcal{E}_R(\tau) - \iint_{\mathbb{T}\times\mathbb{T}} \tau (\mathfrak{D}\varphi)(\zeta, \tau) \, d\mathcal{E}_T(\zeta) \, d\mathcal{E}_R(\tau) \\
&\quad = \iint_{\mathbb{T}\times\mathbb{T}} (\zeta - \tau)(\mathfrak{D}\varphi)(\zeta, \tau) \, d\mathcal{E}_T(\zeta) \, d\mathcal{E}_R(\tau) \\
&\quad = \iint_{\mathbb{T}\times\mathbb{T}} \big(\varphi(\zeta) - \varphi(\tau)\big) \, d\mathcal{E}_T(\zeta) \, d\mathcal{E}_R(\tau) \\
&\quad = \varphi(T) - \varphi(R). \qquad \square
\end{aligned}$$

Let us consider the case where $T - R \in \boldsymbol{S}_2$. The following result establishes formula (4.1) for functions with derivatives in C_A.

THEOREM 4.2. *Let φ be a function analytic in $\mathbb{D}$ such that $\varphi' \in C_A$. If T and R are contractions such that $T - R \in \boldsymbol{S}_2$, then formula* (4.1) *holds.*

PROOF. Let $\{\varphi_n\}_{n\ge1}$ be a sequence of polynomials such that

$$\|\varphi' - \varphi_n'\|_\infty \to 0 \quad \text{as} \quad n\to\infty.$$

It follows from Theorem 4.1 that

$$\varphi_n(T) - \varphi_n(R) = \iint_{\mathbb{T}\times\mathbb{T}} \frac{\varphi_n(\zeta)-\varphi_n(\tau)}{\zeta-\tau}\, d\mathcal{E}_T(\zeta)\,(T-R)\,d\mathcal{E}_R(\tau).$$

By von Neumann's inequality,

$$\lim_{n\to\infty}\|\varphi_n(T)-\varphi(T)\|\to 0 \quad\text{and}\quad \lim_{n\to\infty}\|\varphi_n(R)-\varphi(R)\|\to 0.$$

The result follows now from the trivial observation that

$$\sup_{\zeta,\tau\in\mathbb{T}} |(\mathfrak{D}\varphi_n)(\zeta,\tau)-(\mathfrak{D}\varphi)(\zeta,\tau)|\to 0 \quad\text{as}\quad n\to\infty,$$

which implies that

$$\lim_{n\to\infty}\left\|\iint_{\mathbb{T}\times\mathbb{T}}\Big((\mathfrak{D}\varphi_n)(\zeta,\tau)-(\mathfrak{D}\varphi)(\zeta,\tau)\Big)\,d\mathcal{E}_T(\zeta)\,(T-R)\,d\mathcal{E}_R(\tau)\right\|_{\boldsymbol{S}_2}=0. \quad \square$$

Consider now the more general case where $\varphi'\in H^\infty$.

THEOREM 4.3. *Let $\varphi'\in H^\infty$ and let Φ be the function on $\mathbb{T}\times\mathbb{T}$ defined by*

$$\Phi\Big|(\mathbb{T}\times\mathbb{T}\setminus\boldsymbol{\Delta}) = (\mathfrak{D}\varphi)\Big|(\mathbb{T}\times\mathbb{T}\setminus\boldsymbol{\Delta}) \quad\text{and}\quad \Phi\Big|\boldsymbol{\Delta}=\boldsymbol{0}. \tag{4.2}$$

If T and R are contractions such that $T-R\in\boldsymbol{S}_2$, then

$$\varphi(T)-\varphi(R)=\iint_{\mathbb{T}\times\mathbb{T}}\Phi(\zeta,\tau)\,d\mathcal{E}_T(\zeta)\,(T-R)\,d\mathcal{E}_R(\tau).$$

PROOF. Let φ_n be the nth Cesáro mean of the Fourier series of φ. Then $\varphi_n\in\big(B^1_{\infty1}\big)_+$ and by Theorem 4.1,

$$\begin{aligned}\varphi_n(T)-\varphi_n(R)&=\iint_{\mathbb{T}\times\mathbb{T}}\frac{\varphi_n(\zeta)-\varphi_n(\tau)}{\zeta-\tau}\,d\mathcal{E}_T(\zeta)\,(T-R)\,d\mathcal{E}_R(\tau)\\&=\iint_{\mathbb{T}\times\mathbb{T}}\big(\varphi_n(\zeta)-\varphi_n(\tau)\big)\,d\mathcal{E}_T(\zeta)\,d\mathcal{E}_R(\tau)\\&=\iint_{(\mathbb{T}\times\mathbb{T})\setminus\boldsymbol{\Delta}}\big(\varphi_n(\zeta)-\varphi_n(\tau)\big)\,d\mathcal{E}_T(\zeta)\,d\mathcal{E}_R(\tau).\end{aligned}$$

The same reasoning as in the proof of Theorem 4.1 shows that

$$\begin{aligned}&\iint_{(\mathbb{T}\times\mathbb{T})\setminus\boldsymbol{\Delta}}\frac{\varphi_n(\zeta)-\varphi_n(\tau)}{\zeta-\tau}\,d\mathcal{E}_T(\zeta)\,(T-R)\,d\mathcal{E}_R(\tau)\\&\qquad=\iint_{(\mathbb{T}\times\mathbb{T})\setminus\boldsymbol{\Delta}}\big(\varphi_n(\zeta)-\varphi_n(\tau)\big)\,d\mathcal{E}_T(\zeta)\,d\mathcal{E}_R(\tau).\end{aligned}$$

Clearly, $\|\varphi_n-\varphi\|_\infty\to0$ as $n\to\infty$, and so by von Neumann's inequality,

$$\varphi_n(T)-\varphi_n(R)\to\varphi(T)-\varphi(R)$$

in the operator norm.

On the other hand, since

$$\lim_{n\to\infty}\frac{\varphi_n(\zeta)-\varphi_n(\tau)}{\zeta-\tau}=\frac{\varphi(\zeta)-\varphi(\tau)}{\zeta-\tau},\quad \zeta,\tau\in\mathbb{T},$$

and

$$\sup_n \sup_{\zeta,\tau\in\mathbb{T}} \left|\frac{\varphi_n(\zeta)-\varphi_n(\tau)}{\zeta-\tau}\right| < \infty,$$

it follows from Lemma 3.2 that

$$\left\| \iint_{(\mathbb{T}\times\mathbb{T})\setminus\boldsymbol{\Delta}} \left(\frac{\varphi_n(\zeta)-\varphi_n(\tau)}{\zeta-\tau} - \frac{\varphi(\zeta)-\varphi(\tau)}{\zeta-\tau}\right) d\mathcal{E}_T(\zeta)\,(T-R)\,d\mathcal{E}_R(\tau)\right\|_{\boldsymbol{S}_2} \to 0$$

as $n\to\infty$.

To complete the proof, it remains to observe that

$$\iint_{(\mathbb{T}\times\mathbb{T})\setminus\boldsymbol{\Delta}} \frac{\varphi(\zeta)-\varphi(\tau)}{\zeta-\tau}\, d\mathcal{E}_T(\zeta)\,(T-R)\,d\mathcal{E}_R(\tau)$$
$$= \iint_{\mathbb{T}\times\mathbb{T}} \Phi(\zeta,\tau)\, d\mathcal{E}_T(\zeta)\,(T-R)\,d\mathcal{E}_R(\tau). \qquad \square$$

The following result is an immediate consequence of Theorem 4.3; it was obtained recently in [**KS**] by a completely different method.

COROLLARY 4.4. *Suppose that φ is a function analytic in $\mathbb{D}$ such that $\varphi'\in H^\infty$. If T and R are contractions on Hilbert space such that $T-R\in\boldsymbol{S}_2$, then*

$$\varphi(T)-\varphi(R)\in\boldsymbol{S}_2 \quad \text{and} \quad \|\varphi(R)-\varphi(T)\|_{\boldsymbol{S}_2} \le \|\varphi'\|_{H^\infty}\|T-R\|_{\boldsymbol{S}_2}.$$

Let us now obtain analogs of formula (1.6).

THEOREM 4.5. *Let $\varphi\in\left(B^1_{\infty1}\right)_+$. Then for a contraction T and a bounded linear operator Q on Hilbert space the following formula holds:*

$$\varphi(T)Q-Q\varphi(T) = \iint_{\mathbb{T}\times\mathbb{T}} \frac{\varphi(\zeta)-\varphi(\tau)}{\zeta-\tau}\, d\mathcal{E}_T(\zeta)\,(TQ-QT)\,d\mathcal{E}_T(\tau).$$

The following result is a Hilbert–Schmidt version of Theorem 4.5.

THEOREM 4.6. *Let $\varphi'\in H^\infty$ and let Φ be the function on $\mathbb{T}\times\mathbb{T}$ defined by* (4.2). *Suppose that $\mathbb{T}$ is a contraction and Q is a bounded linear operator such that*

$$TQ-QT\in\boldsymbol{S}_2. \tag{4.3}$$

Then

$$\varphi(T)Q-Q\varphi(T) = \iint_{\mathbb{T}\times\mathbb{T}} \Phi(\zeta,\tau)\, d\mathcal{E}_T(\zeta)\,(TQ-QT)\,d\mathcal{E}_T(\tau).$$

Theorems 4.5 and 4.6 can be proved in the same way as Theorems 4.1 and 4.3.

The following inequality is an immediate consequence of Theorem 4.6; recently it was proved in [**KS**] by a different method.

COROLLARY 4.7. *Suppose that $\varphi'\in H^\infty$, T is a contraction, and Q is a bounded linear operator on Hilbert space such that* (4.3) *holds. Then*

$$\varphi(T)Q-Q\varphi(T)\in\boldsymbol{S}_2$$

and

$$\|\varphi(T)Q-Q\varphi(T)\|_{\boldsymbol{S}_2} \le \|\varphi'\|_{H^\infty}\|TQ-QT\|_{\boldsymbol{S}_2}.$$

5. Differentiability of operator functions in the operator norm

Let T and R be contractions on a Hilbert space $\mathcal{H}$. For $t \in [0,1]$, consider the operator T_t defined by

$$T_t = (1-t)T + tR. \tag{5.1}$$

Clearly, T_t is a contraction.

In this section, for a function $\varphi \in C_A$, we consider the problem of differentiability of the function

$$t \mapsto \varphi(T_t), \quad 0 \le t \le 1, \tag{5.2}$$

and the problem of the existence of higher derivatives. We also compute the derivatives in terms of multiple operator integrals with respect to semispectral measures.

Let $\mathcal{E}_t$ be a semispectral measure of T_t on the unit circle $\mathbb{T}$, i.e.,

$$T_t^n = \int_{\mathbb{T}} \zeta^n \, d\mathcal{E}_t(\zeta), \quad n \ge 0.$$

Put $\mathcal{E} \stackrel{\text{def}}{=} \mathcal{E}_0$.

THEOREM 5.1. *Suppose that $\varphi \in \left(B^1_{\infty 1}\right)_+$. Then the function* (5.2) *is differentiable in the norm and*

$$\frac{d}{ds}\varphi(T_s)\bigg|_{s=t} = \iint_{\mathbb{T}\times\mathbb{T}} \frac{\varphi(\zeta)-\varphi(\tau)}{\zeta-\tau}\, d\mathcal{E}_t(\zeta)\,(R-T)\,d\mathcal{E}_t(\tau).$$

PROOF. We have

$$\begin{aligned}\frac{1}{s}\big(\varphi(T_{t+s})-\varphi(T_t)\big) &= \frac{1}{s}\iint_{\mathbb{T}\times\mathbb{T}} \frac{\varphi(\zeta)-\varphi(\tau)}{\zeta-\tau}\, d\mathcal{E}_{t+s}(\zeta)\,(T_{t+s}-T_t)\,d\mathcal{E}_t(\tau)\\ &= \iint_{\mathbb{T}\times\mathbb{T}} \frac{\varphi(\zeta)-\varphi(\tau)}{\zeta-\tau}\, d\mathcal{E}_{t+s}(\zeta)\,(R-T)\,d\mathcal{E}_t(\tau).\end{aligned}$$

As we mentioned in § 2.3, it was shown in [**Pe2**] that $\mathfrak{D}\varphi$ admits a representation

$$\frac{\varphi(\zeta)-\varphi(\tau)}{\zeta-\tau} = \sum_n f_n(\zeta)g_n(\tau), \tag{5.3}$$

where $f_n, g_n \in C_A$ and

$$\sum_n \|f_n\|_\infty \|g_n\|_\infty < \infty. \tag{5.4}$$

The following identities hold:

$$\begin{aligned}&\iint_{\mathbb{T}\times\mathbb{T}} \frac{\varphi(\zeta)-\varphi(\tau)}{\zeta-\tau}\, d\mathcal{E}_{t+s}(\zeta)\,(R-T)\,d\mathcal{E}_t(\tau)\\ &\qquad= \sum_n \iint_{\mathbb{T}\times\mathbb{T}} f_n(\zeta)g_n(\tau)\, d\mathcal{E}_{t+s}(\zeta)\,(R-T)\,d\mathcal{E}_t(\tau)\\ &\qquad= \sum_n f_n(T_{t+s})(R-T)g_n(T_t).\end{aligned}$$

Clearly,

$$\lim_{s\to 0}\|f_n(T_{t+s}) - f_n(T)\| = 0$$

and in view of (5.4),

$$\lim_{s\to 0}\frac{1}{s}\big(\varphi(T_{t+s})-\varphi(T_t)\big)=\sum_n f_n(T_t)(R-T)g_n(T_t)$$

in the norm. It remains to observe that by (5.3),

$$\begin{aligned}\lim_{s\to 0}\sum_n f_n(T_{t+s})(R-T)g_n(T_t)&=\sum_n f_n(T_t)(R-T)g_n(T_t)\\&=\sum_n\iint_{\mathbb{T}\times\mathbb{T}} f_n(\zeta)g_n(\tau)\,d\mathcal{E}_t(\zeta)\,(R-T)\,d\mathcal{E}_t(\tau)\\&=\iint_{\mathbb{T}\times\mathbb{T}}\frac{\varphi(\zeta)-\varphi(\tau)}{\zeta-\tau}\,d\mathcal{E}_t(\zeta)\,(R-T)\,d\mathcal{E}_t(\tau).\end{aligned}$$

□

In what follows, to simplify the notation, we will not specify the domain of integration: all double or multiple integrals will be taken over unit tori.

THEOREM 5.2. *Let $\varphi\in(B^2_{\infty,1})_+$. Then the function* (5.2) *has second derivative in the norm and*

(5.5) $$\frac{d^2}{ds^2}\varphi(T_s)\Big|_{s=t}=2\iiint(\mathfrak{D}^2\varphi)(\zeta,\tau,\upsilon)\,d\mathcal{E}_t(\zeta)\,(R-T)\,d\mathcal{E}_t(\tau)\,(R-T)\,d\mathcal{E}_t(\upsilon).$$

PROOF. We prove (5.5) for $t=0$. For all other t the proof is the same. By Theorem 5.1,

$$\begin{aligned}&\frac{1}{t}\left(\frac{d}{ds}(\varphi(T_s))\Big|_{s=t}-\frac{d}{ds}(\varphi(T_s))\Big|_{s=0}\right)\\&=\frac{1}{t}\iint(\mathfrak{D}\varphi)(\zeta,\upsilon)\,d\mathcal{E}_t(\zeta)(R-T)\,d\mathcal{E}_t(\upsilon)-\frac{1}{t}\iint(\mathfrak{D}\varphi)(\tau,\upsilon)\,d\mathcal{E}(\tau)\,(R-T)\,d\mathcal{E}(\upsilon)\\&=\frac{1}{t}\iint(\mathfrak{D}\varphi)(\zeta,\upsilon)\,d\mathcal{E}_t(\zeta)(R-T)\,d\mathcal{E}_t(\upsilon)-\frac{1}{t}\iint(\mathfrak{D}\varphi)(\tau,\upsilon)\,d\mathcal{E}(\tau)\,(R-T)\,d\mathcal{E}_t(\upsilon)\\&+\frac{1}{t}\iint(\mathfrak{D}\varphi)(\zeta,\upsilon)\,d\mathcal{E}(\zeta)(R-T)\,d\mathcal{E}_t(\upsilon)-\frac{1}{t}\iint(\mathfrak{D}\varphi)(\zeta,\tau)\,d\mathcal{E}(\zeta)\,(R-T)\,d\mathcal{E}(\tau).\end{aligned}$$

We have

$$\begin{aligned}&\iint(\mathfrak{D}\varphi)(\zeta,\upsilon)\,d\mathcal{E}_t(\zeta)\,(R-T)\,d\mathcal{E}_t(\upsilon)-\iint(\mathfrak{D}\varphi)(\tau,\upsilon)\,d\mathcal{E}(\tau)\,(R-T)\,d\mathcal{E}_t(\upsilon)\\&\quad=\iiint(\mathfrak{D}\varphi)(\zeta,\upsilon)\,d\mathcal{E}_t(\zeta)\,d\mathcal{E}(\tau)\,(R-T)\,d\mathcal{E}_t(\upsilon)\\&\qquad-\iiint(\mathfrak{D}\varphi)(\tau,\upsilon)\,d\mathcal{E}_t(\zeta)\,d\mathcal{E}(\tau)\,(R-T)\,d\mathcal{E}_t(\upsilon)\\&\quad=\iiint(\mathfrak{D}^2\varphi)(\zeta,\tau,\upsilon)(\zeta-\tau)\,d\mathcal{E}_t(\zeta)\,d\mathcal{E}(\tau)\,(R-T)\,d\mathcal{E}_t(\upsilon)\\&\quad=t\iiint(\mathfrak{D}^2\varphi)(\zeta,\tau,\upsilon)\,d\mathcal{E}_t(\zeta)\,(R-T)\,d\mathcal{E}(\tau)\,(R-T)\,d\mathcal{E}_t(\upsilon).\end{aligned}$$

Similarly,

$$\begin{aligned}&\iint(\mathfrak{D}\varphi)(\zeta,\upsilon)\,d\mathcal{E}(\zeta)\,(R-T)\,d\mathcal{E}_t(\upsilon)-\iint(\mathfrak{D}\varphi)(\zeta,\tau)\,d\mathcal{E}(\zeta)\,(R-T)\,d\mathcal{E}(\tau)\\&\quad=t\iiint(\mathfrak{D}^2\varphi)(\zeta,\tau,\upsilon)\,d\mathcal{E}(\zeta)(R-T)\,d\mathcal{E}(\tau)\,(R-T)\,d\mathcal{E}_t(\upsilon).\end{aligned}$$

Thus

$$\frac{1}{t}\left(\frac{d}{ds}(\varphi(T_s))\Big|_{s=t} - \frac{d}{ds}(\varphi(T_s))\Big|_{s=0}\right)$$
$$= \iiint (\mathfrak{D}^2\varphi)(\zeta,\tau,\upsilon)\, d\mathcal{E}_t(\zeta)\,(R-T)\,d\mathcal{E}(\tau)\,(R-T)\,d\mathcal{E}_t(\upsilon)$$
$$+ \iiint (\mathfrak{D}^2\varphi)(\zeta,\tau,\upsilon)\, d\mathcal{E}(\zeta)\,(R-T)\,d\mathcal{E}(\tau)\,(R-T)\,d\mathcal{E}_t(\upsilon).$$

As we mentioned in § 2.4, it follows from the results of [**Pe9**] that $\mathfrak{D}^2\varphi \in C_A\hat{\otimes}C_A\hat{\otimes}C_A$, i.e., there exist sequences $\{f_n\}$, $\{g_n\}$, and $\{h_n\}$ in C_A such that

$$(\mathfrak{D}^2\varphi)(\zeta,\tau,\upsilon) = \sum_{n\ge 0} f_n(\zeta)g_n(\tau)h_n(\upsilon)$$

and

$$\sum_{n\ge 0}\|f_n\|_\infty\|g_n\|_\infty\|h_n\|_\infty < \infty. \tag{5.6}$$

Put

$$Q \stackrel{\text{def}}{=} R - T.$$

We have

$$\iiint (\mathfrak{D}^2\varphi)(\zeta,\tau,\upsilon)\, d\mathcal{E}_t(\zeta)\, Q\, d\mathcal{E}(\tau)\, Q\, d\mathcal{E}_t(\upsilon) = \sum_{n\ge 0} f_n(T_t)Q\, g_n(T)Q\, h_n(T_t).$$

It follows from (5.6) that

$$\lim_{t\to 0}\sum_{n\ge 0} f_n(T_t)Q\, g_n(T)Q\, h_n(T_t) = \sum_{n\ge 0} f_n(T)Q\, g_n(T)Q\, h_n(T)$$

in the operator norm. Thus

$$\lim_{t\to 0}\iiint (\mathfrak{D}^2\varphi)(\zeta,\tau,\upsilon)\, d\mathcal{E}_t(\zeta)\, Q\, d\mathcal{E}(\tau)\, Q\, d\mathcal{E}_t(\upsilon)$$
$$= \sum_{n\ge 0} f_n(T)Q\, g_n(T)Q\, h_n(T)$$
$$= \iiint (\mathfrak{D}^2\varphi)(\zeta,\tau,\upsilon)\, d\mathcal{E}(\zeta)Q\, d\mathcal{E}(\tau)Q\, d\mathcal{E}(\upsilon).$$

Similarly,

$$\lim_{t\to 0}\iiint (\mathfrak{D}^2\varphi)(\zeta,\tau,\upsilon)\, d\mathcal{E}(\zeta)\, Q\, d\mathcal{E}(\tau)\, Q\, d\mathcal{E}_t(\upsilon)$$
$$= \iiint (\mathfrak{D}^2\varphi)(\zeta,\tau,\omega)\, d\mathcal{E}(\zeta)\, Q\, d\mathcal{E}(\tau)\, Q\, d\mathcal{E}(\omega),$$

which proves the result. ☐

The same method allows us to prove the following generalization of Theorem 5.2.

THEOREM 5.3. *Suppose that* $\varphi \in \left(B^n_{\infty 1}\right)_+$. *Then the function* (5.2) *has nth derivative in the norm,*

$$\frac{d^n}{ds^n}\varphi(T_s)\bigg|_{s=t} = n!\underbrace{\int\dots\int}_{n+1}(\mathfrak{D}^n\varphi)(\zeta_1,\dots,\zeta_{n+1})\,d\mathcal{E}_t(\zeta_1)\,(R-T)\cdots(R-T)\,d\mathcal{E}_t(\zeta_{n+1}).$$

6. Differentiability of operator functions in the Hilbert–Schmidt norm

Suppose that T and R are contractions on Hilbert space such that $T - R \in \boldsymbol{S}_2$. In this section we are going to obtain results on the differentiability of the function (5.2) in the Hilbert–Schmidt norm.

THEOREM 6.1. *Let* φ *be a function analytic in* $\mathbb{D}$ *such that* $\varphi' \in C_A$. *Suppose that* T *and* R *are contractions on Hilbert space such that* $T - R \in \boldsymbol{S}_2$ *and* T_t *is defined by* (5.1). *Then the function* (5.2) *is differentiable in the Hilbert–Schmidt norm and*

$$\frac{d}{ds}\varphi(T_s)\bigg|_{s=0} = \iint \frac{\varphi(\zeta)-\varphi(\tau)}{\zeta-\tau}\,d\mathcal{E}(\zeta)\,(R-T)\,d\mathcal{E}(\tau),$$

where $\mathcal{E}$ *is a semispectral measure of* T.

PROOF. Let φ_n be the nth Cesáro mean of the Taylor series of φ. Then $\|\varphi_n' - \varphi'\|_\infty \to 0$ as $n \to \infty$. Let $\mathcal{E}_t$ be a semispectral measure of T_t. We have

$$\begin{aligned}\frac{1}{s}(\varphi(T_s)-\varphi(T)) &= \iint \frac{\varphi(\zeta)-\varphi(\tau)}{\zeta-\tau}\,d\mathcal{E}_s(\zeta)\,(R-T)\,d\mathcal{E}(\tau)\\ &= \iint \frac{\varphi_n(\zeta)-\varphi_n(\tau)}{\zeta-\tau}\,d\mathcal{E}_s(\zeta)\,(R-T)\,d\mathcal{E}(\tau)\\ &+ \iint \frac{(\varphi-\varphi_n)(\zeta)-(\varphi-\varphi_n)(\tau)}{\zeta-\tau}\,d\mathcal{E}_s(\zeta)\,(R-T)\,d\mathcal{E}(\tau).\end{aligned}$$

Let $\varepsilon > 0$. There exists a natural number N such that

$$\sup_{\zeta,\tau}\left|\frac{(\varphi-\varphi_n)(\zeta)-(\varphi-\varphi_n)(\tau)}{\zeta-\tau}\right| < \varepsilon \tag{6.1}$$

whenever $n \ge N$.

Let $n \ge N$. Since φ_n is a polynomial, it is easy to see that

$$\iint \frac{\varphi_n(\zeta)-\varphi_n(\tau)}{\zeta-\tau}\,d\mathcal{E}_s(\zeta)\,(R-T)\,d\mathcal{E}(\tau)$$

tends to

$$\iint \frac{\varphi_n(\zeta)-\varphi_n(\tau)}{\zeta-\tau}\,d\mathcal{E}(\zeta)\,(R-T)\,d\mathcal{E}(\tau)$$

in the Hilbert–Schmidt norm. Let $\delta > 0$ be a number such that

$$\left|\iint \frac{\varphi_n(\zeta)-\varphi_n(\tau)}{\zeta-\tau}\,d\mathcal{E}_s(\zeta)\,(R-T)\,d\mathcal{E}(\tau) - \iint \frac{\varphi_n(\zeta)-\varphi_n(\tau)}{\zeta-\tau}\,d\mathcal{E}(\zeta)\,(R-T)\,d\mathcal{E}(\tau)\right| < \varepsilon$$

whenever $s \le \delta$.

To conclude the proof, we observe that inequalities (6.1) and (3.2) imply that

$$\left\| \iint \frac{(\varphi - \varphi_n)(\zeta) - (\varphi - \varphi_n)(\tau)}{\zeta - \tau} \, d\mathcal{E}_s(\zeta)\,(R-T)\,d\mathcal{E}(\tau) \right\|_{\boldsymbol{S}_2} < \varepsilon$$

and

$$\left\| \iint \frac{(\varphi - \varphi_n)(\zeta) - (\varphi - \varphi_n)(\tau)}{\zeta - \tau} \, d\mathcal{E}(\zeta)\,(R-T)\,d\mathcal{E}(\tau) \right\|_{\boldsymbol{S}_2} < \varepsilon$$

whenever $n \ge N$, and so

$$\left\| \frac{1}{s}(\varphi(T_s) - \varphi(T)) - \iint \frac{\varphi(\zeta) - \varphi(\tau)}{\zeta - \tau} \, d\mathcal{E}(\zeta)\,(R-T)\,d\mathcal{E}(\tau) \right\|_{\boldsymbol{S}_2} < 3\varepsilon$$

whenever $s \le \delta$. □

References

[ABF] J. Arazy, T. Barton, and Y. Friedman, *Operator differentiable functions*, Int. Equat. Oper. Theory **13** (1990), 462–487.

[ACDS] N. A. Azamov, A. L. Carey, P. G. Dodds, and F. A. Sukochev, *Operator integrals, spectral shift and spectral flow*, `arXiv:math/0703442`.

[Be] G. Bennett, *Schur multipliers*, Duke Math. J. **44** (1977), 603–639.

[BS1] M. S. Birman and M. Z. Solomyak, *Double Stieltjes operator integrals*, Problems of Math. Phys., Leningrad. Univ. **1** (1966), 33–67; English transl., Topics Math. Physics **1** (1967), 25–54, Consultants Bureau, Plenum Publishing Corporation, New York.

[BS2] ______, *Double Stieltjes operator integrals.* II, Problems of Math. Phys., Leningrad. Univ. **2** (1967), 26–60; English transl., Topics Math. Physics **2** (1968), 19–46, Consultants Bureau, Plenum Publishing Corporation, New York.

[BS3] ______, *Double Stieltjes operator integrals.* III, Problems of Math. Phys., Leningrad. Univ. **6** (1973), 27–53. (Russian)

[BS4] ______, *Tensor product of a finite number of spectral measures is always a spectral measure*, Integral Equations Operator Theory **24** (1996), 179–187.

[BS5] ______, *Double operator integrals in Hilbert space*, Int. Equat. Oper. Theory **47** (2003), 131–168.

[DK] Yu. L. Daletskiĭ and S. G. Kreĭn, *Integration and differentiation of functions of Hermitian operators and application to the theory of perturbations*, Trudy Sem. Funktsion. Anal., Voronezh. Gos. Univ. **1** (1956), 81–105. (Russian)

[Fa] Yu. B. Farforovskaya, *An estimate of the norm of $|f(B) - f(A)|$ for selfadjoint operators A and B*, Zap. Nauchn. Sem. Leningrad. Otdel. Mat. Inst. Steklov. (LOMI) **56** (1976), 143–162. (Russian)

[FrJ] M. Frazier and B. Jawerth, *A discrete transform and decompositions of distribution spaces*, J. Funct. Anal. **93** (1990), 34–170.

[JTT] K. Jushchenko, I. G. Todorov, and L. Turowska, *Multidimensional operator multipliers*, `arXiv:math/0701645v2`.

[KS] E. Kissin and V. S. Shulman, *On fully operator Lipschitz functions*, J. Funct. Anal. **253** (2007), 711–728.

[MM] M. M. Malamud and S. M. Malamud, *Spectral theory of operator measures in a Hilbert space*, Algebra i Analiz **15** (2003), no. 3, 1–77; English transl., St. Petersburg Math. J. **15** (2004), no. 3, 323–373.

[N] M. A. Naimark, *Spectral functions of symmetric operator*, Izvestia Akad. Nauk SSSR, Ser. Matem. **4** (1940), no. 3, 277–318. (Russian)

[Pee] J. Peetre, *New thoughts on Besov spaces*, Duke Univ. Press., Durham, NC, 1976.

[Pe1] V. V. Peller, *Estimates of functions of power bounded operators on Hilbert space*, J. Oper. Theory **7** (1982), 341–372.

[Pe2] ______, *Hankel operators in the theory of perturbations of unitary and self-adjoint operators*, Funktsional. Anal. i Prilozhen. **19** (1985), no. 2, 37–51; English transl., Funct. Anal. Appl. **19** (1985), 111–123.

[Pe3] ______, *For which f does $A - B \in \mathbf{S}_p$ imply that $f(A) - f(B) \in \mathbf{S}_p$?*, Operators in indefinite metric spaces, scattering theory and other topics (Bucharest, 1985), pp. 289–294, Oper. Theory Adv. Appl. **24**, Birkhäuser, Basel, 1987.

[Pe4] ______, *When is a function of a Toeplitz operator close to a Toeplitz operator?*, Toeplitz operators and spectral function theory, pp. 59–85, Oper. Theory Adv. Appl. **42**, Birkhäuser, Basel, 1989.

[Pe5] ______, *Hankel operators in the perturbation theory of of unbounded self-adjoint operators.* Analysis and partial differential equations, pp. 529–544, Lecture Notes in Pure and Appl. Math. **122**, Dekker, New York, 1990.

[Pe6] ______, *Functional calculus for a pair of almost commuting selfadjoint operators*, J. Funct. Anal. **112** (1993), 325–345.

[Pe7] ______, *Hankel operators and their applications*, Springer-Verlag, New York, 2003.

[Pe8] ______, *An extension of the Koplienko–Neidhardt trace formulae*, J. Funct. Anal. **221** (2005), 456–481.

[Pe9] ______, *Multiple operator integrals and higher operator derivatives*, J. Funct. Anal. **233** (2006), 515–544.

[St] V. V. Sten′kin, *Multiple operator integrals*, Izv. Vyssh. Uchebn. Zaved. Matematika **4 (79)** (1977), 102–115; English transl., Soviet Math. (Iz. VUZ) **21:4** (1977), 88–99.

[SNF] B. Szőkefalvi-Nagy and C. Foiaş, *Harmonic analysis of operators on Hilbert space*, Akadémiai Kiadó, Budapest, 1970.

[T] H. Triebel, *Theory of function spaces*, Monographs in Mathematics, **78**, Birkhäuser, Basel, 1983.

Department of Mathematics, Michigan State University, East Lansing, Michigan 48824, USA

Amer. Math. Soc. Transl.
(2) Vol. **226**, 2009

Asymptotic Behavior of Arguments of Cauchy Integrals

Alexei Poltoratski

To my teacher Victor Petrovich Havin on the occasion of his 75th birthday

ABSTRACT. We study asymptotic behavior of Cauchy integrals of measures on the real line. We describe the sets of measures whose Hilbert transform (conjugate Poisson integral) can grow faster than the Poisson transform, and vice versa.

1. Introduction

In this note we continue the discussion of asymptotic properties of Cauchy integrals started in [**9**]. Our main goal is to identify the classes of measures whose Poisson transform grows faster (or slower) than the conjugate Poisson transform (or the Hilbert transform) near a large set of points. Besides pure curiosity, such questions are motivated by their frequent appearance in complex function theoretic proofs. The idea to estimate the singular Cauchy (Hilbert) transform with a much more manageable Poisson transform lies in the foundation of complex analysis. Such estimates have produced many of the classical results: from Harnack's lemma to Fatou's theorem. The questions considered here may be viewed as an attempt to identify those measures whose Cauchy integrals allow such estimates. Such problems have many interesting connections, like the connection with planar Cauchy transforms via a problem by P. Mattila, which was solved in [**4**], or the connection with mathematical physics via reflectionless measures, which are discussed in the next section.

Let $M(\mathbb{R})$ be the space of all complex measures μ on the real line satisfying

$$\int_{\mathbb{R}} \frac{d|\mu|(t)}{1+t^2} < \infty.$$

2000 *Mathematics Subject Classification.* Primary 30E20.

Key words and phrases. Hilbert transform, Cauchy integral, Poisson integral, boundary behavior.

The author is supported in part by N.S.F. grant DMS 0800300.

We will denote by $P\mu$ and $Q\mu$ the Poisson and the conjugate Poisson integrals of $\mu \in M(\mathbb{R})$ in the upper half-plane $\mathbb{C}_+$, respectively:

$$P\mu(x+iy) = \frac{1}{\pi}\int_{\mathbb{R}} \frac{y}{(x-t)^2+y^2}\,d\mu(t) \qquad \text{and}$$

$$Q\mu(x+iy) = \frac{1}{\pi}\int_{\mathbb{R}} \left(\frac{x-t}{(x-t)^2+y^2} + \frac{t}{1+t^2}\right) d\mu(t).$$

We denote by $C\mu$ the Cauchy integral in $\mathbb{C}_+$:

$$C\mu = i(P\mu + iQ\mu) = \frac{1}{\pi}\int_{\mathbb{R}} \left(\frac{1}{t-z} - \frac{t}{1+t^2}\right) d\mu(x).$$

We say that a measure $\mu \in M(\mathbb{R})$ belongs to Class N ($N =$ I, II, III, or IV) if μ-almost every x satisfies the corresponding condition below. The conditions are:

$$\text{(I)} \quad P\mu(z) = o(Q\mu(z)) \quad \text{as} \quad z \underset{\sphericalangle}{\longrightarrow} x;$$

$$\text{(II)} \quad Q\mu(z) = o(P\mu(z)) \quad \text{as} \quad z \underset{\sphericalangle}{\longrightarrow} x;$$

$$\text{(III)} \quad P\mu(z) = O(Q\mu(z)) \quad \text{as} \quad z \underset{\sphericalangle}{\longrightarrow} x;$$

$$\text{(IV)} \quad Q\mu(z) = O(P\mu(z)) \quad \text{as} \quad z \underset{\sphericalangle}{\longrightarrow} x.$$

We also discuss these conditions with the non-tangential convergence $z \underset{\sphericalangle}{\longrightarrow} x$ replaced by the radial convergence $z = x+iy$, $y \to 0+$. The corresponding classes of measures will be denoted by N_r.

The word "argument" in the title is justified by the fact that, if μ is a real measure, each of the relations (I)–(IV) can be interpreted as a statement about the behavior of $\arg C\mu$. From this point of view, real measures from Class II, for instance, are such that $\arg C\mu$ tends to $\pm\pi$ at μ-a.e. point, etc.

A study of these classes was started in [**9**], where some results on their radial versions were obtained. Many interesting open problems, especially those related to Class II, were also discussed. One of the open problems was later solved in [**4**]; see Section 2 below.

In this note we provide a short review of the new results and connections that appeared since [**9**]. One of such connections concerns so-called reflectionless measures, which recently became an object of active study in complex analysis and mathematical physics; see [**1, 2, 8, 12, 14**]. We discuss reflectionless measures in Section 2. We also give answers to some of the remaining questions and provide several new examples of measures from Classes I–IV. One of the new results is Theorem 3.1, which implies that the set of zeros of a Cauchy integral of a measure μ in $\mathbb{C}_+$ approaches non-tangentially almost every point with respect to the singular component of μ.

Let us start the discussion of Classes I–IV with a remark that the subset of singular measures in each class is spectral. We say that a set of measures is spectral if together with any measure μ it contains the whole equivalence class of μ, i.e., all measures $f\mu$ where $f \in L^1(|\mu|)$. This immediately follows from

THEOREM 1.1 ([**10**]). *Let $\mu, \nu \in M(\mathbb{R})$, $\mu = f\nu + \eta$, where $f \in L^1(|\nu|)$ and $\eta \perp \nu$. Then the limit*

$$\lim_{z \underset{\sphericalangle}{\longrightarrow} x} \frac{C\mu(x)}{C\nu(x)}$$

exists ν-a.e. It is equal to $f(x)$ ν_s-a.e.

Theorem 1.1 also takes care of other questions of similar nature. For instance, the fact that Class II does not contain any singular measures (see below) implies, by Theorem 1.1, that it does not contain any measures with non-trivial singular parts, etc.

Now let us pass to a more detailed discussion of each class. As was proved in [**9**], Classes I and I_r are trivial. Theorem 2.7 in [**9**] actually gives a slightly more general result, where non-tangential convergence is replaced with convergence along a non-tangential path. Similar versions of some of the results discussed below would also be interesting.

2. Class II

Singular measures. As was proved in [**4**], Class II does not contain any singular measures. This follows from a more general result concerning Riesz transforms in $\mathbb{R}^n$.

For each $\mu \in M(\mathbb{R}^{n-1})$ the Riesz transform $R\mu(x)$ is defined on the half-space $\mathbb{R}^n_+ = \{(x_1, \dots, x_n) \mid x_n > 0\}$ as

$$R\mu(x) = \langle R_1\mu(x), \dots, R_n\mu(x)\rangle,$$
$$R_i\mu(x) = \int_{\mathbb{R}^{n-1}} \frac{x_i - y_i}{|x-y|^n}\, d\mu(y), \quad i = 1, 2, \dots, n.$$

It is well known that Riesz transforms can be viewed as generalizations of the Poisson integral (the n-th transform R_n) and the conjugate Poisson integral (all other transforms $R_1, \dots, R_{n-1}$) to higher dimensions.

If $x \in \mathbb{R}^{n-1}$ and $0 < \phi < \pi/2$, we denote by Γ^ϕ_x the truncated cone:

$$\Gamma^\phi_x = \{y = \langle y_1, \dots, y_n\rangle \in \mathbb{R}^n_+ \mid y_n/|y-x| > \sin\phi,\ y_n < 1\}.$$

As usual we write $y \underset{\sphericalangle}{\longrightarrow} x$ if $y \to x$ and there exists $\phi = \phi(x)$ such that $y \in \Gamma^\phi_x$.

THEOREM 2.1 ([**4**]). *Let $\mu \in M(\mathbb{R}^{n-1})$ and let $\Sigma \subset \mathbb{R}^{n-1}$ be such that*

$$|\langle R_1\mu(y), \dots, R_{n-1}\mu(y)\rangle| = o(R_n\mu(y)) \tag{1}$$

as $y \underset{\sphericalangle}{\longrightarrow} x$ for μ-a.e. $x \in \Sigma$. Then the restriction of μ to Σ is absolutely continuous with respect to Lebesgue measure in $\mathbb{R}^{n-1}$.

It would be interesting to see what other problems considered here can be generalized to the settings of Riesz transforms.

Class II_r does not contain any singular continuous measures, but contains all discrete measures:

THEOREM 2.2 ([**4**]). *Let $\mu \in M(\mathbb{R})$ and let Σ be a Borel subset of $\mathbb{R}$. Then the following conditions are equivalent:*

1) $Q\mu(x+iy) = o(P\mu(x+iy))$ *as $y \to 0+$ for μ_s-a.e. $x \in \Sigma$;*
2) *the restriction of μ_s to Σ is discrete.*

This result answers a question raised in [**9**]. It also solves a problem by P. Mattila and some related problems in the theory of planar Cauchy transforms; see [**4**]. For an alternative proof and further developments see also [**16**].

Absolutely continuous measures. It is easy to see that the set of absolutely continuous measures from Class II coincides with the set of measures $\mu \in M(\mathbb{R})$ whose Hilbert transform is equal to 0 μ-a.e. Such measures are called reflectionless; see [**12**]. For absolutely continuous measures, non-tangential and radial convergence are related more closely. In particular, for any $N = \mathrm{I}, \dots, \mathrm{IV}$, the set of absolutely continuous measures belonging to Class N is equal to the analogous subset in Class N_r.

A more general definition of a reflectionless measure pertains to finite complex measures in the plane $\mathbb{C}$. In [**6**], a finite complex measure μ in $\mathbb{C}$ is called reflectionless if it has no point masses and its planar Cauchy transform, defined as the convolution of μ with $\frac{1}{z}$ in the sense of principal value, converges to 0 at μ-almost every point in $\mathbb{C}$.

If μ happens to lie on the real line, this definition is equivalent to our definition with a slightly different kernel for the conjugate Poisson transform, namely, without the term $\frac{t}{1+t^2}\,d\mu(t)$ under the integral. The two definitions are, of course, closely related. In particular, the two different conjugate Poisson transforms coincide for measures such that

$$\int \frac{t}{1+t^2}\,d\mu(t) = 0,$$

which is true, for instance, for all even measures.

Theorem 2.1 above implies that reflectionless measures on the line cannot have non-trivial singular parts. However, many absolutely continuous measures on the line are reflectionless. Examples are given immediately by

THEOREM 2.3 ([**6**]). *Let E be a compact set of positive length, $E \subset \mathbb{R}$. Let ω be a harmonic measure of $\mathbb{C} \setminus E$ with pole at infinity. Suppose that ω is absolutely continuous. Then ω is reflectionless.*

Perhaps the simplest example of a reflectionless measure is the harmonic measure on $[-1,1]$: $\mu = \frac{1}{\pi}(1-x^2)^{-1/2}\,dx$.

A description of the set of reflectionless measures on R seems to be an interesting open problem. In [**6**], a partial description of the class of positive reflectionless measures on $\mathbb{R}$ is given using geometric properties of conformal maps. Here we provide a full description of this set in terms of the arguments of the Cauchy transforms of the measures. A description of all complex reflectionless measures, even in the case of the line, is yet to be found.

Let us mention that positive reflectionless measures on the line naturally appear in Mathematical Physics as spectral measures of Jacobi matrices. In addition to the measures defined as above, in such problems one is also interested in measures reflectionless on a fixed set, i.e., those measures for which condition (II) holds a.e. on a fixed subset of the line. This connection brings up a number of interesting questions on reflectionless measures; see [**2, 8, 12, 14**].

The study of reflectionless measures on the line also seems to be a logical first step towards obtaining a description of all reflectionless measures in the plane. It was proved in [**17**] (see also [**6**] for an alternative proof) that a reflectionless measure in the plane cannot be absolutely continuous with respect to two-dimensional Lebesgue measure. The results of [**4**] imply that a reflectionless measure may not have Hausdorff dimension less than 1. Examples of reflectionless measures on the real line give measures of dimension 1.

Do there exist planar reflectionless measures of Hausdorff dimensions between 1 and 2? This seems to be another intriguing open problem. First significant progress in this direction was recently made by F. Nazarov and A. Volberg [**7**]. It was proved that each planar reflectionless measure corresponds to a Hausdorff gauge function that tends to zero strictly slower than x^2.

Our goal for the rest of this section is to provide a workable description of the set of all positive reflectionless measures on the line that could help answer some of the most basic questions about this set: Does Theorem 2.3 give all possible examples of positive reflectionless measures on $\mathbb{R}$? Is every set of positive Lebesgue measure a minimal support for a reflectionless measure? Is such a measure unique?

As we will see shortly, not all reflectionless measures on the line are harmonic measures. Every set of positive Lebesgue measure is indeed a minimal support of a reflectionless measure. Such a measure is unique if and only if the set is an interval.

Let $\mu \in M(\mathbb{R})$. We say that a Borel set E supports μ (E is a support of μ) if $|\mu|(E^c) = 0$. In the case of μ absolutely continuous, we say that E is a minimal support of μ if E is a support of μ and for any set $G \subset E$, $|G| > 0$, $|\mu|(G) > 0$.

Notice that our definition does not require a support to be closed. Minimal supports of absolutely continuous measures are unique up to sets of Lebesgue measure zero.

Denote by $M(\hat{\mathbb{R}})$ the space of complex measures on $\hat{\mathbb{R}}$: if $\mu \in M(\hat{\mathbb{R}})$ then

$$\mu = \nu + c\delta_\infty,$$

where $\nu \in M(\mathbb{R})$, $c \in \mathbb{R}$, and δ_∞ is the unit point mass at infinity. Then the Cauchy integral of μ, $C\mu$, is defined as

$$C\mu = C\nu + cz.$$

The norm in $M(\hat{\mathbb{R}})$ is defined as

$$\|\mu\| = \int \frac{d|\mu|(x)}{1+x^2} + |c|.$$

Consider the principal branch of the logarithm, $\log z$, defined in $\mathbb{C} \setminus (-\infty, 0]$, with values in the strip $|\Im z| < \pi$. If μ is a positive measure from $M(\hat{\mathbb{R}})$, then $C\mu$ has positive imaginary part in $\mathbb{C}_+$. Therefore $\log C\mu$ is a well-defined function in $\mathbb{C}_+$ with $\Im \log C\mu$ taking values in $(0, \pi)$. Denote

$$u_\mu(x) = \Im \log C\mu(x) \tag{2}$$

for $x \in \mathbb{R}$. That is, u_μ is equal to the boundary values of $\arg C\mu$ on the real line, which exist almost everywhere on $\mathbb{R}$. Here arg stands for the principal branch of the argument in $\mathbb{C} \setminus (-\infty, 0]$ taking values in $(-\pi, \pi)$. Then u_μ is a bounded function on $\mathbb{R}$, $0 \le u_\mu \le \pi$. Denote by Π Poisson measure on the real line, $\Pi = \frac{dx}{\pi(1+x^2)}$. The Poisson integral Pu_μ is equal to $\Im \log C\mu(x)$ in $\mathbb{C}_+$. In particular,

$$Pu_\mu(i) = \int_{\mathbb{R}} u_\mu \, d\Pi = \pi/2$$

because our choice of the conjugate Poisson kernel implies $Q\mu(i) = 0$. Sometimes it is more convenient to rewrite (2) as

$$\exp Cu_\mu(x) = \frac{1}{\|\mu\|} C\mu(x). \tag{3}$$

Moreover, for any bounded function u on $\mathbb{R}$, $0 \le u \le \pi$, $\int_{\mathbb{R}} u\, d\Pi = \pi/2$, one can solve (3) backwards to obtain a unique probability measure $\mu \in M(\hat{\mathbb{R}})$ satisfying

$$\exp Cu(x) = C\mu(x). \tag{4}$$

Indeed, the left hand side is an analytic function in $\mathbb{C}_+$ with a positive imaginary part and, by the representation theorem, it must be equal to a Cauchy integral of a positive measure $\mu \in M(\hat{\mathbb{R}})$, up to a real constant. The constant is zero because the left hand side is imaginary at i. Thus one can consider a one-to-one correspondence between bounded functions u on $\mathbb{R}$, taking values on $[0, \pi]$ and satisfying $\int u\, d\Pi = \pi/2$, and probability measures on $\hat{\mathbb{R}}$.

In perturbation theory, u_μ appears as the so-called Krein–Lifshits shift function corresponding to the self-adjoint operator with the spectral measure μ. The relation $u_\mu \leftrightarrow \mu$ described above is well studied in the theory of rank-one perturbations; see for instance [**3, 11, 15**]. Here we use some of the well-known properties of this relation to establish the following:

LEMMA 2.4. *Let E be a Borel subset of $\mathbb{R}$. A probability measure μ is a reflectionless measure minimally supported on E if and only if the function u_μ defined by (4) satisfies*

1) $u_\mu(x) = \pi/2$ *for almost every $x \in E$ and $u_\mu(x)$ equals either 0 or π for almost every $x \notin E$;*
2) $\int_{\mathbb{R}} u(x)\, d\Pi(x) = \pi/2$;
3) $\Pi(\{x \in \mathbb{R} \mid Qu_\mu(x) > t\}) = o(e^{-t})$ *as $t \to \infty$;*
4) *either* $\int_1^\infty \frac{u_\mu(x)dx}{x} = \infty$ *or* $\int_{-\infty}^{-1} \frac{\pi - u_\mu(x)dx}{x} = -\infty$.

REMARK 2.5. If E is a simple set, the third and fourth conditions can be made much more explicit. For instance, if E is a union of finitely many intervals, these conditions are equivalent to the condition that u_μ does not make any jumps "up" from 0 to π on the complement of E, including at infinity.

The conjugate Poisson transform can be replaced with the Hilbert transform in this statement.

As usual, denote by μ_{ac} the absolutely continuous part, and by μ_s, the singular part of μ.

PROOF. The Poisson integral of μ has non-zero limits a.e. on any minimal support of μ_{ac}, and zero limits a.e. outside of that support. Therefore the argument of the Cauchy integral of μ is equal to 0 or π outside of a minimal support of μ and is strictly between 0 and π a.e. on it. The argument of $C\mu$ is equal to $\pi/2$ a.e. on E if and only if $Q\mu$ is zero a.e. on that set. Hence the first condition is equivalent to the condition that E is a minimal support of μ_{ac} and that $Q\mu$ is zero a.e. on E.

The second condition is equivalent to the condition that μ is a probability measure.

The third condition can be rewritten as

$$\Pi(\{|C\mu| > t\}) = o\left(\frac{1}{t}\right)$$

as $t \to \infty$. This is equivalent to the statement that μ is absolutely continuous on $\mathbb{R}$.

Finally, the fourth condition means that μ does not have a point mass at infinity. $\square$

COROLLARY 2.6. *Let $E \subset \mathbb{R}$ be a Borel subset. Then there exists a reflectionless probability measure μ minimally supported on E. Such a measure is unique if and only if E is an interval.*

PROOF. Without loss of generality, we can assume that $E \neq \mathbb{R}$. To prove the existence, notice that the function $u = \frac{\pi}{2}\chi_E$ satisfies conditions 1), 3), and 4) of the lemma. It remains to "correct" it to satisfy 2) (for u we obviously have $\int_{\mathbb{R}} u(x)\, d\Pi(x) < \pi/2$).

Let u_a, $a \in \mathbb{R}$, be defined as

$$u_a(x) = \begin{cases} \pi/2 & \text{if } x \in E, \\ \pi & \text{if } x \notin E,\ \ x > a, \\ 0 & \text{if } x \notin E,\ \ x < a. \end{cases}$$

Obviously, one can choose $a = a_0$ so that u_{a_0} satisfies 2). It will also satisfy 1) and 4). Moreover, if a_0 is a point of non-zero lower density for E, then u_{a_0} will still satisfy 3). Thus if a_0 is such a point we are done: μ can be chosen so that $u_\mu = u_{a_0}$.

If a_0 is not a point of non-zero lower density for E, then the construction could be finished in the following way. Let b be any point of non-zero lower density for E. Then there are two possibilities:

a) $\displaystyle\int_{\mathbb{R}} u_b(x)\, d\Pi(x) < \pi/2$

or

b) $\displaystyle\int_{\mathbb{R}} u_b(x)\, d\Pi(x) > \pi/2.$

Suppose we are in the case a). Then one can choose a point $a < b$ so that the function u_{ab} defined a.e. on the line as

$$u_{ab}(x) = \begin{cases} \pi/2 & \text{if } x \in E, \\ \pi & \text{if } x \notin E,\ \ x < a \text{ or } x > b, \\ 0 & \text{if } x \notin E,\ \ a < x < b, \end{cases}$$

satisfies the second condition of the lemma. Since the jump at a is a negative jump, it will still satisfy 3) and the rest of the conditions.

Similarly, if we are in the case b), then one can choose a point $c > b$ and the function u_{bc} defined as

$$u_{bc}(x) = \begin{cases} \pi/2 & \text{if } x \in E, \\ 0 & \text{if } x \notin E,\ \ x < b \text{ or } x > c, \\ \pi & \text{if } x \notin E,\ \ b < x < c, \end{cases}$$

to satisfy the conditions of the lemma.

To prove uniqueness for the case where $E = \langle a, b\rangle$ is an interval, notice that if $\int_{\mathbb{R}} u_b(x)\, d\Pi(x) < \pi/2$, then the only function satisfying conditions 1)–4) is the function v_c defined as

$$v_c(x) = \begin{cases} \pi/2 & \text{if } a < x < b, \\ 0 & \text{if } c < x < a, \\ \pi & \text{if } x < c \text{ or } b < x, \end{cases}$$

where c is chosen so that v_c satisfies 2. Similarly, if $\int_{\mathbb{R}} u_b(x)\,d\Pi(x) > \pi/2$, then one can choose $c > b$ and define v_c as

$$v_c(x) = \begin{cases} \pi/2 & \text{if } a < x < b, \\ 0 & \text{if } x < a \text{ or } x > c, \\ \pi & \text{if } b < x < c. \end{cases}$$

From our construction above one can see that when E is a more complicated set, there exist infinitely many functions satisfying the conditions of the lemma. We discuss the case of two intervals below and leave the details of this part of the proof to the reader. □

EXAMPLE 2.7. Suppose that $E = [-2,-1] \cup [1,2]$. Elementary calculations show that the set of all functions satisfying the conditions of Lemma 2.4 can be described as follows. Define u_c, $c \in [0,1/2]$, as

$$u_c(x) = \begin{cases} \pi/2 & \text{if } x \in (-2,-1) \cup (1,2), \\ 0 & \text{if } x < -2 \text{ or } x \in (c,1) \text{ or } x \geq 1/c, \\ \pi & \text{if } x \in (-1,c) \text{ or } x \in (2,1/c) \end{cases}$$

(if $c = 0$ then $1/c$ should be replaced with $+\infty$). For $c \in [-1/2, 0)$, u_c is defined as

$$u_c(x) = \begin{cases} \pi/2 & \text{if } x \in (-2,-1) \cup (1,2), \\ 0 & \text{if } 1/c < x < -2 \text{ or } x \in (c,1), \\ \pi & \text{if } x < 1/c \text{ or } x \in (-1,c) \text{ or } x > 2. \end{cases}$$

Except for such u_c, $c \in [-1/2, 1/2]$, there are no functions satisfying the conditions 1) and 2), that have no jumps from 0 to π at any finite points or at infinity (which, as we remarked earlier, is equivalent to satisfying the conditions 3) and 4)). Thus by Lemma 2.4, all reflectionless probability measures on $[-2,-1] \cup [1,2]$ form a one-parameter family μ_c, $c \in [-1/2, 1/2]$, defined via (4) as $u_{\mu_c} = u_c$.

A description of general complex reflectionless measures on $\mathbb{R}$ as well as numerous problems on reflectionless measures in the plane (some of them mentioned above) remain open.

3. Class III

This class obviously contains all absolutely continuous measures from $M(\mathbb{R})$. Does it contain any singular measures? Theorem 3.1 below shows that the answer is negative if the size of the sector is large enough.

As before, if $x \in \mathbb{R}$ and $0 < \phi < \pi$, denote by Γ_x^ϕ the sector

$$\Gamma_x^\phi = \left\{ u + iv \in \mathbb{C}_+ \mid -\phi/2 < \arctan \frac{v}{u-x} < \phi/2 \right\}.$$

THEOREM 3.1. *Let $\mu \in M(\mathbb{R})$ be a real measure. Define the set $E \subset \mathbb{R}$ as the set of all $x \in \mathbb{R}$ for which there exist $\phi > \pi/2$ and $\epsilon > 0$ such that $Q\mu(z) \neq 0$ for all $z \in \Gamma_x^\phi$, $\Im z < \epsilon$. Then $|\mu_s|(E) = 0$.*

PROOF. The proof utilizes some of the methods from [**4**].

Suppose that $|\mu_s|(E) > 0$. Then there exists a set

$$F \subset E, \quad |\mu_s|(F) > 0,$$

such that there exist uniform $\phi > \pi/2$ and $\epsilon > 0$ such that $Q\mu(z) \neq 0$ for all $z \in \Gamma_x^\phi$, $\Im z < \epsilon$ for any $x \in F$. Note that since $Q\mu$ is a continuous real function in Γ_x^ϕ, one can assume without loss of generality that $Q\mu(z) > 0$ for all $z \in \Gamma_x^\phi$, $\Im z < \epsilon$ for any $x \in F$. Since $P\mu/P|\mu|$ tends to 1 or -1 at μ-a.e. x, we can also assume that $P\mu(z) > 0$ for all $z \in \Gamma_x^\phi$, $\Im z < \epsilon$ for any $x \in F$. Denote

$$\Gamma = \bigcup_{x \in F} \Gamma_x^\phi \cap \{\Im z < \epsilon\}.$$

Then the function $Q\mu P\mu = \Im(C\mu)^2$ is a positive harmonic function in Γ. Therefore it is summable on the boundary $\partial\Gamma$ of Γ with respect to the harmonic measure. Hence its harmonic conjugate function

$$(P\mu)^2 - (Q\mu)^2$$

belongs to L^p for any $p < 1$ with respect to the harmonic measure on $\partial\Gamma$. Since

$$|(Q\mu)^2| \leq (|P\mu Q\mu| + |(P\mu)^2 - (Q\mu)^2|),$$

$(Q\mu)^2$ belongs to L^p for any $p < 1$ with respect to the harmonic measure on $\partial\Gamma$. Therefore $Q\mu$ is in L^p for any $p < 2$ with respect to the harmonic measure on $\partial\Gamma$.

Denote by Ω the lower part of the boundary $\partial\Gamma$:

$$\Omega = \partial\Gamma \cap \{\Im z < \epsilon/2\}.$$

Since Γ is a Lipschitz domain, the harmonic measure on $\partial\Gamma$ is absolutely continuous with respect to the arclength: it is equal to $w\,ds$ for some density w. Since $\phi > \pi/2$, on the lower part of the boundary the density satisfies

$$\int_\Omega w^{-q}\,ds < \infty$$

for some $q > 1$. Choose $1 < r < 2 < s$ so that $1/r + 1/s = 1$, $r < 2$ and $s/r < q$. Then

$$\int_\Omega |Q\mu|\,ds \leq \left(\int_\Omega |Q\mu|^r w\,ds\right)^{1/r} \left(\int_\Omega w^{-s/r}\right)^{1/s} < \infty.$$

Hence $Q\mu$ is summable with respect to the arclength on Ω. Since $Q\mu$ is bounded in $\{\Im z > \epsilon/2\}$, this implies that $Q\mu$ is summable with respect to the arclength on the whole boundary $\partial\Gamma$. By Lemma 3.1 from [**4**], this can only happen when $|\mu_s|(F) = 0$. □

Since for any $\mu \in M(\mathbb{R})$ the Poisson integral $P\mu$ tends to infinity non-tangentially at μ_s-a.e. point, Theorem 3.1 implies that for large sectors, $P\mu(z)$ cannot be equal to $O(Q\mu(z))$ as $z \to x$, $z \in \Gamma_x^\phi$ for a non-zero set of x with respect to μ_s. Can $\phi > \pi/2$ be replaced with $\phi > 0$? in this statement? Furthermore, can non-tangential convergence be strengthened to radial? The methods that allowed a similar transition in [**4**] do not seem to apply here.

To end this section let us remark, that together with results of [**4**], Theorem 3.1 shows that for any $\mu \in M(\mathbb{R})$, $|Q\mu|$ oscillates wildly in μ_s-a.e. sector Γ_x^ϕ from 0 to a value comparable to $|P\mu|$ that tends to infinity. Once again, it would be interesting to see if such an oscillation happens on the μ_s-a.e. vertical line as well.

4. Class IV

All measures from $M(\mathbb{R})$ that do not have a singular continuous part, including all absolutely continuous measures or all pure point measures, belong to this class. Moreover, measures whose singular continuous part is "reasonable", such as the standard Cantor measure for example, do belong to this class as well. Nonetheless, this class does not contain the whole $M(\mathbb{R})$ and even the larger set, Class IV$_r$, does not include some of the measures.

As was mentioned in [**9**], the existence of a measure that does not belong to this class follows from a construction of an inner function given by W. Rudin in [**13**]. This counterexample is quite implicit and does not allow one to study the resulting measure. The existence of the measure follows from the fact that almost every measure from the Clark family corresponding to Rudin's inner function must not satisfy (IV).

Here we give a simpler example of a singular continuous measure not belonging to Class IV. Our construction, once again, utilizes the argument of the Cauchy integral of the measure, the Krein function u_μ. In terms of this function, the example is explicit.

For our construction we will need the following

LEMMA 4.1 ([**5**]). *A probability measure $\mu \in M(\mathbb{R})$ has a point mass at x iff*

$$\int_{x-1}^{x+1} (\pi\chi_{(x;x+1)} - u_\mu(y))\frac{dy}{y-x} < \infty.$$

EXAMPLE 4.2. Our goal is to construct the Krein function u on the real line such that the probability measure μ corresponding to u (i.e., such that $u = u_\mu$) does not belong to Class IV.

Denote $I = [-1, 1]$. For $\epsilon > 0$ consider the intervals

$$J_n = (1 - 2^{-n}, 1 - 2^{-n-1} - \epsilon 2^{-n-1})$$

for $n = 0, 1, 2, \ldots$ and

$$J_n = (-1 + 2^{n-1} + \epsilon 2^n, -1 + 2^n)$$

for $n = -1, -2, -3, \ldots$. Put

$$I^\epsilon = I \setminus \bigcup J_n.$$

For an arbitrary interval $L = [a, b]$ "rescale and shift" the above set and denote

$$L^\epsilon = \left\{ x \mid x = \frac{b-a}{2}(t+1) + a, \ \ t \in I^\epsilon \right\}.$$

The closed set L^ϵ occupies a small portion of L and is thin (has density ϵ) near the endpoints of L. Because of that property one can show that for any $\varepsilon > 0$ there exist ϵ, $\delta > 0$ such that for any $x \in L$

$$1 - \varepsilon < P\chi_{L_\epsilon}(x + i\delta \operatorname{dist}(x, \mathbb{R} \setminus L)) < 1. \tag{5}$$

Now we are ready to define the function u. First let us define it on $[0, 1]$. Choose a sequence of small positive numbers $\epsilon_n \to 0$. Start with the unit interval $I_{0,1} = [0, 1]$. Consider the set $I_{0,1}^{\epsilon_0}$ defined as above. Define u to be π on $[0, 1] \setminus I_{0,1}^{\epsilon_0}$.

Next, consider the disjoint closed intervals $I_{1,n}$, $n = 1, 2, 3, \ldots$, that form $I_{0,1}^{\epsilon_0}$:

$$I_{0,1}^{\epsilon_0} = \bigcup I_{1,n}.$$

On each of these intervals consider the set $I_{1,n}^{\epsilon_1}$. Define u as 0 on each $I_{1,n} \setminus I_{1,n}^{\epsilon_1}$. On the k-th step consider the disjoint closed intervals $I_{k,n}$ that compose the remainder of $[0,1]$ where u is not yet defined. On each of them consider $I_{k,n}^{\epsilon_k}$. If k is even, define u as π on all $I_{k,n} \setminus I_{k,n}^{\epsilon_k}$; if k is odd, define u as 0.

Since $|\bigcup_{n,k} I_{k,n}^{\epsilon_k}| = 1$, the limit function u will be defined a.e. on the unit interval. Assume that $\epsilon_1 < 1/2$. Then outside of $[0,1]$ define u as π on $(1,\infty) \cup (-\infty, c)$ and as 0 on $(c, 0)$, where the negative constant c is chosen so that

$$\int u \, d\Pi = 0.$$

Denote by C the Cantor set $\bigcap_{n,k} I_{k,n}^{\epsilon_k}$. Let μ be a probability measure on $\mathbb{R}$ such that $u_\mu = u$. One can easily show that μ is supported on C. The set C is a closed set that is equal to $[0,1]$ minus the union of countably many complementary open intervals. The measure does not have point masses at the endpoints of the complementary intervals of C. This can be seen from Lemma 4.1 and the definition of the set I^ϵ at the beginning of our construction.

Next, notice that

$$\limsup_{z \underset{\sphericalangle}{\longrightarrow} x} Pu(z) = \pi \quad \text{and} \quad \liminf_{z \underset{\sphericalangle}{\longrightarrow} x} Pu(z) = 0$$

for every $x \in C$ that is not an endpoint of a complementary interval. This also follows from the construction of L^ϵ; see (5). Hence, by (3),

$$\limsup_{z \underset{\sphericalangle}{\longrightarrow} x} \arg C\mu(z) = \pi \quad \text{and} \quad \liminf_{z \underset{\sphericalangle}{\longrightarrow} x} \arg C\mu(z) = 0$$

for μ-a.e. x. Therefore the condition (IV) is not satisfied and μ does not belong to Class IV.

It would be interesting to describe Class IV and to understand its relations with properties of the measure such as dimension, symmetry, smoothness, etc.

References

[1] F. Gesztesy and P. Yuditskii, *Spectral properties of a class of reflectionless Schrödinger operators*, J. Funct. Anal. **241** (2006), 486–527.

[2] F. Gesztesy and M. Zinchenko, *Local spectral properties of reflectionless Jacobi, CMV, and Schrödinger operators*, preprint.

[3] D. Hundertmark and B. Simon, *An optimal L^p-bound on the Krein spectral shift function*, J. d'Anal. Math. **87** (2002), no. 1, 199–208.

[4] P. W. Jones and A. Poltoratski, *Asymptotic Behavior of Cauchy Transforms*, Ann. Acad. Sci. Fenn. Math. **29** (2004), no. 1, 99–120.

[5] M. Martin and M. Putinar, *Lectures on Hyponormal Operators*, Operator Theory: Advances and Applications **39**, Birkhäuser, Basel, 1989.

[6] M. Melnikov, A. Poltoratski, and A. Volberg, *Uniqueness theorems for Cauchy integrals*, to appear in Publ. Math.

[7] F. Nazarov and A. Volberg, *A remark on reflectionless measures*, preprint.

[8] F. Nazarov, A. Volberg, and P. Yuditskii, *Reflectionless measures with a point mass and singular continuous component*, preprint.

[9] A. Poltoratski, *Images of non-tangential sectors under Cauchy transforms*, J. d'Anal. Math. **89** (2003), 385–395.

[10] ———, *On the boundary behavior of pseudocontinuable functions*, St. Petersburg Math. J. **5** (1994), 389–406.

[11] ———, *The Krein spectral shift and rank one perturbations of spectra*, St. Petersburg Math. J. **10** (1999), 833–859.

[12] A. Poltoratski and C. Remling, *Reflectionless Herglotz functions and Jacobi matrices*, to appear in Comm. Math. Phys.

[13] W. Rudin, *Inner function images of radii*, Proc. Camb. Phil. Soc. **85** (1979), 357–372.

[14] M. Sodin and P. Yuditskii, *Almost periodic Jacobi matrices with homogeneous spectrum, infinite-dimensional Jacobi inversion, and Hardy spaces of character-automorphic functions*, J. Geom. Anal. **7** (1997), 387–435.

[15] B. Simon, *Spectral analysis of rank one perturbations and applications*, CRM Proc. and Lect. Notes, 8, 1995.

[16] X. Tolsa, *Growth estimates for Cauchy integrals of measures and rectifiability*, Geom. Funct. Anal. **17** (2007), no. 2, 605–643.

[17] X. Tolsa and J. Verdera, *May the Cauchy transform of a non-trivial finite measure vanish on the support of the measure?* Ann. Acad. Sci. Fenn. Math. **31** (2006), no. 2, 479–494.

Department of Mathematics, Texas A&M University, College Station, TX 77843, USA

E-mail address: alexei@math.tamu.edu

Amer. Math. Soc. Transl.
(2) Vol. **226**, 2009

Free Interpolation in the Nevanlinna Class

Donald Sarason

To Victor Havin, in admiration of his fundamental contributions, spanning 50 years, to functional analysis and complex analysis

ABSTRACT. Given a Blaschke sequence in the unit disk, one can arbitrarily specify the value of a function of Nevanlinna class, and the values of its derivatives to the appropriate order, at each point of the sequence. A proof is presented, and a connection with operator theory discussed.

1. Introduction

By the Nevanlinna class, denoted by N, is meant here the class of functions of bounded characteristic in the unit disk $\mathbb{D}$ — equivalently, the class of functions in $\mathbb{D}$ writable as quotients ψ/χ, where ψ and χ are in H^∞ and χ is not the zero function. The functions in N are meromorphic in $\mathbb{D}$; the set of poles of such a function, if nonempty, forms a Blaschke sequence. (The notation here deviates from a more common one, according to which N denotes the class of holomorphic functions of bounded characteristic.)

Roughly speaking, a set is called a set of free interpolation for a class of functions if every function on the set satisfying "natural conditions" can be interpolated by a function in the class. Extensive comments on the use of the term "free interpolation" can be found in the book [**4**] of Nikolai Nikol'skiĭ. The premier example concerns the class H^∞; a subset of $\mathbb{D}$ is termed a set of free interpolation for H^∞ if any bounded function on the set can be interpolated by a function in H^∞. Such sets, so-called uniformly separated sets, are characterized by the famous theorem of L. Carleson (see for example [**4**]). Another, more elementary, example is often given as an exercise in complex analysis courses: every sequence in $\mathbb{C}$ tending to ∞ is a set of free interpolation for the class of entire functions, which is to say, the values of an entire function can be arbitrarily specified on the sequence.

The zero set of a nonzero function in N is a Blaschke sequence, making it clear that any set of free interpolation for N must be a Blaschke sequence. Here the

2000 *Mathematics Subject Classification.* Primary 30D50, 30E05, 47B35, 47A20.
Key words and phrases. Functions of bounded characteristic, Nevanlinna class, free interpolation, Toeplitz operator, restricted backward shift.

converse will be established: every Blaschke sequence is a set of free interpolation for N. Theorem 2.1 is the precise statement.

Theorem 2.1 was suggested by a result from [**5**] in operator theory (and can be deduced from that result). The operator theory connection is discussed in Sections 3 and 4.

References [**1**]–[**3**], kindly supplied by the referee, are related to the present study. Over 50 years ago, in [**3**], A. G. Naftalevič considered various interpolation problems for Nevanlinna functions and for holomorphic Nevanlinna functions. Among them is a variant for holomorphic Nevanlinna functions of the problem considered here for Nevanlinna functions. In [**2**], A. Hartmann and coauthors obtain definitive results on free interpolation for the holomorphic Nevanlinna class and for the Smirnov class (the class of functions in N with outer denominators). Their interpretation of what free interpolation means in their setting is given a detailed justification. The paper also contains a discussion of the result of Naftalevič alluded to above. In [**1**], K. M. Dyakonov and A. Nicolau characterize the sets of free interpolation for the class of nonvanishing functions in H^∞.

Notation

- For λ a point of $\mathbb{D}$, the kernel function for the evaluation functional at λ on H^2 is denoted by $k_\lambda : k_\lambda(z) = 1/(1-\overline{\lambda}z)$.
- For λ in $\mathbb{D}$, the (unnormalized) Blaschke factor b_λ is defined by $b_\lambda(z) = (z-\lambda)/(1-\overline{\lambda}z)$.
- The domain and graph of an operator T are denoted, respectively, by $\mathcal{D}(T)$ and $\mathcal{G}(T)$.
- For θ a function in L^∞ of $\partial\mathbb{D}$, the Toeplitz operator on H^2 with symbol θ, the operator of multiplication by θ followed by orthogonal projection onto H^2, will be denoted, as usual, by T_θ.

2. Free interpolation in N

THEOREM 2.1. *Let u be an infinite Blaschke product, with zeros $z_1, z_2, \ldots$, enumerated without multiplicities, and with respective multiplicities $m_1, m_2, \ldots$. Let $w_{n,j}$, $j = 0, \ldots, m_n - 1$, $n = 1, 2, \ldots$, be complex numbers. Then there is a function φ in N such that $\varphi^{(j)}(z_n) = w_{n,j}$ for each n and j.*

Some preparation is needed prior to the proof. For λ a point of $\mathbb{D}$, we let $e_{\lambda,j} = \sqrt{1-|\lambda|^2}\, b_\lambda^j k_\lambda$, $j = 0, 1, \ldots$. The functions $e_{\lambda,j}$ form an orthonormal basis for H^2, in fact, the image of the monomial basis under the unitary map $f \mapsto \sqrt{1-|\lambda|^2}(f \circ b_\lambda)k_\lambda$.

For $l = 0, 1, \ldots$, the kernel function for the functional on H^2 of evaluation of the lth Taylor coefficient at λ will be denoted by $k_\lambda^{[l]}$; it is given by $k_\lambda^{[l]}(z) = z^l/(1-\overline{\lambda}z)^{l+1}$, as one can verify, for example, on the basis of the residue theorem.

LEMMA 2.2. *For each l, the function $k_\lambda^{[l]}$ is a linear combination of $e_{\lambda,0}, \ldots, e_{\lambda,l}$.*

PROOF. Fix l. We consider the vector space of all functions pk_λ^{l+1} with p a polynomial of degree at most l. It has dimension $l+1$. The functions $e_{\lambda,0}, \ldots, e_{\lambda,l}$ belong to this space and are linearly independent, hence form a basis. The function $k_\lambda^{[l]}$ also belongs and so is a linear combination of $e_{\lambda,0}, \ldots, e_{\lambda,l}$. □

REMARK. The same reasoning shows that for each j the function $e_{\lambda,j}$ is a linear combination of $k_\lambda^{[0]}, \dots, k_\lambda^{[j]}$.

We shall work in the space $K_u^2 = H^2 \ominus uH^2$ (with u as in Theorem 2.1). It carries a natural symmetry C, an antiunitary involution given by $(Cf)(z) = u(z)\overline{zf(z)}$ $(z \in \partial\mathbb{D})$.

We simplify the notation a bit by writing $e_{n,j}$ in place of $e_{z_n,j}$. For each n the functions $e_{n,0}, \dots, e_{n,m_n-1}$ belong to K_u^2, i.e., they are annihilated by the Toeplitz operator $T_{\bar{u}}$. In fact, for $0 \le j \le m_n - 1$, the function $u/b_{z_n}^j$ is holomorphic and vanishes at z_n, so that its corresponding coanalytic Toeplitz operator annihilates k_{z_n}. Since that operator equals $T_{\bar{u}}T_{b_{z_n}^j}$ and $e_{n,j}$ is a constant times $T_{b_{z_n}^j}k_{z_n}$, the equality $T_{\bar{u}}e_{n,j} = 0$ follows.

We let $\widetilde{e}_{n,j} = Ce_{n,j}$ $(j = 0, \dots, m_n - 1)$. For z in $\partial\mathbb{D}$,

$$
\begin{aligned}
\frac{\widetilde{e}_{n,j}(z)}{\sqrt{1-|z_n|^2}} &= u(z)\overline{z b_{z_n}(z)^j k_{z_n}(z)} \\
&= \frac{u(z)\bar{z}(\bar{z}-\bar{z}_n)^j}{(1-z_n\bar{z})^{j+1}} = \frac{(1-\bar{z}_n z)^j u(z)}{(z-z_n)^{j+1}}.
\end{aligned} \tag{2.1}
$$

We see that, for $n' \neq n$, the function $\widetilde{e}_{n,j}$ has a zero of order $m_{n'}$ at $z_{n'}$, while for $j < m_n - 1$ it has a zero of order $m_n - j - 1$ at z_n. For $j = m_n - 1$ we have $\widetilde{e}_{n,m_n-1}(z_n) = (1-|z_n|^2)^{m_n-1/2}u^{(m_n)}(z_n)/m_n!$, which is nonzero.

LEMMA 2.3. *The inner product $\langle \widetilde{e}_{n,l}, e_{n,j}\rangle$ vanishes for $j + l < m_n - 1$ and is nonzero for $j + l = m_n - 1$.*

PROOF. We have $e_{n,j} = b_{z_n}^j e_{n,0}$ and, by (2.1), $\widetilde{e}_{n,l} = \widetilde{e}_{n,0}/b_{z_n}^l$. Hence

$$
\begin{aligned}
\langle \widetilde{e}_{n,l}, e_{n,j}\rangle &= \langle \widetilde{e}_{n,0}/b_{z_n}^l, b_{z_n}^j e_{n,0}\rangle \\
&= \langle \widetilde{e}_{n,0}/b_{z_n}^{j+l}, k_{z_n}\rangle.
\end{aligned}
$$

For $j + l \le m_n - 1$, by (2.1) again,

$$
\widetilde{e}_{n,0}/b_{z_n}^{j+l} = \widetilde{e}_{n,j+l},
$$

and (see above) $\widetilde{e}_{n,j+l}(z_n)$ is zero for $j + l < m_n - 1$ and is nonzero for $j + l = m_n - 1$. □

PROOF OF THEOREM 2.1. By Lemma 2.3, for each n the (Hankel) matrix

$$
\Big(\langle \widetilde{e}_{n,l}, e_{n,j}\rangle\Big)_{j,l=0}^{m_n-1}
$$

is nonzero at each entry on the principal cross diagonal, and is zero at each entry above it. The rows of the matrix are thus linearly independent, i.e., the matrix is invertible. We can conclude that there is a linear combination of $\widetilde{e}_{n,0}, \dots, \widetilde{e}_{n,m_n-1}$ whose inner products with $e_{n,0}, \dots, e_{n,m_n-1}$ take any preassigned values.

Combining the last conclusion with Lemma 2.2 and its following remark, we see that, for each n, there is a linear combination ψ_n of $\widetilde{e}_{n,0}, \dots, \widetilde{e}_{n,m_n-1}$ such that $\psi_n^{(j)}(z_n) = w_{n,j}$, $j = 0, \dots, m_n - 1$. Similarly, there is a linear combination χ_n of $\widetilde{e}_{n,0}, \dots, \widetilde{e}_{n,m_n-1}$ such that $\chi_n^{(j)}(z_n)$ equals 1 for $j = 0$ and equals 0 for $j = 1, \dots, m_n - 1$. The functions ψ_n and χ_n are bounded because each $\widetilde{e}_{n,l}$ is, and for $n' \neq n$ they vanish at $z_{n'}$ at least to order $m_{n'}$.

Let us choose positive numbers $\rho_1, \rho_2, \ldots$ such that $\sum_1^\infty \rho_n \|\psi_n\|_\infty < \infty$ and $\sum_1^\infty \rho_n \|\chi_n\|_\infty < \infty$. The series $\sum_1^\infty \rho_n \psi_n$ and $\sum_1^\infty \rho_n \chi_n$ then converge uniformly in $\mathbb{D}$, say to the respective functions ψ and χ, which belong to $K_u^2 \cap H^\infty$. For each n we have

$$\psi^{(j)}(z_n) = \rho_n w_{n,j}, \; j = 0, \ldots, m_n - 1,$$

$$\chi^{(j)}(z_n) = \begin{cases} \rho_n, & j = 0, \\ 0, & j = 1, \ldots, m_n - 1. \end{cases}$$

In particular $\chi(z_n)$ is nonzero for all n. The function $\varphi = \psi/\chi$ is in N and satisfies $\varphi^{(j)}(z_n) = w_{n,j}$ for $j = 0, \ldots, m_n - 1$ and all n. □

Holomorphic functions in N obey a growth restriction near $\partial\mathbb{D}$ (see for example [**5**]), so the interpolation in Theorem 2.1 need not be achievable by a holomorphic function in N.

The referee has pointed out that there is a more direct proof of Theorem 2.1, one that by-passes the space K_u^2 and its structure. I have retained the original proof here because it reflects how the theorem first arose, as an outgrowth of the theorem in operator theory to be discussed below. The referee's elegant argument, though, deserves to be presented, as it clearly reveals the elementary nature of the theorem. It goes as follows.

The notations in the statement of the theorem are retained. The crux of the matter is to find, for each n, a function ψ_n in H^∞ satisfying $\psi_n^{(j)}(z_n) = w_{n,j}$ for $j = 0, \ldots, m_n - 1$, and vanishing at least to order $m_{n'}$ at $z_{n'}$ for $n' \neq n$. Once that is done the proof can be completed as in the last paragraph of the proof above.

Let $v_n = u/b_n^{m_n}$, and note that v_n vanishes to the appropriate order at $z_{n'}$ for $n' \neq n$. We seek ψ_n in the form $\psi_n = v_n g_n$, where g_n is an H^∞ function to be determined. Using Leibniz's differentiation formula we see that the values $g_n^{(j)}(z_n)$, $j = 0, \ldots, m_n - 1$, must satisfy a system of linear equations, namely

$$\sum_{l=0}^{j} \binom{j}{l} v_n^{(l)}(z_n) g_n^{(j-l)}(z_n) = w_{n,j}, \quad j = 0, \ldots, m_n - 1.$$

The matrix for the system is lower triangular, each diagonal entry having the same nonzero value $v_n(z_n)$. The matrix is therefore invertible, so the values $g_n^{(j)}(z_n)$, $j = 0, \ldots, m_n - 1$, are well defined in terms of the given values $w_{n,j}$, $j = 0, \ldots, m_n - 1$, and one can attain them by taking g_n to be a polynomial of degree at most $m_n - 1$.

3. Operator theory connection: Introduction

For the time being, let u denote any nonconstant inner function. The compression to the subspace $K_u^2 = H^2 \ominus uH^2$ of the unilateral shift operator S will be denoted by S_u; its adjoint S_u^* is the restriction of S^* to K_u^2. For θ in L^∞, the compression of the Toeplitz operator T_θ to K_u^2 will be denoted by A_θ.

The prototype of the commutant lifting theorem states that the general bounded operator on K_u^2 that commutes with S_u^* is the operator $A_{\overline{\varphi}}$ with φ in H^∞. For the case where u is not a finite Blaschke product, i.e., the case where $\dim K_u^2$ is infinite, Daniel Suárez in his paper [**6**] studied the closed, densely defined operators on K_u^2

that commute with S_u^*. Suárez's analysis was extended by the author in [**5**]. The end result is a sort of commutant lifting theorem for unbounded operators.

Theorem 2.1 enables one to treat the case where u is an infinite Blaschke product without recourse to Suárez's analysis. This was done in [**5**] for the case where u has simple zeros. It will be done in Section 4 for the general infinite Blaschke product, by an extension of the analysis in [**5**].

For future use we note that, for θ in L^∞, the operator $A_{\overline{\theta}}$ is both the adjoint of A_θ and, by a simple argument, the transform of A_θ under the symmetry $C : A_{\overline{\theta}} = CA_\theta C$.

4. Operator theory connection: Details

Throughout the section, let u be as in Theorem 2.1, namely, an infinite Blaschke product with zeros $z_1, z_2, \ldots$, enumerated without multiplicities, with respective multiplicities $m_1, m_2, \ldots$. The other notations introduced in Section 2 will be retained.

We let N_u^+ denote the set of functions in N not having a pole at any of the zeros of u. It will be shown that each function φ in N_u^+ induces a coanalytic Toeplitz operator $A_{\overline{\varphi}}$ on K_u^2, and that these are the closed, densely defined operators on K_u^2 that commute with S_u^*.

For each n we let E_n denote the span of the functions $e_{n,0}, \ldots, e_{n,m_n-1}$. These are invariant subspaces of S^*, and they span K_u^2. (In fact, by Lemma 2.2, any function in H^2 orthogonal to every E_n has a zero at each z_n of order at least m_n, so it lies in uH^2.) We let E denote the linear span of the subspaces $E_1, E_2, \ldots$.

LEMMA 4.1. *If A is a closed, densely defined operator on K_u^2 that commutes with S_u^*, then the domain $\mathcal{D}(A)$ of A contains E, and A is the closure of its restriction to E.*

PROOF. To establish the inclusion $E \subset \mathcal{D}(A)$ it will suffice to show that $E_n \subset \mathcal{D}(A)$ for every n. Fix a positive integer n_0. The graph $\mathcal{G}(A)$ is invariant under $S^* \oplus S^*$, the backward shift of multiplicity 2, so it is invariant under the weakly closed operator algebra generated by $S^* \oplus S^*$, which consists of the operators $T_{\overline{\psi}} \oplus T_{\overline{\psi}}$ with ψ in H^∞. In particular, $\mathcal{D}(A)$ is $T_{\overline{\psi}}$-invariant (equivalently, $A_{\overline{\psi}}$-invariant) for ψ in H^∞.

We apply the preceding remark with $\psi(z) = u(z)/(z - z_{n_0})^{m_{n_0}}$. We consider the operator A_ψ, the C-transform of $A_{\overline{\psi}}$. As the subspaces E_n are $A_{\overline{\psi}}$-invariant, the subspaces $\widetilde{E}_n = CE_n$ are A_ψ-invariant. Clearly, a function in K_u^2 is annihilated by A_ψ if and only if it vanishes at z_{n_0} to order at least m_{n_0}. This applies to every function in $\widetilde{E}_n$ for $n \neq n_0$, so A_ψ annihilates each of those subspaces, and $A_{\overline{\psi}}$ annihilates E_n for $n \neq n_0$. By equality (2.1),

$$\frac{\psi(z)\widetilde{e}_{n_0,j}(z)}{\sqrt{1 - |z_{n_0}|^2}} = \frac{u(z)^2(1 - \overline{z}_{n_0} z)^j}{(z - z_{n_0})^{m_{n_0}+j+1}}, \quad j = 0, \ldots, m_{n_0} - 1,$$

and these functions are linearly independent modulo uH^2. Thus $A_\psi \widetilde{E}_{n_0} = \widetilde{E}_{n_0}$, and $A_{\overline{\psi}} E_{n_0} = E_{n_0}$. Because the subspaces E_n span K_u^2, we can conclude that $A_{\overline{\psi}} K_u^2 = E_{n_0}$.

Since the domain $\mathcal{D}(A)$ is dense in K_u^2, its image under $A_{\overline{\psi}}$ is dense in $A_{\overline{\psi}}K_u^2$. But $A_{\overline{\psi}}K_u^2$ is the finite-dimensional subspace E_{n_0}, so $A_{\overline{\psi}}\mathcal{D}(A) = E_{n_0}$, giving the desired inclusion $E_{n_0} \subset \mathcal{D}(A)$, and hence the inclusion $E \subset \mathcal{D}(A)$.

It remains to prove that A is the closure of its restriction to E. For each n, let b_n be the normalized Blaschke factor for the point z_n: $b_n = -|z_n|b_{z_n}/z_n$ if $z_n \neq 0$, $b_n(z) = z$ if $z_n = 0$. We can assume without loss of generality that the Blaschke product u is normalized, meaning that $u = \prod_1^\infty b_n^{m_n}$. For $\nu = 1, 2, \ldots$ let $u_\nu = u/\prod_1^\nu b_n^{m_n}$. Then $u_\nu \to 1$ locally uniformly in $\mathbb{D}$ as $\nu \to \infty$.

Let $f \oplus g$ be a vector in $\mathcal{G}(A)$. Since $\mathcal{G}(A)$ is invariant under $A_{\overline{u}_\nu} \oplus A_{\overline{u}_\nu}$ for each ν, each vector $A_{\overline{u}_\nu} f \oplus A_{\overline{u}_\nu} g$ is in $\mathcal{G}(A)$. But $A_{\overline{u}_\nu}$ annihilates E_n for $n > \nu$ and maps E_n into itself for $n \leq \nu$. We can conclude that the range of $A_{\overline{u}_\nu}$ is contained in the span of $E_1, \ldots, E_\nu$, hence is contained in E. Thus, $A_{\overline{u}_\nu} f \oplus A_{\overline{u}_\nu} g$ belongs to $\mathcal{G}(A \mid E)$ for each ν.

Standard reasoning shows that, because $u_\nu \to 1$ locally uniformly in $\mathbb{D}$, the operators $A_{\overline{u}_\nu}$ converge in the weak operator topology as $\nu \to \infty$ to the identity operator on K_ν^2. In particular, $A_{\overline{u}_\nu} f \oplus A_{\overline{u}_\nu} g \to f \oplus g$ weakly. □

Consider now a nonzero function φ in N_u^+. We construct the associated coanalytic Toeplitz operator $A_{\overline{\varphi}}$ on K_u^2. The function φ can be written as $\varphi = \psi/\chi$, where ψ and χ are in H^∞, the function χ is nonzero at each z_n, and the inner factors of ψ and χ are relatively prime.

That χ is nowhere zero on the zero set of u implies that the operator A_χ is injective. Hence $A_{\overline{\chi}}$, the adjoint and C-transform of A_χ, has a dense range and is injective. As the first step in constructing $A_{\overline{\varphi}}$, we let $A_{\overline{\varphi}}^0$ be the operator with domain $\mathcal{D}(A_{\overline{\varphi}}^0) = A_{\overline{\chi}}K_u^2$ defined by $A_{\overline{\varphi}}^0 A_{\overline{\chi}} h = A_{\overline{\psi}} h$ ($h \in K_u^2$). Thus $A_{\overline{\varphi}}^0$ is densely defined. We also let A_φ^0 be the C-transform of $A_{\overline{\varphi}}^0$:

$$\begin{aligned}\mathcal{D}(A_\varphi^0) &= C\mathcal{D}(A_{\overline{\varphi}}^0) = A_\chi K_u^2,\\ A_\varphi^0 A_\chi h &= CA_{\overline{\varphi}}^0 CA_\chi h = CA_{\overline{\varphi}}^0 A_{\overline{\chi}} Ch\\ &= CA_{\overline{\psi}}Ch = A_\psi h \quad (h \in K_u^2).\end{aligned}$$

We determine the adjoint of A_φ^0. The vector $f \oplus g$ in $K_u^2 \oplus K_u^2$ belongs to the graph of the adjoint of A_φ^0 if and only if $g \oplus -f$ is orthogonal to $\mathcal{G}(A_\varphi^0)$. That happens if and only if, for all h in K_u^2,

$$\begin{aligned}0 &= \langle g \oplus -f, A_\chi h \oplus A_\psi h\rangle\\ &= \langle g, A_\chi h\rangle - \langle f, A_\psi h\rangle\\ &= \langle A_{\overline{\chi}} g - A_{\overline{\psi}} f, h\rangle,\end{aligned}$$

in other words, if and only if $A_{\overline{\psi}} f = A_{\overline{\chi}} g$. Hence

$$\mathcal{G}((A_\varphi^0)^*) = \{f \oplus g \in K_u^2 \oplus K_u^2 : A_{\overline{\psi}} f = A_{\overline{\chi}} g\}. \tag{4.1}$$

From this it is clear that $(A_\varphi^0)^*$ commutes with S_u^*.

We note that the adjoint $(A_\varphi^0)^*$ is independent of the representation of φ used in its definition. In fact, suppose $\varphi = \psi_1/\chi_1$ is another such representation, and suppose $A_{\overline{\psi}}f = A_{\overline{\chi}}g$. From the equality $\psi\chi_1 = \psi_1\chi$ we conclude that $A_{\overline{\chi}_1}A_{\overline{\psi}} = A_{\overline{\chi}}A_{\overline{\psi}_1}$. Hence

$$A_{\overline{\chi}}A_{\overline{\chi}_1}g = A_{\overline{\chi}_1}A_{\overline{\psi}}f = A_{\overline{\chi}}A_{\overline{\psi}_1}f,$$

which because of the injectivity of $A_{\overline{\chi}}$ implies that $A_{\overline{\psi}_1}f = A_{\overline{\chi}_1}g$, giving the asserted independence.

From (4.1) we see that $A^0_{\overline{\varphi}} \subset (A^0_{\varphi})^*$. As $A^0_{\overline{\varphi}}$ has a closed extension it is closable; we let $A_{\overline{\varphi}}$ be the closure of $A^0_{\overline{\varphi}}$. Thus $A_{\overline{\varphi}} \subset (A^0_{\varphi})^*$, and $A_{\overline{\varphi}}$ commutes with S_u^* because $A^0_{\overline{\varphi}}$ does. By Lemma 4.1, both operators $A_{\overline{\varphi}}$ and $(A^0_{\varphi})^*$ are the closures of their restrictions to E, so they are in fact equal. This completes our construction.

THEOREM 4.2. *Let A be a closed, densely defined operator on K_u^2 that commutes with S_u^*. Then there is a function φ in N_u^+ such that $A = A_{\overline{\varphi}}$.*

PROOF. By Lemma 4.1, $\mathcal{D}(A)$ contains each subspace E_n, and by the same lemma it will suffice to show that there is a φ in N_u^+ such that $A_{\overline{\varphi}}$ agrees with A on each E_n.

The restriction of A to E_n is an operator on E_n that commutes with $S^* \mid E_n$. Each of the operators S^* and $T_{\overline{b}_{z_n}}$ can be written as a norm-convergent power series in the other. Their restrictions to E_n thus have the same commutant. With respect to the orthonormal basis $e_{n,0}, \ldots, e_{n,m_n-1}$, the operator $T_{\overline{b}_{z_n}} \mid E_n$ is a standard nilpotent Jordan block. The operators commuting with it are the polynomials of degrees at most $\dim(E_n)-1 = m_n-1$ in it. Hence, for each n, there is a polynomial p_n, of degree at most $m_n - 1$, such that $A \mid E_n = p_n(T_{\overline{b}_{z_n}}) \mid E_n$. We define the conjugate polynomial p_n^* by $p_n^*(z) = \overline{p_n(\overline{z})}$, so that $p_n(T_{\overline{b}_{z_n}})^* = p_n^*(T_{b_{z_n}}) = T_{p_n^*(b_{z_n})}$.

By Theorem 2.1, there is a function φ in N_u^+ whose value at z_n, for each n, agrees with that of $p_n^*(b_{z_n})$, and whose derivatives at z_n agree with those of $p_n^*(b_{z_n})$ up to order $m_n - 1$. Thus, we can write $\varphi = p_n^*(b_{z_n}) + \varphi_n$, where φ_n is divisible by $(z - z_n)^{m_n}$. Then, the compression of A_φ to E_n $(= H^2 \ominus b_{z_n}^{m_n}H^2)$ is the same as the compression of $T_{p_n^*(b_{z_n})}$, so the restriction of $A_{\overline{\varphi}}$ to E_n is the same as the restriction of $p_n(T_{\overline{b}_{z_n}})$, hence equal to $A \mid E_n$. □

Theorems 2.1 and 4.2 show that, for a closed, densely defined operator A on K_u^2 that commutes with S_u^*, one may choose the restrictions $A \mid E_n$ arbitrarily, subject only to the condition that $A \mid E_n$ commute with $S_u^* \mid E_n$ for all n.

Acknowledgements

I am grateful to the referee for giving the paper a careful reading (including spotting a few notational errors), for supplying references [**1**]–[**3**], and for the very nice alternative proof of Theorem 2.1.

References

[1] K. Dyakonov and A. Nicolau, *Free interpolation by nonvanishing functions*, Trans. Amer. Math. Soc. **359** (2007), no. 9, 4449–4465; MR2309193 (2008g:46087).

[2] A. Hartman, X. Massaneda, A. Nicolau, and P. Thomas, *Interpolation in the Nevanlinna and Smirnov classes and harmonic majorants*, J. Funct. Anal. **217** (2004), no. 1, 1–37; MR2097605 (2005h:30065).

[3] A. G. Naftalevič, *On interpolation by functions of bounded characteristic* (Russian), Vilniaus Valst. Univ. Mokslų Darbai. Mat. Fiz. Chem. Mokslų Ser. **5** (1956), 5–27; MR0120387 (22#11141).

[4] N. K. Nikol'skiĭ, *Treatise on the shift operator*, Grundlehren der Mathematischen Wissenschaften, 273, Springer-Verlag, Berlin, 1986; MR0827223 (87i:47042).

[5] D. Sarason, *Unbounded operators commuting with restricted backward shifts*, Operators and Matrices, forthcoming.

[6] D. Suárez, *Closed commutants of the backward shift operator*, Pacific J. Math. **179** (1997), 371–396; MR1452540 (99a:47050).

University of California, Department of Mathematics, Berkeley, CA 94720-3840, USA

E-mail address: `sarason@math.berkeley.edu`

Amer. Math. Soc. Transl.
(2) Vol. **226**, 2009

Interpolation by Dirichlet Series in H^∞

Kristian Seip

ABSTRACT. We present some basic results about interpolating sequences for $\mathscr{H}^\infty$, the algebra of Dirichlet series in H^∞ of the right half-plane. As a consequence of a theorem of B. Berndtsson, S.-Y. A. Chang, and K.-C. Lin on interpolating sequences in polydiscs, we obtain a description of bounded interpolating sequences for $\mathscr{H}^\infty$. We show how interpolating sequences for $\mathscr{H}^\infty$ can be described via interpolating sequences for bounded analytic functions in polydiscs. Finally, we deduce a necessary density condition from a theorem on the existence of a linear operator of interpolation.

1. Introduction

This note presents a few basic results about interpolating sequences for the algebra $\mathscr{H}^\infty$, which consists of those bounded analytic functions f on $\mathbb{C}_+ = \{s = \sigma + it : \sigma > 0\}$ that can be represented by an ordinary Dirichlet series

$$f(s) = \sum_{n=1}^{\infty} a_n n^{-s} \tag{1}$$

in some half-plane. We cannot expect to obtain a complete geometric description of the interpolating sequences in this case, but even partial results can give interesting function-theoretic information about $\mathscr{H}^\infty$.

We note that $\mathscr{H}^\infty$ is a closed subalgebra of $H^\infty(\mathbb{C}_+)$, the algebra of bounded analytic functions, assuming both spaces are equipped with the supremum norm:

$$\|f\|_\infty = \sup_{\sigma>0} |f(\sigma + it)|.$$

In general, if A is an algebra of bounded analytic functions on some domain Ω, a sequence $S = (s_j)$ of distinct points $s_j = \sigma_j + it_j$ in Ω (possibly finite) is said to be an interpolating sequence for A if every interpolation problem

$$f(s_j) = a_j \tag{2}$$

has a solution f in A whenever (a_j) is a bounded sequence of complex numbers. Our first theorem is the following local result.

2000 *Mathematics Subject Classification.* Primary 30B50, 30E05, 30H05.

Key words and phrases. Dirichlet series, bounded analytic functions, interpolating sequences, linear operator of interpolation.

The author is supported by the Research Council of Norway grant 185359/V30.

THEOREM 1. *A bounded sequence S in $\mathbb{C}_+$ is an interpolating sequence for $\mathscr{H}^\infty$ if and only if it is an interpolating sequence for $H^\infty(\mathbb{C}_+)$.*

In other words, for bounded sequences, L. Carleson's characterization of interpolating sequences for $H^\infty(\mathbb{C}_+)$ [**5**] applies to $\mathscr{H}^\infty$ as well. The result is similar to the main result of [**13**]. It is quite illuminating because it somehow suggests that, locally, functions in $\mathscr{H}^\infty$ look like functions in $H^\infty(\mathbb{C}_+)$.

Theorem 1 is a direct consequence of a theorem of Berndtsson, Chang, and Lin [**3**] concerning interpolating sequences in the unit polydisc. This means that Theorem 1 is really a result about a much smaller algebra than $\mathscr{H}^\infty$. Indeed, for m a positive integer, let $\mathscr{H}_m^\infty$ be the closed subalgebra of $\mathscr{H}^\infty$ consisting of Dirichlet series of the form (1) involving only integers n generated by the first m primes 2, 3, ..., p_m. We will see below that in Theorem 1 we may replace $\mathscr{H}^\infty$ by $\mathscr{H}_2^\infty$.

The next theorem shows how to connect the finite-dimensional case (corresponding to $\mathscr{H}_m^\infty$) to the infinite-dimensional one (corresponding to $\mathscr{H}^\infty$). To state that result, we need to introduce what is known as the constant of interpolation. To this end, let A be any of the algebras considered so far. Using the open mapping theorem, we find that if S is an interpolating sequence for A, then we can always solve (2) with an f in A such that

$$\|f\|_\infty \le C\|(a_j)\|_\infty$$

for some $C < \infty$ depending only on S. The constant of interpolation $M = M(S)$ is the smallest such C; we set $M(S) = \infty$ if S is not an interpolating sequence for A. Also, for $\delta > 0$, let S_δ be the subsequence of S consisting of points from the half-plane $\Re s > \delta$. We now have:

THEOREM 2. *The sequence S is an interpolating sequence for $\mathscr{H}^\infty$ if and only if there exist numbers $M > 1$ and $c > 2$ such that, for every $\delta > 0$ and $m > c^{1/\delta}$, the constant of interpolation for S_δ, considered as an interpolating sequence for $\mathscr{H}_m^\infty$, is at most M.*

It is a rather simple fact that each of the sequences S_δ in Theorem 2 is finite. The following theorem gives a much more precise quantitative result.

THEOREM 3. *If $S = (s_j)$ is an interpolating sequence for $\mathscr{H}^\infty$ with constant of interpolation M, then*

$$\sum_j M^{-c/\sigma_j} < \infty$$

whenever $c > 1/2$.

We will exhibit an example showing that the theorem fails if we replace c/σ_j by $\alpha/(\sigma_j|\log \sigma_j|)$ for any positive constant α. We will also see that there is a huge gap between the condition of Theorem 3 and the weaker one that we get from the mere fact that the sequence S has to be separated with respect to Gleason distance (to be defined below). Here the contrast to the situation for $H^\infty(\mathbb{C}_+)$ or indeed any of the spaces $\mathscr{H}_m^\infty$ is quite striking.

The proofs of the three theorems stated above are relatively short, but somewhat elaborate preparations are required. It is essential that we associate $\mathscr{H}^\infty$ with H^∞ of the infinite-dimensional polydisc and $\mathscr{H}_m^\infty$ with H^∞ of the m-disc, following a fundamental idea of H. Bohr. We explain this viewpoint in the next section. Section 3 presents a well-known perturbation argument that (1) shows that interpolating sequences are stable under small perturbations and (2) leads to a proof of Theorem 2. In Section 4, we present the theorem of Berndtsson–Chang–Lin, and we show how it gives Theorem 1. In Section 5, we obtain Theorem 3 as a consequence of a general result about the existence of a linear operator of interpolation. Finally, Section 6 contains the discussion around Theorem 3 alluded to the above.

2. $\mathscr{H}_m^\infty$ and $\mathscr{H}^\infty$ as spaces on polydiscs

Following Bohr [**4**], we set

$$z_1 = 2^{-s},\quad z_2 = 3^{-s},\quad \ldots,\quad z_m = p_m^{-s},\quad \ldots,$$

where p_m denotes the m-th prime number. Thus, in view of the fundamental theorem of arithmetic, Dirichlet series in $\mathscr{H}_m^\infty$ and $\mathscr{H}^\infty$ can be viewed as power series in m variables and infinitely many variables, respectively. Indeed, if the prime factorization of the positive integer n is written as $n = p_1^{\alpha_1} p_2^{\alpha_2} \cdots p_r^{\alpha_r}$, we obtain from (1) the formal power series

$$F(\mathbf{z}) = \sum_{n=1}^{\infty} a_n z_1^{\alpha_1} z_2^{\alpha_2} \cdots z_r^{\alpha_r},$$

where $\mathbf{z} = (z_1, z_2, \ldots, z_m)$ or $\mathbf{z} = (z_1, z_2, \ldots)$ depending on whether f is a function in $\mathscr{H}_m^\infty$ or in $\mathscr{H}^\infty$. The key point is that, thanks to Kronecker's theorem on Diophantine approximation [**10**, Ch. XXIII], this is not just a formal correspondence:

LEMMA 1. *Let $2, 3, \ldots, p_m$ be the first m primes ($m \geq 2$). Then the map*

$$t \mapsto (2^{-it}, 3^{-it}, \ldots, p_m^{-it})$$

from $\mathbb{R}^+$ to $\mathbb{T}^m$ has dense range in $\mathbb{T}^m$.

It is immediate from Lemma 1 and the maximum principle that

$$\|f\|_\infty = \|F\|_\infty, \tag{3}$$

where the norm on the right-hand side is the $H^\infty(\mathbb{D}^m)$ norm. The result is the same in the infinite-dimensional case, but some care has to be taken when defining the norm in the polydisc. First, we restrict point evaluation to the set $\mathbb{D}^\infty \cap c_0$, where $c_0 = c_0(\mathbb{N})$ is the space of sequences tending to zero. Indeed, for a point $\mathbf{z} = (z_1, z_2, \ldots, z_m, \ldots)$ in $\mathbb{D}^\infty \cap c_0$ we put

$$\mathbf{z}^{(m)} = (z_1, z_2, \ldots, z_m, 0, 0, \ldots),$$

meaning that all coordinates z_j for $j > m$ are replaced by zeros. Plugging $\mathbf{z}^{(m)}$ into F, we obtain a function $F(\mathbf{z}^{(m)})$ in the polydisc $\mathbb{D}^m$, the "m-te Abschnitt", as Bohr called it. We declare F to be in $H^\infty(\mathbb{D}^\infty)$ if the H^∞ norms of these functions are uniformly bounded, and denote the smallest uniform bound on these norms by $\|f\|_\infty$. By the Schwarz lemma for the polydisc, we have for $m < l$

$$|F(\mathbf{z}^{(m)}) - F(\mathbf{z}^{(l)})| \leq 2\|F\|_\infty \max\{|z_j| : m < j \leq l\},$$

so that we may define

$$F(\mathbf{z}) = \lim_{m\to\infty} F(\mathbf{z}^{(m)}).$$

We may view $\mathbb{C}_+$ as a one-dimensional complex variety in $\mathbb{D}^\infty \cap c_0$ via the map

$$s \mapsto (2^{-s}, 3^{-s}, \ldots, p_n^{-s}, \ldots).$$

It was proved in [11] that (3) remains true in the infinite-dimensional case. In other words, we may associate $\mathscr{H}^\infty$ with $H^\infty(\mathbb{D}^\infty)$. In what follows, we will call F the Bohr extension of f.

The considerations above imply that our interpolation problems are special versions of interpolation problems on finite or infinite polydiscs. This does not mean that our interpolation problems are much easier than their counterparts on the larger domains, but we have at least the option of applying methods from either setting. For instance, as already mentioned, Theorem 1 is really a result for polydiscs, while the one-dimensional Phragmén–Lindelöf principle is an essential ingredient in the proof of Theorem 3.

3. Perturbation arguments and proof of Theorem 2

The existence of a constant of interpolation gives a way of solving interpolation problems that are somehow small perturbations of problems for which solutions are known to exist. We will now give two versions of this kind of reasoning, one of which constitutes the proof of Theorem 2.

We begin with distortions with respect to Gleason distance. We declare the Gleason distance between two points $\mathbf{z}$ and $\mathbf{w}$ in $\mathbb{D}^m$ to be the supremum of $|f(\mathbf{w})|$ when f is a function in the unit ball of $H^\infty(\mathbb{D}^m)$ such that $f(\mathbf{z}) = 0$. It follows from the Schwarz lemma that the Gleason distance between $\mathbf{z} = (z_1, \dots, z_m)$ and $\mathbf{w} = (w_1, \dots, w_m)$ in $\mathbb{D}^m$ is

$$d_m(\mathbf{z}, \mathbf{w}) = \max_{1\le j\le m} \left| \frac{z_j - w_j}{1 - \overline{w}_j z_j} \right|.$$

Likewise, the Gleason distance between two points $\mathbf{z}$ and $\mathbf{w}$ in $\mathbb{D}^\infty \cap c_0$ is the supremum of $|f(\mathbf{w})|$ when f is a function in the unit ball of $H^\infty(\mathbb{D}^\infty)$ such that $f(\mathbf{z}) = 0$. Using again the Schwarz lemma, we find that the Gleason distance between $\mathbf{z} = (z_1, z_2, \dots)$ and $\mathbf{w} = (w_1, w_2, \dots)$ in $\mathbb{D}^\infty \cap c_0$ is

$$d_\infty(\mathbf{z}, \mathbf{w}) = \max_{j} \left| \frac{z_j - w_j}{1 - \overline{w}_j z_j} \right|.$$

If $Z = (\mathbf{z}_j)$ is an interpolating sequence for $H^\infty(\mathbb{D}^m)$ ($m \ge 1$ or $m = \infty$) with constant of interpolation M, then it follows from the definition of Gleason distance that

$$d_m(\mathbf{z}_j, \mathbf{z}_l) \ge M^{-1}$$

whenever $j \ne l$. When we have such a uniform lower estimate for the Gleason distance between distinct points, we say that Z is separated with respect to Gleason distance.

The following lemma is folklore. A proof is included for the sake of completeness and to demonstrate another usage of the iteration scheme appearing in the proof of Theorem 2.

LEMMA 2. *If $Z = (\mathbf{z}_j)$ is an interpolating sequence for $H^\infty(\mathbb{D}^m)$ ($m \ge 1$ or $m = \infty$) with constant of interpolation M, then any sequence $Z' = (\mathbf{z}'_j)$ satisfying*

$$\sup_j d_m(\mathbf{z}_j, \mathbf{z}'_j) < \frac{1}{M+1}$$

is also an interpolating sequence for $H^\infty(\mathbb{D}^m)$.

PROOF. Set $\varepsilon = \sup_j d_m(\mathbf{z}_j, \mathbf{z}'_j)$. We want to solve $f(\mathbf{z}'_j) = a_j$. We first solve $f_0(\mathbf{z}_j) = a_j$ with $\|f_0\|_\infty \leq M\|(a_j)\|_\infty$. We set $a_j^{(1)} = a_j - f_0(\mathbf{z}'_j)$ and note that we have $\|(a_j^{(1)})\|_\infty \leq (M+1)\varepsilon\|(a_j)\|_\infty$. We next solve $f_1(\mathbf{z}_j) = a_j^{(1)}$ with $\|f_1\|_\infty \leq M(M+1)\varepsilon\|(a_j)\|_\infty$. We iterate this procedure: at the k-th step, we set $a_j^{(k)} = a_j^{(k-1)} - f_{(k-1)}(\mathbf{z}'_j)$ and solve the interpolation problem $f_k(\mathbf{z}_j) = a_j^{(k)}$ with the bound $\|f_k\|_\infty \leq M(M+1)^k\varepsilon^k\|(a_j)\|_\infty$. Therefore,

$$f = \sum_{k=0}^{\infty} f_k$$

solves the problem $f(\mathbf{z}'_j) = a'_j$ whenever $\varepsilon < 1/(M+1)$. □

We mentioned in Section 2 that $F(\mathbf{z})$ could be defined on $\mathbb{D}^\infty \cap c_0$ as a limit of the "m-te Abschnitt" $F(\mathbf{z}^{(m)})$ when $m \to \infty$. This construction relies on the Schwarz lemma. The starting point for the proof of Theorem 2 is the restriction of this reasoning to $\mathbb{C}_+$, which should now be viewed as a subset of $\mathbb{D}^\infty \cap c_0$.

For the rest of this section we let f_m ($m \geq 2$) denote the element in $\mathscr{H}_m^\infty$ whose Bohr extension is $F(\mathbf{z}^{(m)})$. The Schwarz lemma implies that

$$|f_m(s) - f(s)| \leq 2p_{m+1}^{-\sigma}\|f\|_\infty.$$

A loose interpretation of this inequality is that when we restrict attention to a smaller half-plane than $\mathbb{C}_+$, say $\sigma \geq \delta > 0$, then only a finite number of the "variables" 2^{-s}, 3^{-s}, ... play a "significant" role. Theorem 2 can be viewed as a precise quantitative version of this statement.

PROOF OF THEOREM 2. Sufficiency is an immediate consequence of Montel's theorem. To prove necessity, we begin by applying the Schwarz lemma in the following way. Assuming S is an interpolating sequence with constant of interpolation L, we solve $f(s_j) = a_j$ for s_j in S_δ. Then

$$|f(s) - f_m(s)| \leq 2p_{m+1}^{-\sigma}L\|(a_j)\|_\infty$$

and $\|f_m\|_\infty \leq L\|(a_j)\|_\infty$. We define a new sequence $a_j^{(1)} = a_j - f_m(s_j)$ and solve $f^{(1)}(s_j) = a_j^{(1)}$, which then satisfies

$$\|f^{(1)}\|_\infty \leq 2p_{m+1}^{-\delta}L^2\|(a_j)\|_\infty.$$

We take again the m-te Abschnitt and obtain the corresponding Dirichlet series $f_m^{(1)}$ for which we have $\|f_m^{(1)}\|_\infty \leq 2p_{m+1}^{-\delta}L^2\|(a_j)\|_\infty$. We iterate this procedure: at the k-th step, we set $a_j^{(k)} = a_j^{(k-1)} - f_m^{(k-1)}(s_j)$ and solve the interpolation problem $f^{(k)}(s_j) = a_j^{(k-1)}$ with the bound

$$\|f^{(k)}\|_\infty \leq 2^k p_{m+1}^{-k\delta}L^{k+1}\|(a_j)\|_\infty.$$

If $p_{m+1}^{\delta} \geq (2+\varepsilon)L$, then the interpolation problem $g(s_j) = a_j$ can be solved in $\mathscr{H}_m^\infty$ by the function

$$g = f_m + \sum_{k=1}^{\infty} f_m^{(k)}.$$

For the constant of interpolation M we get the estimate

$$M \leq L(2+\varepsilon)/\varepsilon.$$ □

The main obstacle for applying Theorem 2 is of course that it is hard to control the constant of interpolation in nontrivial situations.

4. The Berndtsson–Chang–Lin condition and proof of Theorem 1

Our starting point is the following theorem of Berndtsson, Chang, and Lin [**3**].

LEMMA 3 (Berntsson–Chang–Lin theorem). *If a sequence $Z = (\mathbf{z}_j)$ in $\mathbb{D}^m$ $(1 \le m < \infty)$ satisfies*

$$\inf_j \prod_{l \neq j} d_m(\mathbf{z}_j, \mathbf{z}_l) > 0,$$

then $Z = (\mathbf{z}_j)$ is an interpolating sequence for $H^\infty(\mathbb{D}^m)$.

We note that the proof relies on a one-variable result of P. Jones [**12**]. It would be interesting to know whether this theorem remains valid for $m = \infty$ when Z is a sequence of distinct points from $\mathbb{D}^\infty \cap c_0$.

We will now switch to the half-plane $\mathbb{C}_+$. We set

$$\varrho_2(s_1, s_2) = d_2((2^{-s_1}, 3^{-s_1}), (2^{-s_2}, 3^{-s_2}))$$

and

$$\varrho(s_1, s_2) = \left| \frac{s_1 - s_2}{s_1 + \overline{s}_2} \right|,$$

which is the Gleason distance for $H^\infty(\mathbb{C}_+)$. Clearly,

$$\varrho_2(s_1, s_2) \le \varrho(s_1, s_2).$$

Carleson's theorem [**5**] says that

$$\inf_j \prod_{l \neq j} \varrho(s_j, s_l) > 0 \tag{4}$$

is a necessary and sufficient condition for $S = (s_j)$ to be an interpolating sequence for $H^\infty(\mathbb{C}_+)$. Consequently, in view of Lemma 3, Theorem 1 will follow once we have established the following lemma.

LEMMA 4. *For every $M > 0$, there exists a positive constant γ such that if s_1 and s_2 are points in $\mathbb{C}_+$ satisfying $|s_1|, |s_2| \le M$, then*

$$\varrho_2(s_1, s_2) \ge [\varrho(s_1, s_2)]^\gamma.$$

PROOF. It suffices to show that there exist positive constants c and C such that

$$\varrho_2(s_1, s_2) \ge c\varrho(s_1, s_2) \quad \text{and} \quad 1 - \varrho_2(s_1, s_2) \le C(1 - \varrho(s_1, s_2)). \tag{5}$$

To prove the first inequality in (5), we set $K = \{s = \sigma + it : \ \sigma \ge 0, \ |s| \le M\}$ and define the following function $\psi(s_1, s_2)$ on the compact set $K \times K$. We set $\psi(s_1, s_2) = \varrho(s_1, s_2)/\varrho_2(s_1, s_2)$ if $s_1 \neq s_2$ and both s_1 and s_2 are in $\mathbb{C}_+$. If at least one of the points s_1 and s_2 lies on the imaginary axis, we set $\psi(s_1, s_2) = 1$. Finally, if s is in $\mathbb{C}_+$, we put

$$\psi(s, s) = \frac{2^\sigma - 2^{-\sigma}}{2\sigma \log 2}.$$

We observe that ψ is a continuous function, and so we may choose for c the reciprocal of its maximum on $K \times K$.

To prove the second inequality in (5), we begin by noting that

$$1-[\varrho_2(s_1,s_2)]^2=\frac{(1-p^{-2\sigma_1})(1-p^{-2\sigma_2})}{(1-p^{-(\sigma_1+\sigma_2)})^2+2p^{-(\sigma_1+\sigma_2)}(1-\cos((\log p)(t_1-t_2)))},$$

where p is either 2 or 3. Hence

$$1-[\varrho_2(s_1,s_2)]^2\le\frac{C\,\sigma_1\sigma_2}{(\sigma_1+\sigma_2)^2+(t_1-t_2)^2}$$

for a constant C depending on M. The result follows since

$$1-[\varrho(s_1,s_2)]^2=\frac{4\sigma_1\sigma_2}{(\sigma_1+\sigma_2)^2+(t_1-t_2)^2}.\qquad\square$$

Defining $\varrho_m(s_1,s_2)$ for $m>2$ or $m=\infty$ in the obvious way, we note that for $m<\infty$ we also have the reverse estimate

$$1-[\varrho_m(s_1,s_2)]^2\ge\frac{1}{4}(1-p_m^{-2\sigma_1})(1-p_m^{-2\sigma_2}).$$

This means that if $S=(\sigma_j+\mathrm{i}t_j)$, viewed as a sequence in $\mathbb{D}^m$ via the map $s\mapsto(2^{-s},3^{-s},\dots,p_m^{-s})$, meets the Berndtsson–Chang–Lin condition, then

$$\sum_j\sigma_j<\infty.$$

This summability condition is not even necessary for S to be an interpolating sequence for $\mathscr{H}_2^\infty$ (see Section 6).

5. P. Beurling functions and proof of Theorem 3

A theorem of Pehr Beurling [**6**] says that if $S=(s_j)$ is an interpolating sequence for $H^\infty(\mathbb{C}_+)$ with constant of interpolation M, then there are functions f_j in $H^\infty(\mathbb{C}_+)$ such that $f_j(s_k)=\delta_{j,k}$ and

$$\sum_j|f_j(s)|\le M$$

holds for all s in $\mathbb{C}_+$. In other words, we may solve the interpolation problem $f(s_j)=a_j$ by means of the linear operator

$$f(s)=\sum_j a_jf_j(s),$$

and the norm of this operator is optimal.

We cannot expect such linear operators of interpolation to exist for general Banach algebras. However, for the spaces $\mathscr{H}_m^\infty$ and $\mathscr{H}^\infty$, linear operators of interpolation are known to exist, thanks to a general theorem from the theory of uniform algebras. (See J. B. Garnett's book [**9**, p. 288], where the result is attributed to N. Th. Varopoulos [**14**] and A. Bernard [**2**].) Explicitly, we have the following version of this result:

LEMMA 5. *Let A be any of the algebras $\mathscr{H}^\infty$ or $\mathscr{H}_m^\infty$. Then if S is an interpolating sequence for A with constant of interpolation M, then there are functions f_j in A satisfying $f_j(s_k)=\delta_{j,k}$, $\|f_j\|_\infty\le M$, and*

$$\sum_j|f_j(s)|\le M^2.$$

The proof relies on normal families along with a general argument valid for uniform algebras on compact sets. See [**9**, pp. 288–289] for details. We see that we get a poorer bound on the norm than in the P. Beurling theorem. We will still refer to the functions f_j of Lemma 5 as P. Beurling functions.

To place our application of Lemma 5 in context, we begin with some remarks on how it applies to $H^\infty(\mathbb{C}_+)$ and $\mathscr{H}_m^\infty$. Suppose first that S is an interpolating sequence for $H^\infty(\mathbb{C}_+)$ with constant of interpolation M. Let I be an arbitrary interval on the imaginary axis, and let $Q(I)$ be the square in $\mathbb{C}_+$ such that I is one of its sides. Also, for an arbitrary positive number ξ, let ξI be the interval of length $\xi|I|$ concentric with I. Then starting with the Poisson formula

$$1 = f_j(s_j) = \frac{1}{\pi}\int_{-\infty}^{\infty} f_j(\mathrm{i}t)\frac{\sigma_j}{(t-t_j)^2+\sigma_j^2}\,\mathrm{d}t,$$

we get the following estimate for $s_j = \sigma_j + \mathrm{i}t_j$ in $Q(I)$:

$$\sigma_j \le \int_{3MI} |f_j(\mathrm{i}t)|\,\mathrm{d}t.$$

Summing over j and applying Lemma 5, we get

$$\sum_{s_j\in Q(I)} \sigma_j \le 3M^3|I|,$$

which is Carleson's condition. (Using P. Beurling's theorem, we could replace M^3 by M^2 in the latter estimate.)

We may act similarly when we consider $\mathscr{H}_m^\infty$, but we need to switch to the polydisc. To this end, let $Z = (\mathbf{z}_j)$ be an interpolating sequence for $H^\infty(\mathbb{D}^m)$, and let R_j be the "rectangular base" of $\mathbf{z}_j$ on the distinguished boundary $\mathbb{T}^m$ of $\mathbb{D}^m$. Then using Poisson integrals for each f_j and summing as above, we get

$$\sum_{R_j\subset\Omega} |R_j| \le C|\Omega|$$

for every open subset of the distinguished boundary. Here the constant C depends only on M. This result is stated in [**3**] and is obtained there as a consequence of the work of Varopoulos [**15**].

Returning to the half-plane and taking Ω to be the entire distinguished boundary, we get the necessary condition

$$\sum_j \sigma_j^m < \infty \tag{6}$$

for S to be an interpolating sequence for $\mathscr{H}_m^\infty$. This condition is only slightly more restrictive than the one we get from the fact that the sequence has to be separated with respect to Gleason distance. In particular, the latter condition implies that

$$\sum_j \sigma_j^{m+\varepsilon} < \infty$$

for every $\varepsilon > 0$.

We now turn to $\mathscr{H}^\infty$. Instead of using Poisson integrals as above, we will make use of the link between $\mathscr{H}^\infty$ and the Banach space $\mathscr{H}^1$, which we choose to define in the following way. For Dirichlet polynomials

$$f(s) = \sum_{n=1}^{N} a_n n^{-s},$$

we set

$$\|f\|_1 = \lim_{T\to\infty} \frac{1}{T}\int_0^\infty |f(\mathrm{i}t)|\mathrm{d}t,$$

where the limit exists thanks to a classical theorem of F. Carlson [**7**]. The space $\mathscr{H}^1$ is obtained by taking the closure of the set of Dirichlet polynomials with respect to this norm. For the details of this construction, we refer to [**1**].

We will need the following fact from [**8**]. A function f represented by a Dirichlet series in $\mathscr{H}^1$ is analytic in the half-plane $\sigma > 1/2$. Indeed, we have the following estimate for point evaluation:

$$|f(\sigma + \mathrm{i}t)| \le C\frac{\|f\|_1}{\sigma - 1/2}$$

for a constant C independent of f. We will also need the fact that we can compute $\|f\|_1$ for f in $\mathscr{H}^\infty$ in the following way:

$$\|f\|_1 = \lim_{\sigma\to 0}\lim_{T\to\infty} \frac{1}{T}\int_0^T |f(\sigma + \mathrm{i}t)|\,\mathrm{d}t. \tag{7}$$

Thus, in particular,

$$\|f\|_1 \le \|f\|_\infty$$

for functions f in $\mathscr{H}^\infty$. This inequality can be seen as a consequence of a much stronger result, proved in [**1**], which says that $\mathscr{H}^\infty$ is in fact the multiplier algebra for $\mathscr{H}^1$.

We are now ready to prove Theorem 3.

Proof of Theorem 3. Set $M(\sigma) = \sup_t |f(\sigma + \mathrm{i}t)|$. Then the Phragmén–Lindelöf principle gives for $\sigma_0 \ge \sigma$

$$|f(\sigma + \mathrm{i}t)| \le \|f\|_\infty^{1-\sigma/\sigma_0}[M(\sigma_0)]^{\sigma/\sigma_0}.$$

If $\sigma_0 > 1/2$, we can use the bound for point evaluation in $\mathscr{H}^1$ to obtain from this

$$|f(\sigma + \mathrm{i}t)| \le [C/(\sigma_0 - 1/2)]^{\sigma/\sigma_0}\|f\|_\infty^{1-\sigma/\sigma_0}\|f\|_1^{\sigma/\sigma_0}.$$

Applying this inequality to $f = f_j$ at the point s_j, we get

$$\|f_j\|_1 \ge [(\sigma_0 - 1/2)/C]\, M^{-\sigma_0/\sigma_j + 1}.$$

But in view of Lemma 5 and formula (7), we also have

$$\sum_j \|f_j\|_1 \le M^2,$$

and so we conclude that

$$\sum_j M^{-c/\sigma_j} < \infty$$

whenever $c > 1/2$. □

6. Remarks on Theorem 3

The following result on separation of points on a vertical line shows that the condition of Theorem 3 is extremely restrictive compared with what we get from the mere fact that an interpolating sequence is separated with respect to Gleason distance.

PROPOSITION 1. *Suppose* $0 < \varepsilon < 1$ *and a positive number* $\gamma < 2$ *are given. For sufficiently small* σ *we can find at least* $2^{(1/\varepsilon)^{\gamma/\sigma}}$ *distinct points* $\sigma + \mathrm{i}t_j$ *such that*

$$\varrho_\infty(\sigma + \mathrm{i}t_j, \sigma + \mathrm{i}t_k) \geq \varepsilon$$

for $j \neq k$.

PROOF. A rough estimate will suffice. We begin by finding the smallest radius r such that there are two distinct points on the circle $|z| = r$ with the pseudohyperbolic distance at least ε between them. We obtain the equation

$$\frac{2r}{1+r^2} = \varepsilon,$$

which gives $r = 1 - \sqrt{1-\varepsilon^2}$. We now estimate the number of primes p for which

$$p^{-\sigma} > 1 - \sqrt{1-\varepsilon^2},$$

or, in other words, for which

$$\log p < \frac{1}{\sigma} \log \frac{1+\sqrt{1-\varepsilon^2}}{\varepsilon^2}.$$

For sufficiently small σ there are at least $(1/\varepsilon)^{\gamma/\sigma}$ such primes in view of the prime number theorem. The claim now follows by an application of Lemma 1. □

The proposition could have been stated in a slightly sharper form, but we note that it would not hold if we allowed γ to be any number larger than 2. We may see this by essentially repeating the argument above, this time starting with the observation that the maximum number of points pseudohyperbolically separated by ε on the circle $|z| = 2^{-\sigma}$ is bounded by a constant times $1/(\varepsilon\sigma)$. We then get $(c/\varepsilon\sigma)^{(1/\varepsilon)^{2/\sigma}}$ as an upper bound for the number of t_j that can be chosen.

To see to what extent (6) and Theorem 3 are sharp, we consider two examples. Set

$$\Sigma_l = \{s = 2^{-l} + \mathrm{i}2\pi k \log 2,\ k \text{ an integer}\}$$

for l a positive integer and look for an interpolating sequence S that is a subsequence of $\bigcup_l \Sigma_l$. Then the problem reduces to the problem of solving the interpolation problem $f_l(s_j) = a_j$ for s_j in $\Sigma_l \cap S$ with uniform control of norms, because the global interpolation problem can be solved as

$$f(s) = \sum_l g_l(2^{-s}) f_l(s),$$

with g_l being P. Beurling functions for the sequence $2^{-2^{-l}}$ in the unit disc. By Lemma 1, we may, for each l, choose 2^l points s_j from Σ_l such that the points 3^{-s_j} are uniformly separated with respect to pseudohyperbolic distance on the unit disc. This means that the f_l can be chosen as functions of the "variable" 3^{-s}, where the functions in question are the P. Beurling functions associated with the finite sequence (3^{-s_j}). Thus we have an example showing that we may have

$$\sum \sigma_j = \infty$$

even in $\mathscr{H}_2^\infty$. This argument can be further developed to a proof that an interpolating sequence S for $\mathscr{H}_m^\infty$ may have

$$\sum \sigma_j^{m-1} = \infty.$$

Let us now turn to $\mathscr{H}^\infty$. We begin by considering the two points $z_{n,0} = p_n^{-\sigma}$ and $z_{n,1} = -p_n^{-\sigma}$ on each of the circles $|z_n| = p_n^{-\sigma}$. Then by Pick's theorem (see [**9**, p. 7]), the constant of interpolation for the two-point sequence $z_{n,0}$, $z_{n,1}$ is p_n^σ. In the $(m-1)$-disk, we take the cross product of these sequences,

$$Z_m(\sigma) = (\mathbf{z_r}),$$

where $\mathbf{z_r} = (z_{2,r_2}, z_{3,r_3}, \ldots, z_{m,r_m})$ and $\mathbf{r} = (r_2, r_3, \ldots, r_m)$ ranges over $\{0,1\}^{m-1}$. Thus $Z_m(\sigma)$ is a sequence consisting of 2^{m-1} points. The interpolation problem $f(\mathbf{z_r}) = a_\mathbf{r}$ for $Z_m(\sigma)$ can be solved by the function

$$f(z) = \sum_\mathbf{r} a_\mathbf{r} f_{r_2}(z_2) f_{r_3}(z_3) \cdots f_{r_m}(z_m),$$

where the functions f_{r_n} are the P. Beurling functions for the respective two-point problems. The constant of interpolation M_m therefore satisfies

$$M_m \le \prod_{n=2}^m p_n^\sigma = e^{\sigma \sum_{n=2}^m \log p_m}.$$

This expression is uniformly bounded if

$$m \le \frac{c}{\sigma \log(1/\sigma)}$$

for some positive constant c. By Lemma 1, we can find points on Σ_l that are arbitrarily close to the corresponding points on $Z_m(2^{-l})$. By Lemma 2, this means that we can find a sequence of 2^m points from Σ_l with a constant of interpolation arbitrarily close to M_m. In other words, we may place at least

$$2^{-\frac{c}{\sigma_l \log \sigma_l}}$$

points on Σ_l, where $\sigma_l = 2^{-l}$. We have therefore constructed a sequence which satisfies

$$\sum_j 2^{\frac{c}{\sigma_j \log \sigma_j}} = \infty.$$

Acknowledgement

I am grateful to Eero Saksman for a pertinent remark that reduced the bound in the proof of Theorem 3 from $c > 1$ to $c > 1/2$.

References

[1] F. Bayart, *Hardy spaces of Dirichlet series and their composition operators*, Monatsh. Math. **136** (2002), 203–236.

[2] A. Bernard, *Algèbres quotients d'algèbres uniformes*, C. R. Acad. Sci. Paris, Sér. A **272** (1971), 1101–1104.

[3] B. Berndtsson, S.-Y. A. Chang, and K.-C. Lin, *Interpolating sequences in the polydisc*, Trans. Amer. Math. Soc. **302** (1987), 161–169.

[4] H. Bohr, *Über die Bedeutung der Potenzreihen unendlich vieler Variabeln in der Theorie der Dirichletschen Reihen* $\sum a_n/n^s$, Nachr. Akad. Wiss. Göttingen Math.-Phys. Kl., 1913, 441–488.

[5] L. Carleson, *An interpolation problem for bounded analytic functions*, Amer. J. Math. **80** (1958), 921–930.

[6] ———, *Interpolations by bounded analytic functions and the Corona problem*, Proc. Internat. Congr. Mathematicians (Stockholm, 1962), Inst. Mittag-Leffler, Djursholm, 1963, 314–316.

[7] F. Carlson, *Contributions à la théorie des séries de Dirichlet*, Note I, Ark. Mat. **16** (1922), no. 18 , 1–19.

[8] B. Cole and T. Gamelin, *Representing measures and Hardy spaces for the infinite polydisk algebra*, Proc. London Math. Soc. **53** (1986), 112–142.

[9] J. B. Garnett, *Bounded Analytic Functions.* Revised first edition, Graduate Texts in Mathematics, 236, Springer, New York, 2007.

[10] G. H. Hardy and E. M. Wright, *An Introduction to the Theory of Numbers.* 3rd ed., Clarendon Press, Oxford, 1954.

[11] H. Hedenmalm, P. Lindqvist, and K. Seip, *A Hilbert space of Dirichlet series and systems of dilated functions in* $L^2(0,1)$, Duke Math. J. **86** (1997), 1–37.

[12] P. W. Jones, L^∞ *estimates for the* $\overline{\partial}$ *problem in a half-plane*, Acta Math. **150** (1983), 137–152.

[13] J.-F. Olsen and K. Seip, *Local interpolation in Hilbert spaces of Dirichlet series*, Proc. Amer. Math. Soc. **136** (2008), 203–212.

[14] N. Th. Varopoulos, *Ensembles pics et ensembles d'interpolation pour les algèbre uniforme*, C. R. Acad. Sci. Paris, Sér. A **272** (1971), 866–867.

[15] ———, *Sur un problème d'interpolation*, C. R. Acad. Sci. Paris, Sér. A **274** (1972), 1539–1542.

Department of Mathematical Sciences, Norwegian University of Science and Technology (NTNU), NO-7491 Trondheim, Norway

E-mail address: seip@math.ntnu.no

Amer. Math. Soc. Transl.
(2) Vol. **226**, 2009

Remarks on Counting Negative Eigenvalues of the Schrödinger Operator on Regular Metric Trees

Michael Solomyak

To Victor Petrovich Havin, a friend and colleague, on his 75th birthday

ABSTRACT. We discuss estimates on the number $N_-(\mathbf{H}_{\alpha V})$ of negative eigenvalues of the Schrödinger operator $\mathbf{H}_{\alpha V} = -\Delta - \alpha V$ on regular metric trees, as depending on the properties of the potential $V \geq 0$ and on the value of the large parameter α. For any given $p \geq 1/2$, we obtain conditions on V guaranteeing the behavior $N_-(\mathbf{H}_{\alpha V}) = O(\alpha^p)$. For a special class of trees we show that these conditions are not only sufficient but also necessary. For $p > 1/2$ the order-sharp estimates involve a (quasi-)norm of V in some 'weak' L_p- or $\ell_p(L_1)$-space. We show that the results obtained can be easily derived from the ones of [**8**].

The results considerably improve the estimates presented in the recent paper [**5**].

1. Introduction

1.1. Preliminaries. In this note we discuss estimates for the number $N_-(\mathbf{H}_V)$ of the negative eigenvalues of the Schrödinger operator $\mathbf{H}_V = -\Delta - V$ on regular metric trees. The classical Birman–Schwinger principle allows one to reduce such estimates to the study of the spectrum of a certain compact, self-adjoint operator acting in an appropriate Hilbert space; see Section 5 below for an explanation of this reduction. The detailed study of this spectrum was initiated by Naimark and the author [**8**], and some estimates for $N_-(\mathbf{H}_V)$ immediately follow from there.

Further results on the behavior of $N_-(\mathbf{H}_V)$ were obtained recently by Ekholm, Frank, and Kovařík [**5**]. The main novelty in [**5**] is an analysis of the connection between the estimates for $N_-(\mathbf{H}_V)$ and the *global dimension* of a regular tree. This is an important step in the understanding of spectral properties of the Schrödinger operator on trees.

The proofs in [**5**] are based upon a scheme developed by Naimark and Solomyak [**8, 9**], and independently by Carlson [**4**]. Our main goal in this note is to show that the estimates obtained in [**5**] can be considerably improved by the direct reduction to the results of [**8**]. We also show that the estimates in terms of the 'weak L_p-spaces'

2000 *Mathematics Subject Classification.* Primary 34L40, 34B45.
Key words and phrases. Metric tree, Schrödinger operator, negative spectrum.

obtained by this approach, are order-sharp in the strong coupling limit. This means that when applied to the operator $\mathbf{H}_{\alpha V}$, where $\alpha > 0$ is a large parameter, these estimates are of the type

$$N_-(\mathbf{H}_{\alpha V}) = O(\alpha^p), \qquad \alpha \to \infty, \tag{1.1}$$

where 'O' cannot be replaced by 'o'. On the contrary, in the estimates obtained in [**5**] in terms of the standard L_p, such replacement is always possible.

We also discuss (in Theorem 4.3) estimates of a different nature, where the function $N_-(\mathbf{H}_{\alpha V})$ is evaluated in terms of a certain number sequence associated with the potential. In Theorem 4.4 we single out a subclass of regular trees for which the conditions guaranteeing the order $N_-(\mathbf{H}_{\alpha V}) = O(\alpha^p)$ with $p > 1/2$ are not only sufficient, but also necessary.

Any metric tree has local dimension one. As a consequence, for the rapidly decaying potentials V one always has $N_-(\mathbf{H}_{\alpha V}) = O(\alpha^{1/2})$. For the slowly decaying potentials such that (1.1) is satisfied with some $p > 1/2$, the estimates in L^p-terms are never order-sharp: any order-sharp estimate must involve a norm, or a quasi-norm, of some non-separable function space. In this connection, see a discussion in [**11**]. The results of the present paper once more substantiate this claim.

1.2. Geometry of a tree. Let Γ be a rooted metric tree, with the root o. We assume that the number of vertices and the number of edges are infinite, and that each edge has finite length. The distance $\rho(x,y)$ and the partial ordering $x \preceq y$ on Γ are introduced in a natural way, and we write $x \prec y$ if and only if $x \preceq y$ and $x \neq y$. We denote $|x| = \rho(o,x)$. On each edge $e \subset \Gamma$ the partial ordering turns into the standard ordering defined by the inequality $|x| \le |y|$. With each vertex $v \in \Gamma$ we associate its *generation* $\mathrm{Gen}(v)$, i.e., the number of vertices w such that $o \preceq w \prec v$. In particular, $\mathrm{Gen}(o) = 0$. The *branching number* $b(v)$ of a vertex v is the number of edges emanating from v. We assume that $b(o) = 1$ and that $1 < b(v) < \infty$ for any vertex $v \neq o$.

In this paper we consider the regular trees; see [**8, 9**] for more detail. A tree Γ is said to be *regular* (or *symmetric*, or *radial*), if all vertices of the same generation have equal branching numbers and lie at the equal distance from the root. Each regular tree is completely determined by specifying two number sequences, $\{b_n\}$ and $\{t_n\}$, where $b_n = b(v)$ and $t_n = |v|$ for any vertex v of generation n, $n = 0, 1, \dots$. We assume that the sequence $\{t_n\}$ is unbounded. The *branching function* of Γ, defined as

$$g_0(t) = b_0 b_1 \cdots b_n \quad \text{for } t_n < t \le t_{n+1}, \quad n = 1, 2, \dots, \tag{1.2}$$

is an important characteristic of a regular tree.

The natural measure dx on Γ is induced by Lebesgue measure on the edges. The spaces $L_p(\Gamma)$ are understood as the Lebesgue spaces with respect to this measure. The norm in $L_p(\Gamma)$ is denoted by $\|\cdot\|_p$; for $p = 2$, we write simply $\|\cdot\|$.

1.3. Sobolev spaces on a tree. Differentiation, in the direction compatible with the ordering on Γ, is well defined at any point $x \in \Gamma$ except for the vertices. The Sobolev space $H^1(\Gamma)$ consists of all continuous functions u on Γ such that $u \restriction e \in H^1(e)$ on each edge $e \subset \Gamma$, and the condition

$$\|u\|^2_{H^1(\Gamma)} := \int_\Gamma (|u'(x)|^2 + |u(x)|^2)\, dx < \infty$$

is satisfied. It is easy to show that any function $u \in H^1(\Gamma)$ vanishes as $|x| \to \infty$; see, e.g., Lemma 3.1 in [**12**]. It immediately follows that the functions $u \in H^1(\Gamma)$ with bounded support are dense in $H^1(\Gamma)$. The boundary condition $u(o) = 0$ selects the subspace

$$H^{1,0}(\Gamma) \subset H^1(\Gamma).$$

The functions $u \in H^{1,0}(\Gamma)$ with bounded support are dense in this subspace.

Along with $H^{1,0}(\Gamma)$, we also need the *homogeneous* Sobolev space $\mathcal{H}^{1,0}(\Gamma)$, formed by all continuous functions u on Γ such that $u(o) = 0$, $u \restriction e \in H^1(e)$ on each edge $e \subset \Gamma$, and $u' \in L_2(\Gamma)$. By definition,

$$\|u\|_{\mathcal{H}^1(\Gamma)} = \|u'\|.$$

In contrast with the space $H^{1,0}(\Gamma)$, the functions $u \in \mathcal{H}^{1,0}(\Gamma)$ with bounded support are not always dense in $\mathcal{H}^{1,0}(\Gamma)$. As a consequence, the set $H^{1,0}(\Gamma)$, which is evidently contained in $\mathcal{H}^{1,0}(\Gamma)$, is not always dense in the latter space. This property depends on the geometry of the tree. A tree such that $H^{1,0}(\Gamma)$ is dense in $\mathcal{H}^{1,0}(\Gamma)$ is called *recurrent*; otherwise it is called *transient*. We shall denote the closure of $H^{1,0}(\Gamma)$ in $\mathcal{H}^{1,0}(\Gamma)$ by $\mathcal{H}^\circ(\Gamma)$. So, $\mathcal{H}^\circ(\Gamma) = \mathcal{H}^{1,0}(\Gamma)$ if and only if the tree Γ is recurrent.

A simple necessary and sufficient condition of transiency is known for the regular trees; see, e.g., [**9**], Theorem 3.2. Namely, such a tree is transient if and only if its *reduced height*

$$l(\Gamma) := \int_\Gamma \frac{dt}{g_0(t)} \tag{1.3}$$

is finite.

2. Schrödinger operator on Γ

Our main object in this paper is the Schrödinger operator on Γ,

$$\mathbf{H}_{\alpha V} = -\Delta_D - \alpha V, \tag{2.1}$$

with the non-negative potential $V(x)$ and a large parameter (the *coupling constant*) $\alpha > 0$. In (2.1) the symbol Δ_D stands for the Dirichlet Laplacian on Γ. In Section 6 we discuss also an analogue of (2.1) for the Neumann Laplacian Δ_N.

The operator (2.1) is defined via its quadratic form, which is

$$\mathbf{h}_{\alpha V}[u] = \int_\Gamma (|u'|^2 - \alpha V|u|^2)\, dx, \qquad u \in H^{1,0}(\Gamma). \tag{2.2}$$

We are interested in estimating the number of negative eigenvalues of $\mathbf{H}_{\alpha V}$ in terms of the potential V and the coupling constant α.

Assume that the potential $V \geq 0$ is such that the inequality

$$\int_\Gamma V|u|^2 dx \leq C_V \int_\Gamma |u'|^2\, dx, \qquad \forall u \in H^{1,0}(\Gamma), \tag{2.3}$$

is satisfied. By continuity, it extends to all $u \in \mathcal{H}^\circ(\Gamma)$. We assume that C_V is the least possible constant in (2.3). We call any such potential V a *Hardy weight on* $\mathcal{H}^\circ(\Gamma)$, and we say that the Hardy weight is *normalized* if $C_V = 1$. If V is a Hardy weight, the operator (2.1) is well defined at least for small values of α. Some further assumptions about V that we impose later imply the finiteness of the negative spectrum of this operator.

In general, let $\mathbf{H}$ be a self-adjoint operator in a Hilbert space. Suppose the negative spectrum of $\mathbf{H}$ is finite. Then we denote by $N_-(\mathbf{H})$ the number of its negative eigenvalues counted according to their multiplicities.

For estimating the quantity $N_-(\mathbf{H}_{\alpha V})$ it is important to have a description of the Hardy weights on $\mathcal{H}^\circ(\Gamma)$. Such description, for general trees and the Hardy weights on a wider class $\mathcal{H}^{1,0}(\Gamma)$, was obtained in [**6**]. A similar result for $\mathcal{H}^\circ(\Gamma)$ seems to be unknown so far. The situation changes if we restrict ourselves to the regular trees and to the *symmetric* weights

$$V(x) = v(|x|), \qquad \forall x \in \Gamma,$$

where $v \geq 0$ is a measurable function on $[0, \infty)$. For the regular trees an exhaustive description of all symmetric Hardy weights can be easily derived from the classical Hardy-type inequalities; see, e.g., [**10**], or Section 1.3 in the book [**7**]. The result of such reduction was presented in [**9**], Theorem 5.2, but only for the transient regular trees. We reproduce it below, including also the recurrent case. The proof for this case remains the same.

PROPOSITION 2.1. *Let Γ be a regular tree, and let $g_0(t)$ be its branching function* (1.2). *Suppose $V(x) = v(|x|) \geq 0$ is a symmetric weight function on Γ.*

1° *Let Γ be recurrent. Then $V(x)$ is a Hardy weight on $\mathcal{H}^\circ(\Gamma)$ if and only if*

$$B_0(v) := \sup_{t>0} \left(\int_t^\infty v(s) g_0(s)\, ds \cdot \int_0^t \frac{ds}{g_0(s)} \right) < \infty.$$

Moreover, the least possible constant C_V in (2.3) *satisfies*

$$B_0(v) \leq C_V \leq 4B_0(v).$$

2° *Let Γ be transient. Then $V(x)$ is a Hardy weight on $\mathcal{H}^\circ(\Gamma)$ if and only if the following two conditions are satisfied:*

$$B_1(v) := \sup_{t>t_1} \left(\int_{t_1}^t v(s) g_0(s)\, ds \cdot \int_t^\infty \frac{ds}{g_0(s)} \right) < \infty, \tag{2.4}$$

$$B_2(v) := \sup_{t<t_1} \left(t \int_t^{t_1} v(s)\, ds \right) < \infty. \tag{2.5}$$

Moreover,

$$C_V \leq 4(B_1(v) + B_2(v)); \qquad B_1(v) \leq C_V; \; B_2(v) \leq \left(1 + \frac{b_1 t_1}{t_2 - t_1}\right) C_V.$$

3. Classes of functions, of number sequences, and of compact operators

As we shall see, the sharp estimates of the function $N_-(\mathbf{H}_{\alpha V})$ involve the *weak L_p-spaces* on the tree, rather than the classical Lebesgue spaces L_p. Here we recall their definitions. See, e.g., [**1**], Section 1.3, for more detail.

Let $(\mathcal{X}, \mu)$ be a measure space with σ-finite measure. A measurable function f on $\mathcal{X}$ belongs to the class $L_{p,w}(\mathcal{X}, \mu)$, $0 < p < \infty$, if and only if

$$\|f\|_{L_{p,w}(\mathcal{X},\mu)} := \sup_{t>0} \left(t \cdot (\mu\{x \in \mathcal{X} : |f(x)| > t\})^{1/p} \right) < \infty. \tag{3.1}$$

The spaces $L_{p,w}$ are linear. The functional (3.1) defines a *quasi-norm* on $L_{p,w}$, and the space is complete with respect to this quasi-norm. If $p > 1$, and only in this case, a norm equivalent to the quasi-norm (3.1) does exist. However, in various estimates it is more convenient to use the quasi-norm (3.1).

If the measure μ does not reduce to the sum of a finite number of atoms, the space $L_{p,w}(\mathcal{X},\mu)$ is non-separable. The condition

$$\mu\{x\in\mathcal{X}:|f(x)|>t\}=o(t^{-p})\qquad\text{as } t+t^{-1}\to\infty$$

singles out a separable subspace in $L_{p,w}(\mathcal{X},\mu)$, which we denote by $L^{\circ}_{p,w}(\mathcal{X},\mu)$. It is well known (and easy to check) that

$$\|f\|^p_{L_{p,w}(\mathcal{X},\mu)}\le\int_{\mathcal{X}}|f|^p\,d\mu,\qquad L_p(\mathcal{X},\mu)\subset L^{\circ}_{p,w}(\mathcal{X},\mu).$$

In particular, let $\mathcal{X}=\Gamma$ be a metric tree, and let μ be the measure generated by a weight function $\Phi(x)\ge 0$ on Γ:

$$d\mu=\Phi(x)\,dx.$$

We denote the corresponding space by $L_{p,w}(\Gamma,\Phi)$ and the functional (3.1) by $\|f\|_{p,w;\Phi}$. So,

$$\|f\|^p_{p,w;\Phi}=\sup_{t>0}\left(t^p\int_{|f(x)|>t}\Phi(x)\,dx\right).$$

We also need the weak ℓ_p-spaces of number sequences. These are a particular case of the spaces $L_{p,w}(\mathcal{X},\mu)$, where $\mathcal{X}=\mathbb{N}$ or $\mathcal{X}=\mathcal{N}_0:=\{0,1,\dots\}$. The measure μ is either the standard counting measure, or it is given by a sequence of positive weights $\Phi_n=\mu(\{n\})$. For a sequence $\boldsymbol{f}=\{f_n\}_{n\in\mathcal{X}}$ we have

$$\|\boldsymbol{f}\|^p_{p,w;\Phi_n}=\sup_{t>0}\left(t^p\sum_{|f(n)|>t}\Phi_n\right).$$

The case of the counting measure is especially simple; here we drop the symbol Φ_n in the notation. Consider the sequence $\{f^*_n\}$ whose terms are the numbers $|f_n|$ rearranged in the non-increasing order. Then $\boldsymbol{f}\in\ell_{p,w}$ if and only if $f^*_n=O(n^{-1/p})$, and $\boldsymbol{f}\in\ell^{\circ}_{p,w}$ if and only if $f^*_n=o(n^{-1/p})$. Moreover,

$$\|\boldsymbol{f}\|_{p,w}=\sup_n(n^{1/p}f^*_n).$$

Now, let $\mathbf{A}$ be a compact operator on a Hilbert space $\mathfrak{H}$, and let $\{s_n(\mathbf{A})\}$ be the sequence of its singular numbers, counted according to their multiplicities. Recall that for a non-negative self-adjoint operator its singular numbers coincide with the eigenvalues. By definition, $\mathbf{A}$ belongs to the weak Schatten class $\mathcal{C}_{p,w}$ if $\{s_n(\mathbf{A})\}\in\ell_{p,w}$ and

$$\|\mathbf{A}\|_{p,w}:=\|\{s_n(\mathbf{A})\}\|_{p,w}=\sup_n\left(n^{1/p}s_n(\mathbf{A})\right).$$

The class $\mathcal{C}^{\circ}_{p,w}$ is defined as the set of all compact operators such that $s_n(\mathbf{A})=o(n^{-1/p})$.

For a compact operator $\mathbf{A}$ we denote

$$n(s;\mathbf{A})=\#\{n:s_n(\mathbf{A})>s\},\qquad s>0.$$

Evidently, the inclusion $\mathbf{A}\in\mathcal{C}_{p,w}$ is equivalent to the inequality

$$n(s;\mathbf{A})\le Cn^{-p},\qquad\forall n\in\mathbb{N},$$

and moreover, the best possible constant here is $C = \|\mathbf{A}\|^p_{p,w}$. Also, $\mathbf{A} \in C^\circ_{p,w}$ if and only if $n(s;\mathbf{A}) = o(n^{-p})$.

4. Main results on the behavior of $N_-(\mathbf{H}_{\alpha V})$

Here we present the basic estimates for the number of negative eigenvalues of the operator $\mathbf{H}_{\alpha V}$. Their proofs are given in Section 5.

THEOREM 4.1. *Let Γ be a regular metric tree and let a normalized and symmetric Hardy weight $\Psi(x) = \psi(|x|) > 0$ on $\mathcal{H}^\circ(\Gamma)$ be chosen. Define the weight function $\Phi(x) = |x|\Psi(x)$. Suppose V is a non-negative potential on Γ such that $V\Psi^{-1} \in L_{p,w}(\Gamma, \Phi)$ for some $p > 1$. Then*

$$N_-(\mathbf{H}_{\alpha V}) \le C\alpha^p \sup_{t>0} \left(t^p \int_{V(x)>t\Psi(x)} \Phi(x)\,dx \right), \qquad \forall \alpha > 0, \tag{4.1}$$

and

$$\limsup_{\alpha\to\infty} \alpha^{-p} N_-(\mathbf{H}_{\alpha V}) \le C \limsup_{t+t^{-1}\to\infty} \left(t^p \int_{V(x)>t\Psi(x)} \Phi(x)\,dx \right). \tag{4.2}$$

In particular, if $\int_{V(x)>t\Psi(x)} \Phi(x)\,dx = o(t^{-p})$ as $t + t^{-1} \to \infty$, then

$$N_-(\mathbf{H}_{\alpha V}) = o(\alpha^p). \tag{4.3}$$

If $V\Psi^{-1}$ belongs to the narrower space $L_p(\Gamma, \Phi)$, then

$$N_-(\mathbf{H}_{\alpha V}) \le C\alpha^p \int_\Gamma |x| V^p \Psi^{1-p}\,dx, \tag{4.4}$$

and (4.3) is satisfied.

4.1. Global dimension of a regular tree. Suppose that the branching function $g_0(t)$ satisfies the two-sided inequality

$$c_1(1+t)^{d-1} \le g_0(t) \le c_2(1+t)^{d-1}, \quad 0 < c_1 < c_2, \qquad \forall t > 0,$$

with some $d \ge 1$. Following [**5**], we shall say that the number d is the *global dimension* of the regular tree Γ. This is an important invariant of a tree. The existence of a global dimension is itself a serious restriction on the structure of the tree.

It is more convenient for us to use an equivalent definition of global dimension:

$$c_1' t^{d-1} \le g_0(t) \le c_2' t^{d-1}, \quad 0 < c_1' \le c_2', \qquad \forall t > t_1. \tag{4.5}$$

A direct inspection shows that the function $|x|^{-2}$ is a symmetric Hardy weight on $\mathcal{H}^\circ(\Gamma)$ provided that Γ has global dimension $d \ne 2$. Taking $\Psi(x) = c|x|^{-2}$ with an appropriate c (and hence, $\Phi(x) = c|x|^{-1}$) and applying Theorem 4.1, we come to the following result.

THEOREM 4.2. *Let Γ be a regular metric tree of global dimension $d \ne 2$. Suppose the potential $V \ge 0$ is such that $V|x|^2 \in L_{p,w}(\Gamma, |x|^{-1})$ for some $p > 1$. Then*

$$N_-(\mathbf{H}_{\alpha V}) \le C\alpha^p \sup_{t>0} \left(t^p \int_{V(x)|x|^2>t} |x|^{-1}\,dx \right) \tag{4.6}$$

and

$$\limsup_{\alpha\to\infty} \alpha^{-p} N_-(\mathbf{H}_{\alpha V}) \le C \limsup_{t+t^{-1}\to\infty} \left(t^p \int_{V(x)|x|^2>t} |x|^{-1}\,dx \right). \tag{4.7}$$

In particular, if $\int_{V(x)|x|^2>t} |x|^{-1}\,dx = o(t^{-p})$ *as* $t+t^{-1}\to\infty$, *then* (4.3) *is satisfied. If* $V|x|^2 \in L_p(\Gamma, |x|^{-1})$, *then along with* (4.3) *we have*

$$N_-(\mathbf{H}_{\alpha V}) \le C\alpha^p \int_\Gamma V^p |x|^{2p-1}\,dx. \tag{4.8}$$

Any metric tree has local dimension one, and hence, for the fast decaying potentials one must have $N_-(\mathbf{H}_{\alpha V}) = O(\alpha^{1/2})$. At the same time, both Theorems 4.1 and 4.2 describe only the situation where $N_-(\mathbf{H}_{\alpha V}) = O(\alpha^p)$ with $p > 1$, which corresponds to the potentials that decay rather slowly. The next result concerns the case $1/2 < p \le 1$. On the other hand, it applies only to the symmetric potentials $V(x) = v(|x|)$. We also assume that $V \in L_1(\Gamma)$. Actually, the latter is a restriction on the behavior of V only near the root. It is necessary in the case of the Neumann boundary condition (see Section 6) but not in the case of the Dirichlet condition.

The properties of the potential in the next two theorems are expressed in terms different from those in Theorems 4.1 and 4.2. With any symmetric potential $V(x) = v(|x|) \ge 0$ from $L_1(\Gamma)$ we associate the sequence $\boldsymbol{\eta}(V) = \{\eta_n(V)\}$, where

$$\eta_n(V) = t_{n+1} \int_{t_n}^{t_{n+1}} v(t)\,dt, \qquad n = 0, 1, \dots .$$

THEOREM 4.3. *Let* $V \in L_1(\Gamma)$ *be a symmetric, non-negative potential on a regular tree* Γ.

1° *Suppose the sequence* $\boldsymbol{\eta}(V) = \{\eta_n(V)\}$ *belongs to the space* $\ell_{1/2}$ *with respect to the weight sequence* $\{g_0(t_{n+1})\}$. *Then the estimate*

$$N_-(\mathbf{H}_{\alpha V}) \le C\alpha^{1/2} \sum_n \eta_n^{1/2}(V) g_0(t_{n+1}) \tag{4.9}$$

and the Weyl type asymptotic formula

$$\lim_{\alpha\to\infty} \alpha^{-1/2} N_-(\mathbf{H}_{\alpha V}) = \frac{1}{\pi} \int_\Gamma V^{1/2}(x)\,dx = \frac{1}{\pi} \int_0^\infty v^{1/2}(t) g_0(t)\,dt \tag{4.10}$$

are satisfied.

2° *Suppose* $\boldsymbol{\eta}(V) \in \ell_{p,w}(\mathbb{N}_0; g_0(t_{n+1}))$ *for some* $p \in (1/2, 1)$. *Then*

$$N_-(\mathbf{H}_{\alpha V}) \le C\alpha^p \sup_{t>0} \left(t^p \sum_{\eta_n(V)>t} g_0(t_{n+1}) \right) \tag{4.11}$$

and

$$\limsup_{\alpha\to\infty} \alpha^{-p} N_-(\mathbf{H}_{\alpha V}) \le C \limsup_{t\to 0} \left(t^p \sum_{\eta_n(V)>t} g_0(t_{n+1}) \right). \tag{4.12}$$

In particular, if $\sum_{\eta_n(V)>t} g_0(t_{n+1}) = o(t^{-p})$ *as* $t \to 0$, *then* (4.3) *is satisfied.*

If $\boldsymbol{\eta}(V) \in \ell_p(\mathbb{N}_0; g_0(t_{n+1}))$ *for some* $p \in (1/2, 1]$, *then along with* (4.3) *we have*

$$N_-(\mathbf{H}_{\alpha V}) \le C\alpha^p \sum_{n=0}^\infty \eta_n^p(V) g_0(t_{n+1}). \tag{4.13}$$

4.2. Estimates for the b-regular trees. More advanced results can be obtained if we suppose that for all the vertices $v \neq o$ the branching number is the same integer $b > 1$. In other words, we assume that

$$b_n = b, \qquad \forall n > 0.$$

In the paper [8], Section 7.2, such trees were called *b-regular.* For a b-regular tree one has

$$g_0(t) = b^n, \qquad t_n < t \le t_{n+1}. \tag{4.14}$$

If, besides, Γ has global dimension d, then (4.14) and (4.5) imply (as $t \to t_n+$) that

$$c_1' t_n^{d-1} \le b^n < c_2' t_n^{d-1}. \tag{4.15}$$

Hence, the sequence $\{t_n\}$ grows as a geometric progression.

For the b-regular trees the result of Theorem 4.3 can be considerably strengthened: it extends to arbitrary values of $p > 1/2$ and, most importantly, the condition

$$\boldsymbol{\eta}(V) \in \ell_{p,w}(\mathbb{N}_0, g_0(t_{n+1})) = \ell_{p,w}(\mathbb{N}_0, b^n)$$

turns out to be not only sufficient, but also necessary for $N_-(\mathbf{H}_{\alpha V}) = O(\alpha^p)$.

THEOREM 4.4. *Let $V \in L_1(\Gamma)$ be a symmetric, non-negative potential on a b-regular tree Γ. Suppose that Γ has global dimension $d \neq 2$.*

1° *Suppose $\boldsymbol{\eta}(V) \in \ell_{p,w}(\mathbb{N}_0; b^n)$ for some $p > 1/2$. Then*

$$N_-(\mathbf{H}_{\alpha V}) \le C\alpha^p \sup_{t>0} \left(t^p \sum_{\eta_n(V)>t} b^n \right) \tag{4.16}$$

and

$$\limsup_{\alpha\to\infty} \alpha^{-p} N_-(\mathbf{H}_{\alpha V}) \le C \limsup_{t\to 0} \left(t^p \sum_{\eta_n(V)>t} b^n \right). \tag{4.17}$$

In particular, if $\boldsymbol{\eta}(V) \in \ell^\circ_{p,w}(\mathbb{N}_0, b^n)$, then (4.3) is satisfied.

If $\boldsymbol{\eta}(V) \in \ell_p(\mathbb{N}_0, b^n)$, then along with (4.3) we have

$$N_-(\mathbf{H}_{\alpha V}) \le C\alpha^p \sum_{n=0}^{\infty} \eta_n^p(V) b^n. \tag{4.18}$$

2° *Conversely, if $N_-(\mathbf{H}_{\alpha V}) = O(\alpha^p)$ with some $p \ge \frac{1}{2}$, then $\boldsymbol{\eta}(V) \in \ell_{p,w}(\mathbb{N}_0; b^n)$. If $N_-(\mathbf{H}_{\alpha V}) = o(\alpha^p)$ with some $p > 1/2$, then $\boldsymbol{\eta}(V) \in \ell^\circ_{p,w}(\mathbb{N}_0; b^n)$.*

5. Proofs

5.1. Using the Birman–Schwinger principle. The proofs of all Theorems 4.1–4.4 are based upon the Birman–Schwinger principle; see [**2, 3**] for its exposition most convenient for our purposes. In order to formulate it for the particular case we need, let us consider the quadratic form

$$\mathbf{a}_V[u] = \int_\Gamma V(x)|u(x)|^2\,dx, \tag{5.1}$$

where $V \ge 0$ is a Hardy weight on $\mathcal{H}^\circ(\Gamma)$. Then the quadratic form $\mathbf{a}_V$ is bounded in this space, and therefore, it generates in $\mathcal{H}^\circ(\Gamma)$ a bounded, self-adjoint and non-negative operator, which we denote $\mathbf{A}_V^\circ$. The notation $A^\circ_{\Gamma,V}$ was used for this operator in [**8**]; see formula (3.7) there.

The following statement expresses the Birman–Schwinger principle for the case of operators on trees.

PROPOSITION 5.1. *Let a Hardy weight V on $\mathcal{H}^\circ(\Gamma)$ be such that the operator $\mathbf{A}_V^\circ$ is compact. Then for any $\alpha > 0$ the quadratic form (2.2) is bounded below and closed in $L_2(\Gamma)$, the negative spectrum of the corresponding operator $\mathbf{H}_{\alpha V}$ is finite, and moreover,*

$$N_-(\mathbf{H}_{\alpha V}) = n(\alpha^{-1}; \mathbf{A}_V^\circ), \qquad \forall \alpha > 0. \tag{5.2}$$

This proposition reduces estimates of $N_-(0; \mathbf{H}_{\alpha V})$ to the spectral analysis of the compact operator $\mathbf{A}_V^\circ$. Such analysis was carried out in [**8**].

Formula (5.2) implies that for any $p > 0$,

$$\sup_{\alpha>0} \alpha^{-p} N_-(\mathbf{H}_{\alpha V}) = \sup_{s>0} s^p\, n(s; \mathbf{A}_V^\circ) = \|\mathbf{A}_V^\circ\|_{p,w}^p \tag{5.3}$$

and

$$\limsup_{\alpha\to\infty} \alpha^{-p} N_-(\mathbf{H}_{\alpha V}) = \limsup_{s\to 0} s^p\, n(s; \mathbf{A}_V^\circ) = \inf_{\mathbf{T}\in\mathcal{C}_{p,w}^\circ} \|\mathbf{A}_V^\circ - \mathbf{T}\|_{p,w}^p.$$

5.2. Proof of Theorems 4.1–4.3. Due to (5.3), inequality (4.1) turns into the estimate (3.14) in Theorem 3.4 of the paper [**8**]. By inequality (3.13) in the same theorem, the finiteness of the integral in (4.4) implies $\mathbf{A}_V^\circ \in \mathcal{C}_p$. Since $\mathcal{C}_p \subset \mathcal{C}_{p,w}$, this immediately yields (4.4).

Inequality (4.2) follows from (4.1) automatically; see the proof of a similar statement in [**2**], Section 4.3. Note also that below we give a detailed proof of inequality (4.12) in Theorem 4.3. The proof of (4.2) can be reconstructed by analogy. Theorem 4.2 is a particular case of Theorem 4.1 for a special choice of the weight function, and it does not need a separate proof.

Theorem 4.3, except for the estimate (4.12), is a direct consequence (actually, just a reformulation in the equivalent terms) of Theorem 4.1 in [**8**], for the regular trees and the symmetric weights. One only has to take into account that in [**8**] the operators $\mathbf{A}_V$ were considered, rather than $\mathbf{A}_V^\circ$. The operator $\mathbf{A}_V$ corresponds to the quadratic form (5.1) on the space $\mathcal{H}^{1,0}(\Gamma)$. If the tree is transient, this space is larger than $\mathcal{H}^\circ(\Gamma)$. For this reason,

$$\|\mathbf{A}_V^\circ\|_{\mathcal{C}} \le \|\mathbf{A}_V\|_{\mathcal{C}},$$

where $\|\cdot\|_{\mathcal{C}}$ stands for the (quasi-)norm in any class of operators we are dealing with. It immediately follows that all the estimates derived in [**8**] for $\mathbf{A}_V$, hold also for the operator $\mathbf{A}_V^\circ$.

To justify (4.12), we note that for any $V \in L_1(\Gamma)$ with compact support the estimate (4.9) holds. For the corresponding operator $\mathbf{A}_V^\circ$ this implies

$$n(s; \mathbf{A}_V^\circ) = O(s^{-1/2}) = o(s^{-p})$$

for any $p > 1/2$. Given an $\epsilon > 0$, we can find a finite subset $E_\epsilon \subset \mathbb{N}$ such that

$$\sup_{t>0}\left(t^p \sum_{n\notin E_\epsilon:\eta_n(V)>t} g_0(t_{n+1})\right) \le \limsup_{t>0}\left(t^p \sum_{n\in\mathbb{N}_0:\eta_n(V)>t} g_0(t_{n+1})\right) + \epsilon.$$

Let χ_ϵ stand for the characteristic function of the set $\bigcup_{n\in E_\epsilon}(t_n, t_{n+1})$. Take $V_\epsilon(x) = V(x)\chi_\epsilon(|x|)$ and $V^\epsilon(x) = V(x) - V_\epsilon(x)$, then $\mathbf{A}_V^\circ = \mathbf{A}_{V_\epsilon}^\circ + \mathbf{A}_{V^\epsilon}^\circ$ and $n(s; \mathbf{A}_{V_\epsilon}^\circ) = o(s^{-p})$. It follows from (4.11) that

$$\limsup_{s\to 0} s^p n(s; \mathbf{A}_V^\circ) = \limsup_{s\to 0} s^p n(s; \mathbf{A}_{V^\epsilon}^\circ)$$

$$\leq \sup_{t>0}\left(t^p \sum_{\eta_n(V^\epsilon)>t} g_0(t_{n+1})\right) \leq \limsup_{t>0}\left(t^p \sum_{n\in\mathbb{N}_0:\eta_n(V)>t} g_0(t_{n+1})\right) + \epsilon.$$

Since $\epsilon > 0$ is arbitrary, (4.12) is justified.

5.3. Proof of Theorem 4.4. The estimate (4.16) for $1/2 < p < 1$ is a part of Theorem 4.3 above. Now we prove that if $\boldsymbol{\eta}(V) \in \ell_\infty$, then the operator $\mathbf{A}_V^\circ$ is bounded, with the estimate

$$\|\mathbf{A}_V^\circ\| \leq C \sup_n \eta_n(v).$$

To check this, it is sufficient to show that $V(x)$ is a Hardy weight on $H^\circ(\Gamma)$. Suppose first that $d < 2$, so that the tree is recurrent. The condition (2.4) is equivalent to

$$\int_t^\infty v(s)s^{d-1}ds \leq Ct^{d-2}, \qquad \forall t > 0.$$

Since the numbers t_n grow as a geometric progression, it is sufficient to check this inequality for the values $t = t_n,\ n \in \mathbb{N}$. We have

$$\int_{t_n}^\infty v(s)s^{d-1}\,ds = \sum_{j=n}^\infty \int_{t_j}^{t_{j+1}} v(s)s^{d-1}\,ds \leq \sum_{j=n}^\infty t_{j+1}^{d-2}\eta_j(V) \leq \sup_j\left(\eta_j(V)\sum_{j=n}^\infty t_{j+1}^{d-2}\right).$$

By (4.15), the series in the last expression converges and its sum is controlled by t_n^{d-2}. This gives the desired result in the recurrent case. For the transient case $d > 2$ the argument is similar.

The estimates (4.16) and (4.18) follow from the estimates for $p = 1/2$ and $p = \infty$ by the same interpolation argument as the one used in [**8**] for the proof of Theorem 4.1. For interpolation, it is necessary to extend the definition of the operator $\mathbf{A}_V^\circ$ to the case where $V(x)$ is an arbitrary measurable function on Γ such that $|V(x)|$ is a Hardy weight. This extension is evident and we give no detail; cf. Section 3.1 in [**8**].

We interpolate the mapping $\Pi : v \mapsto \mathbf{A}_V^\circ,\ V(x) = v(|x|)$. In both cases, one of the basic results for interpolation is the one for $p = 1/2$, that is, the estimate

$$\|\mathbf{A}_V^\circ\|_{1/2,w} \leq C\|\boldsymbol{\eta}(V)\|_{\ell_{1/2}}.$$

Recall that this is equivalent to (4.9). Under the assumptions of Theorem 4.4, due to the special character of the tree and the function V, another basic result concerns the case $p = \infty$, while for the general trees and arbitrary potentials we could use only Theorem 3.3 in [**8**], which corresponds to the case $p = 1$. This argument justifies statement 1° of Theorem 4.4.

For justifying statement 2°, we use Theorem 4.6 and Remark to Corollary 4.7 in [**8**]. Note that in this theorem the properties of the operators $\mathbf{A}_V$ on $\mathcal{H}^{1,0}(\Gamma)$ were discussed. However, the construction in the proof of Theorem 4.6, suggested in [**8**], uses only functions with bounded support, and hence the result applies also to the operators $\mathbf{A}_V^\circ$. Note also that Corollary 4.7 in [**8**] summarizes the results of

Theorems 4.1 and 4.6 of that paper. For the b-regular trees the first of these results extends to all $p > 1/2$, and this automatically leads to extension of Corollary 4.7 to all such p.

6. Neumann boundary condition at the root

Here we discuss the changes in the above results, appearing if in (2.1) we replace the Dirichlet Laplacian $\Delta_{\mathcal{D}}$ by the Neumann Laplacian $\Delta_{\mathcal{N}}$. In order to indicate the difference between these two cases, below we denote the corresponding operators by $\mathbf{H}_{\alpha V;\mathcal{D}}$ and $\mathbf{H}_{\alpha V;\mathcal{N}}$. It immediately follows from the variational principle that

$$N_-(\mathbf{H}_{\alpha V;\mathcal{D}}) \le N_-(\mathbf{H}_{\alpha V;\mathcal{N}}) \le N_-(\mathbf{H}_{\alpha V;\mathcal{D}}) + 1,$$

so that both quantities behave in a similar way as $\alpha \to \infty$. The behavior in the weak coupling limit (i.e., as $\alpha \to 0$) may differ. Indeed, if the tree Γ is recurrent, then for any non-trivial potential $V \ge 0$ the operator $\mathbf{H}_{\alpha V;\mathcal{N}}$ has at least one negative eigenvalue. Hence, no estimate homogeneous in α is possible for $N_-(\mathbf{H}_{\alpha V;\mathcal{N}})$ in the recurrent case.

Such estimates of $N_-(\mathbf{H}_{\alpha V;\mathcal{N}})$ are possible for the transient trees. They are based upon an analogue of Proposition 5.1 for the $\mathcal{N}$-case. In order to define an appropriate substitute for the operator $\mathbf{A}_V^\circ$, we need one more version of the homogeneous Sobolev space on Γ.

Let Γ be a regular transient tree, and let $u \in H^1(\Gamma)$ be a function with bounded support. Suppose also that u is symmetric, that is, $u(x) = \phi(|x|)$, where $\phi(t)$ is some function on $[0, \infty)$. Then

$$u(o) = -\int_0^\infty \phi'(t)\,dt = -\int_0^\infty \phi'(t)\sqrt{g_0(t)}\frac{dt}{\sqrt{g_0(t)}},$$

whence

$$|u(o)|^2 \le l(\Gamma)\int_0^\infty |\phi'(t)|^2 g_0(t)\,dt = l(\Gamma)\int_\Gamma |u'|^2\,dx. \tag{6.1}$$

Here $l(\Gamma)$ is the reduced height of the tree Γ; see (1.3). The resulting inequality extends to all $u \in H^1(\Gamma)$ with bounded support, since for the symmetric component u_{sym} of u we always have $\|u'_{sym}\| \le \|u'\|$ by [**12**], Theorem 3.2 (applied to the function u').

Define the space $\mathcal{H}(\Gamma)$ as the completion of $H^1(\Gamma)$ in the norm

$$\|u\|_{\mathcal{H}} := \|u'\|_{L_2(\Gamma)}.$$

The inequality (6.1) extends by continuity to all $u \in \mathcal{H}(\Gamma)$, and it shows that $\mathcal{H}(\Gamma)$ can be regarded as a function space on Γ. This is no more true if the tree is recurrent: then the elements of the completion are the classes of functions, differing by an arbitrary constant.

Below we assume that Γ is a transient regular tree. We have shown that then the linear functional $\gamma : u \mapsto u(o)$ on the space $\mathcal{H}(\Gamma)$ is continuous, and

$$\|\gamma\| \le \sqrt{l(\Gamma)}. \tag{6.2}$$

The space $\mathcal{H}^\circ(\Gamma)$, introduced in Subsection 1.3, can be realized as the subspace $\{u \in \mathcal{H}(\Gamma) : \gamma u = 0\}$. Clearly,

$$\dim \mathcal{H}(\Gamma)/\mathcal{H}^\circ(\Gamma) = 1.$$

By the Riesz theorem, there is a unique function $h \in \mathcal{H}(\Gamma)$ such that

$$u(o) = (u, h)_{\mathcal{H}(\Gamma)} = \int_\Gamma u'\overline{h'}\, dx, \qquad \forall u \in \mathcal{H}(\Gamma).$$

It is easy to see that this function is symmetric, and is given by

$$h(x) = h_0(|x|), \quad \text{where} \quad h_0(t) = \int_t^\infty \frac{ds}{g_0(s)}.$$

Actually, $h(t)$ is the harmonic function on Γ, vanishing at infinity and such that $h(o) = h_0(0) = l(\Gamma)$. Its norm in $\mathcal{H}(\Gamma)$ is given by

$$\|h\|^2_{\mathcal{H}} = \int_\Gamma |h'(x)|^2\, dx = \int_0^\infty h_0'(t)^2 g_0(t)\, dt = l(\Gamma).$$

This shows that in (6.2) we actually have equality.

Let a function $V(x) \ge 0$ on a transient regular tree Γ be such that the quadratic form (5.1) is bounded in the space $\mathcal{H}(\Gamma)$. Then it defines a bounded operator in this space, say, $\widetilde{\mathbf{A}}_V$. It also defines the bounded operator $\mathbf{A}_V^\circ$ in its subspace $\mathcal{H}^\circ(\mathfrak{H})$; see Subsection 5.1. Our next goal is to compare their (quasi-)norms in various spaces of operators.

THEOREM 6.1. *Let Γ be a transient regular tree, and let $V \in L_1(\Gamma)$, $V \ge 0$. Suppose that the operator $\mathbf{A}_V^\circ$ belongs to some space $\mathcal{C}$, where $\mathcal{C} = \mathcal{C}_p, \mathcal{C}_{p,w}$, or $\mathcal{C}^\circ_{p,w}$. Then the operator $\widetilde{\mathbf{A}}_V$ belongs to the same space, and, moreover,*

$$\|\widetilde{\mathbf{A}}_V\|_{\mathcal{C}} \le C\left(\|\mathbf{A}_V^\circ\|_{\mathcal{C}} + t_1 \int_{e_0} V dx\right), \tag{6.3}$$

where e_0 stands for the edge of Γ emanating from the root, and the constant C depends only on the reduced height $l(\Gamma)$ and on the length of e_0 (that is, on the number t_1).

PROOF. Define a function ϕ on Γ in the following way. Let $\phi(x) = 0$ everywhere outside the edge e_0. We identify e_0 with the segment $[0, t_1]$ and take $\phi(x) = t_1^{-1}(t_1 - x)$, $x \in e_0$. Given a function $u \in \mathcal{H}^1(\Gamma)$, we decompose it as

$$u(x) = u_0(x) + u(o)\phi(x);$$

then $u_0 \in \mathcal{H}^\circ(\Gamma)$. We have

$$\begin{aligned} \int_\Gamma V(x)|u(x)|^2\, dx &\le 2\left(\int_\Gamma V(x)|u_0(x)|^2\, dx + |u(o)|^2 \int_{e_0} V(x)|\phi(x)|^2\, dx\right) \\ &\le 2\left(\int_\Gamma V(x)|u_0(x)|^2\, dx + |u(o)|^2 \int_{e_0} V(x)\, dx\right). \end{aligned} \tag{6.4}$$

Furthermore, we have $\|\phi'\|^2 = t_1^{-1}$, and hence,

$$\|u_0'\| \le \|u'\| + |u(o)|\, \|\phi'\| \le \|u'\|\left(1 + \sqrt{l(G)t_1^{-1}}\right).$$

It follows that

$$\|u_0'\|^2 + |u(o)|^2\|\phi'\|^2 \le \|u'\|^2(2 + 3l(\Gamma)t_1^{-1}). \tag{6.5}$$

Inequalities (6.4) and (6.5) show that the (quasi-)norm $\|\widetilde{\mathbf{A}}_V\|_{\mathcal{C}}$ is controlled by the sum of two terms. The first is $\|\mathbf{A}_V^\circ\|_{\mathcal{C}}$ and the second is the (quasi-)norm in $\mathcal{C}$ of the operator of multiplication by the number $M = \int_{e_0} V\, dx\, \|\phi'\|^{-2} = t_1 \int_{e_0} V\, dx = \eta_0(V)$, acting in the one-dimensional space generated by the function $\phi(x)$. This results in (6.3). □

An analogue of Proposition 5.1 for the N-case shows that

$$N_-(\mathbf{H}_{\alpha V;\mathcal{N}}) = n(\alpha^{-1}; \widetilde{\mathbf{A}}_V).$$

As a result of this equality and Theorem 6.1, we come to the following general conclusions.

THEOREM 6.2. *Let Γ be a regular, transient metric tree and let $V \in L_1(\Gamma)$, $V \geq 0$. Then*

$1°$ *Each estimate* (4.1), (4.4), (4.6), (4.8), (4.11), (4.13), (4.16), (4.18), *with an additional term $C\alpha^p t_1 \int_{e_0} V dx$ on the right-hand side, holds for the operator $\mathbf{H}_{\alpha V;\mathcal{N}}$. If $p = 1/2$, the same is true for the estimate* (4.9), *and the asymptotic formula* (4.10) *remains valid.*

$2°$ *Inequalities* (4.2), (4.7), (4.12), *and* (4.17) *remain valid for the operator $\mathbf{H}_{\alpha V;\mathcal{N}}$. In particular, the finiteness of the right-hand sides in* (4.4), (4.8), (4.13), *and* (4.18) *implies* (4.3).

$3°$ *Statement $2°$ of Theorem 4.4 remains valid for the operator $\mathbf{H}_{\alpha V;\mathcal{N}}$.*

We also present a simple but useful result which follows from Theorem 6.1 and Theorem 3.3 in [**8**].

THEOREM 6.3. *Let Γ be a transient regular tree, and let $V \geq 0$ be a measurable function on Γ. The operator $\widetilde{\mathbf{A}}_V$ is trace class if and only if $\int_\Gamma |x|V(x)\,dx < \infty$, and*

$$\|\widetilde{\mathbf{A}}_V\|_1 = \operatorname{Tr} \widetilde{\mathbf{A}}_V \leq C\left(\int_\Gamma |x|V(x)\,dx + t_1 \int_{e_0} V\,dx\right), \tag{6.6}$$

where C is the constant from (6.3).

It follows from this theorem that under its assumptions we have

$$N_-(\mathbf{H}_{\alpha V;\mathcal{N}}) \leq C\alpha\left(\int_\Gamma |x|V(x)\,dx + t_1 \int_{e_0} V\,dx\right), \tag{6.7}$$

and also $N_-(\mathbf{H}_{\alpha V;\mathcal{N}}) = o(\alpha)$.

7. Concluding remarks

7.1. On the sharpness of Theorem 4.2. Below we present an example in which both Theorems 4.2 and 4.4 apply. Then, based upon statement $2°$ of the latter, we derive the sharpness of the first theorem.

EXAMPLE 7.1. Let Γ be a b-regular tree with some $b > 1$, and let $t_n = b^{\frac{n}{d-1}}$. Then Γ has global dimension d, and we assume that $d > 2$. On Γ we consider the symmetric potential $V(x) = v(|x|)$, where

$$v(t) = t_{n+1}^{-2} b^{-n/p}, \qquad t_n < t \leq t_{n+1}.$$

Then

$$\eta_n(V) = t_{n+1}^{-2}(t_{n+1} - t_n)b^{-n/p} = cb^{-n/p}, \qquad c = 1 - b^{-1/(d-1)}.$$

Let us check the assumptions of Theorem 4.4, $1°$. For small $t > 0$ the sum

$$\sum_{cb^{-n/p} > t} b^n = \sum_{b^n < (t/c)^{-p}} b^n$$

has the order $O(t^{-p})$ but not $o(t^{-p})$. It follows from Theorem 4.4 that $N_-(\mathbf{H}_{\alpha V}) = O(\alpha^p)$, where O cannot be replaced by o.

Now, assuming $p > 1$, we apply Theorem 4.2 to the same potential. A simple calculation shows that the supremum on the right-hand side of (4.6) is finite, and hence, $N_-(\mathbf{H}_{\alpha V}) = O(\alpha^p)$. The reference to Theorem 4.4 shows that the estimate is sharp. As a consequence, we see that the estimate (4.7) in Theorem 4.2 is order-sharp.

7.2. Comparison with the results of [5]. We do not discuss here the general Theorem 2.4 in [**5**] and restrict ourselves to comparison between its Corollary 2.6 and our Theorem 4.2 — more exactly, its analogue for the operator $\mathbf{H}_{\alpha V;\mathcal{N}}$. For the symmetric potentials and $p > 1$, the corollary (where one should take $a = 2p - 1$) is equivalent to our estimate (4.4), within the value of the constant factor. The authors of [**5**] overlooked the possibility to extend their result to the general (non-symmetric) potentials. For $p = 1$ the result is covered by our Theorem 6.3. They also did not discuss the important problem of the sharpness of the estimate in the strong coupling limit, and they have nothing similar to our inequalities (4.6) and (4.7), where the weak L_p-spaces are involved.

The results of [**5**] concern only the case $p \geq 1$, while our Theorem 4.3 covers the case $1/2 \leq p < 1$, and for the b-regular trees Theorem 4.4 gives a necessary and sufficient conditions for $N_-(\mathbf{H}_{\alpha V}) = O(\alpha^p)$ with an arbitrary $p > 1/2$.

We would like to emphasize that this criticism does not concern other parts of the paper [**5**], including the material on the so-called Lieb–Thirring estimates on trees.

References

[1] J. Bergh and J. Löfström, *Interpolation spaces. An introduction*, Springer-Verlag, Berlin, 1976.

[2] M. Sh. Birman and M. Solomyak, *Estimates for the number of negative eigenvalues of the Schrödinger operator and its generalizations*, Estimates and asymptotics for discrete spectra of integral and differential equations (Leningrad, 1989–90), pp. 1–55, Adv. Soviet Math., vol. 7, Amer. Math. Soc., Providence, RI, 1991.

[3] ———, *Schrödinger operator. Estimates for number of bound states as function-theoretical problem*, Spectral theory of operators (Novgorod, 1989), pp. 1–54, Amer. Math. Soc. Transl. Ser. 2, vol. 150, Amer. Math. Soc., Providence, RI, 1992.

[4] R. Carlson, *Nonclassical Sturm–Liouville problems and Schrödinger operators on radial trees*, Electron. J. Differential Equations 2000, no. 71, 24 pp. (electronic).

[5] T. Ekholm, R. Frank, and H. Kovařík, *Eigenvalue estimates for Schrödinger operators on metric trees*, arXive:0710.5500.

[6] W. D. Evans, D. J. Harris, and L. Pick, *Weighted Hardy and Poincaré inequalities on trees*, J. London. Math. Soc. (2) **52** (1995), 121–136.

[7] V. G. Maz'ya, *Sobolev spaces*, Izdat. Leningr. Univ., Leningrad, 1985; English transl., Springer-Verlag, Berlin, 1985.

[8] K. Naimark and M. Solomyak, *Eigenvalue estimates for the weighted Laplacian on metric trees*, Proc. London Math. Soc. (3) **80** (2000), 690–724.

[9] ———, *Geometry of Sobolev spaces on regular trees and the Hardy inequalities*, Russ. J. Math. Phys. **8** (2001), no. 3, 322–335.

[10] B. Opic and A. Kufner, *Hardy-type inequalities*, Pitman Research Notes in Mathematics, **219**, Longman Scientific & Technical, Harlow, 1990.

[11] G. Rozenblum and M. Solomyak, *Counting Schrödinger boundstates: semiclassics and beyond*, arXive:0803.3138.

[12] M. Solomyak, *On the spectrum of the Laplacian on regular metric trees*, Waves in Random Media **14** (2004), 155–171.

Department of Mathematics, The Weizmann Institute of Science, Rehovot 76100, Israel

E-mail address: michail.solomyak@weizmann.ac.il

Amer. Math. Soc. Transl.
(2) Vol. **226**, 2009

H^1 and Dyadic H^1

Sergei Treil

To Victor Petrovich Havin

ABSTRACT. In this paper we give a simple proof of the fact that the average over all dyadic lattices of the dyadic H^1-norm of a function gives an equivalent H^1-norm. The proof we present works for both one-parameter and multi-parameter Hardy spaces.

The results of such type are known; cf. [**2**] for the one-parameter case. Also, by duality, such results are equivalent to the "BMO from dyadic BMO" statements proved in [**4**] (one-parameter case) and [**8**] (two-parameter case).

While the paper generalizes these results to the multi-parameter setting, this is not its main goal. The purpose of the paper is to present an approach leading to a simple proof, which works in both one-parameter and multi-parameter cases.

The main idea of treating the square function as a Calderón–Zygmund operator is a commonplace in harmonic analysis; the main observation, on which the paper is based, is that one can treat the random dyadic square function this way. After that, all is proved by using the standard and well-known results about Calderón–Zygmund operators in the Hilbert-space-valued setting.

As an added bonus, we get a simple proof of the (equivalent by duality) inclusion BMO $\subset$ BMO$_\mathrm{d}$, $H^1_\mathrm{d} \subset H^1$ in the multi-parameter case. Note that, unlike the one-parameter case, the inclusions in the general situation are far from trivial.

Notation

$\mathcal{H} \otimes \mathcal{K}$	tensor product of Hilbert spaces $\mathcal{H}$ and $\mathcal{K}$; we consider it endowed with the "canonical" norm, which makes $\mathcal{H} \otimes \mathcal{K}$ a Hilbert space. If one of the spaces is a function space, for example if $\mathcal{H} = L^2$, then $\mathcal{H} \otimes \mathcal{K}$ can be interpreted as L^2 with values in $\mathcal{K}$.
$L^p(X; \mathcal{H})$	L^p-space of functions on X with values in $\mathcal{H}$; X is usually $\mathbb{R}^N$ with Lebesgue measure. The symbols X and/or $\mathcal{H}$ can be omitted if they are clear from the context.

2000 *Mathematics Subject Classification.* Primary 42B30, 42B20; Secondary 42B35.

Key words and phrases. H^1, dyadic H^1, dyadic BMO, random dyadic lattice, square function, Calderón–Zygmund operator.

This research is partially supported by the NSF grant DMS-0800876.

dx	Lebesgue measure in $\mathbb{R}^N$ if $x \in \mathbb{R}^N$.
$\mathcal{D}$	Dyadic lattice in $\mathbb{R}^N$; see below.
H^1, $H^1_{\mathcal{D}}$	real variable Hardy space on $\mathbb{R}^N$ and its dyadic counterpart, associated with a dyadic lattice $\mathcal{D}$. We will also consider Hilbert-space-valued spaces, and the notation like $H^1(X;\mathcal{H})$ will be used.
$\ell(Q)$, $\lvert Q\rvert$	sidelength and volume of a cube $Q \subset \mathbb{R}^N$.
$\mathbb{E}\xi$	expectation of a random variable ξ, $\mathbb{E}\xi = \int_\Omega \xi(\omega)dP(\omega)$; sometimes, to distinguish the random variable in a formula, the notation $\mathbb{E}\xi(\omega)$ or $\mathbb{E}_\omega \xi(\omega)$ will be used instead of $\mathbb{E}\xi$.

Cubes and dyadic lattices. Throughout the paper we will often speak about dyadic cubes and dyadic lattices, so let us first fix some terminology. A *cube* in $\mathbb{R}^N$ is an object obtained from the *standard cube* $[0,1)^N$ by dilations and shifts.

For a cube Q we denote by $\ell(Q)$ its *size*, i.e., the length of its side. Given a cube Q one can split it by dividing each side in halves into 2^N cubes Q_k of size $\ell(Q)/2$: we will call such cubes Q_k the *children of* Q.

For a cube Q and $\lambda > 0$ we denote by λQ the cube Q dilated λ times with respect to its center.

Now, let us define the *standard dyadic lattice* $\mathcal{D}_0$: for each $k \in \mathbb{Z}$, consider the cube $[0, 2^k)^N$ and all its shifts by elements of $\mathbb{R}^N$ with coordinates of the form $j \cdot 2^k$, $j \in \mathbb{Z}$. The collection of all such cubes (union over all k) is called the *standard dyadic lattice.*

A *dyadic lattice* $\mathcal{D}$ is just a shift of the standard dyadic lattice $\mathcal{D}_0$. A collection of all cubes from a dyadic lattice $\mathcal{D}$ of a fixed size 2^k is called a *layer* of the lattice.

Random dyadic lattice. Our random lattice will contain the dyadic cubes of standard size 2^k ($k \in \mathbb{Z}$), but will be "randomly shifted" with respect to the standard dyadic lattice $\mathcal{D}_0$. The simplest idea would be to pick up a random variable x uniformly distributed over $\mathbb{R}^N$ and to define the random lattice as $x + \mathcal{D}_0$. This idea works for the torus $\mathbb{T}^N$, but unfortunately, there exists no such random variable x in $\mathbb{R}^N$, so we have to act in a little more sophisticated way.

We construct a random lattice of dyadic intervals on the real line $\mathbb{R}$, and then define a random lattice in $\mathbb{R}^N$ as the product of the lattices of intervals.

Let Ω_1 be a probability space and let $x(\omega)$ be a random variable uniformly distributed over the interval $[0, 1)$.

Let $\xi_j(\omega)$ be random variables satisfying $\mathbb{P}\{\xi_j = +1\} = \mathbb{P}\{\xi_j = -1\} = 1/2$. Assume also that $x(\omega)$, $\xi_j(\omega)$, $j \in \mathbb{N}$ are independent. Define the random lattice $\mathcal{D}(\omega)$ as follows:

(1) We require that $I_0(\omega) := [x(\omega)-1, x(\omega)] \in \mathcal{D}(\omega)$; this gives us all intervals in $\mathcal{D}(\omega)$ of length 2^k, $k \le 0$.

(2) To determine the rest of the intervals, it is enough to know dyadic intervals $I_k(\omega) \supset I_0(\omega)$, of length 2^k, $k \ge 0$. The intervals $I_k(\omega)$ are determined inductively: if $I_{k-1}(\omega) \in \mathcal{D}(\omega)$ is already known (and thus all intervals of length 2^{k-1} in $\mathcal{D}(\omega)$), then $I_k(\omega)$ is the union of $I_{k-1}(\omega)$ and its right neighbor if $\xi_k(\omega) = +1$, and $I_k(\omega)$ is the union of $I_{k-1}(\omega)$ and its left neighbor if $\xi_k(\omega) = -1$.

To get a random dyadic lattice in $\mathbb{R}^N$ we just take N independent random dyadic lattices $\mathcal{D}_1, \mathcal{D}_2, \dots, \mathcal{D}_N$ in $\mathbb{R}$ and consider all cubes $Q = I_1 \times I_2 \times \cdots \times I_N$, $I_k \in \mathcal{D}_k$.

1. Introduction and main results

1.1. One-parameter case. Let $H^1 = H^1(\mathbb{R}^N)$ be the usual *real variable* Hardy space on $\mathbb{R}^N$, and let $H^1_{\mathcal{D}}$ be its dyadic counterpart, defined as follows.

Consider a dyadic lattice $\mathcal{D}$ in $\mathbb{R}^N$. Let $E_k = E_k^{\mathcal{D}}$ be the averaging operator over cubes $Q \in \mathcal{D}$ of size 2^k, $E_k f(x) = |Q|^{-1} \int_Q f$, where Q is the cube in $\mathcal{D}$ of size 2^k containing x. If $Q \in \mathcal{D}$ is a cube of size 2^k, define E_Q by $E_Q f = (|Q|^{-1} \int_Q f)\mathbf{1}_Q = \mathbf{1}_Q(E_k f)$.

Define the martingale differences $\Delta_k = \Delta_k^{\mathcal{D}} := E_{k-1}^{\mathcal{D}} - E_k^{\mathcal{D}}$, and again for a cube $Q \in \mathcal{D}$ of size 2^k, define Δ_Q by $\Delta_Q f = \mathbf{1}_Q(E_k f)$.

Define the dyadic square function $S = S_{\mathcal{D}}$ by

$$(1.1) \qquad (S_{\mathcal{D}} f)(x) = \left(\sum_{k \in \mathbb{Z}} \left|\Delta_k^{\mathcal{D}} f(x)\right|^2 \right)^{1/2} = \left(\sum_{Q \in \mathcal{D}, Q \ni x} |\Delta_Q f(x)|^2 \right)^{1/2}.$$

One can also consider a slightly different square function $\widetilde{S} f = \widetilde{S}_{\mathcal{D}} f$,

$$(1.2) \quad (\widetilde{S}_{\mathcal{D}} f)(x) = \left(\sum_{k \in \mathbb{Z}} \left(E_k^{\mathcal{D}} \left|\Delta_k^{\mathcal{D}} f\right|^2 \right)(x) \right)^{1/2} = \left(\sum_{Q \in \mathcal{D},\, Q \ni x} \left(E_Q^{\mathcal{D}} \left|\Delta_Q^{\mathcal{D}} f\right|^2 \right)(x) \right)^{1/2};$$

in other words, each term in the sum on the right equals the average of $|\Delta_Q f|^2$ if $x \in Q$ and 0 otherwise.

It is not hard to show and will be explained later that $C^{-1}\|Sf\|_1 \le \|\widetilde{S} f\|_1 \le C\|Sf\|_1$, where the constant C depends only on the dimension N.

DEFINITION. A function $f \in L^1_{\text{loc}}$ is in the dyadic Hardy space $H^1_{\mathcal{D}}$ (with respect to the dyadic lattice $\mathcal{D}$) if $\|f\|_{H^1_{\mathcal{D}}} := \|S_{\mathcal{D}} f\|_1 < \infty$.

One can use the square function $\widetilde{S}_{\mathcal{D}} f$ here and get an equivalent norm.

Now we state one of the main results of the paper. Let $\mathcal{D}(\omega)$, $\omega \in \Omega$, be the random dyadic lattice, as described above, and recall that $\mathbb{E} = \mathbb{E}_\omega$ denotes the expectation (average with respect to ω).

THEOREM 1.1. *A function $f \in L^1_{\text{loc}}(\mathbb{R}^n)$ belongs to H^1 if and only if*

$$\int_{\mathbb{R}^N} \left[\mathbb{E}(|S_{\mathcal{D}(\omega)} f(x)|^2)\right]^{1/2} dx < \infty.$$

Moreover, the latter quantity gives an equivalent norm on H^1.

The same result is true with the square function $\widetilde{S}_{\mathcal{D}(\omega)}$ in place of $S_{\mathcal{D}(\omega)}$.

1.2. Multi-parameter case. The above results can be generalized to the case of multi-parameter Hardy spaces. Let $X_k = \mathbb{R}^{N_k}$, $k = 1, 2, \dots, n$, and let $H^1(X_1 \times X_2 \times \cdots \times X_n)$ be the n-parameter Hardy space; see Subsection 3.1 below for the definition.

Define the dyadic Hardy space as follows. Let $\mathcal{D}_k$ be a dyadic lattice on X_k, $k = 1, 2, \dots, n$, and let $\mathcal{D} = \mathcal{D}_1 \times \mathcal{D}_2 \times \cdots \times \mathcal{D}_n$ be the product dyadic lattice on

$X = X_1 \times X_2 \times \cdots \times X_n$; the elements on $\mathcal{D}$ are the "rectangles" (parallelepipeds) $R = Q_1 \times Q_2 \times \cdots \times Q_n$, $Q_k \in \mathcal{D}_k$.

For a multi-index $\mathbf{k} = (k_1, k_2, \ldots, k_n)$ define the average $E_{\mathbf{k}} := E^1_{k_1} E^2_{k_2} \cdots E^n_{k_n}$ on $X = X_1 \times X_2 \times \cdots \times X_n$ and the martingale differences $\Delta_{\mathbf{k}} := \Delta^1_{k_1} \Delta^2_{k_2} \cdots \Delta^n_{k_n}$, where $E^j_{k_j}$ and $\Delta^j_{k_j}$ are the "one-variable" averages E_{k_j}, and the martingale difference Δ_{k_j} as defined in Subsection 1.1 and taken in the variable $x_j \in X_j$.

For a "rectangle" $R = Q_1 \times \cdots \times Q_n$, $Q_k \in \mathcal{D}_k$ define $\Delta_R := \Delta^1_{Q_1} \Delta^2_{Q_2} \cdots \Delta^n_{Q_n}$, where again $\Delta^j_{Q_j}$ is the operator Δ_{Q_j} defined by (1.2) and taken in the variable x_j.

Define the multi-parameter square function $S = S_{\mathcal{D}}$ by

$$(1.3) \qquad (S_{\mathcal{D}} f)(x) = \left(\sum_{\mathbf{k} \in \mathbb{Z}^n} |\Delta_{\mathbf{k}} f(x)|^2 \right)^{1/2} = \left(\sum_{R \in \mathcal{D}, R \ni x} |\Delta_R f(x)|^2 \right)^{1/2}.$$

One can also define the square function $\widetilde{S} = \widetilde{S}_{\mathcal{D}}$:

$$(1.4) \quad \widetilde{S} f(x) := \left(\sum_{k \in \mathbb{Z}} \left(E_k |\Delta_k f|^2 \right)(x) \right)^{1/2} = \left(\sum_{R \in \mathcal{D}, R \ni x} \left(E_R |\Delta_R f|^2 \right)(x) \right)^{1/2}.$$

The definitions look very similar to (1.1), (1.2), only here the sums on the right hand side are taken over all "rectangles" R, while in (1.1), (1.2) they are taken over all cubes.

We use the same notation for the one-parameter and multi-parameter square function, but since we will treat these cases in different sections, we hope to avoid confusion.

DEFINITION. Let $\mathcal{D} = \mathcal{D}_1 \times \mathcal{D}_2 \times \cdots \times \mathcal{D}_n$ be a product dyadic lattice on $X = X_1 \times X_2 \times \cdots \times X_n$. We say that a function $f \in L^1_{\text{loc}}(X)$ belongs to the dyadic Hardy space $H^1_{\mathcal{D}}$ if $\|f\|_{H^1_{\mathcal{D}}} := \|S_{\mathcal{D}} f\|_1 < \infty$.

Now let $\mathcal{D}_1(\omega), \mathcal{D}_2(\omega), \ldots, \mathcal{D}_n(\omega)$ be independent random dyadic lattices on the spaces $X_1, X_2, \ldots, X_n$, respectively, and let $\mathcal{D}(\omega) = \mathcal{D}_1(\omega) \times \mathcal{D}_2(\omega) \times \cdots \times \mathcal{D}_n(\omega)$ be the multi-parameter random dyadic lattice.

THEOREM 1.2. *A function $f \in L^1_{\text{loc}}(\mathbb{R}^n)$ belongs to $H^1(X_1 \times X_2 \times \cdots \times X_n)$ if and only if*

$$\int_{\mathbb{R}^N} \left[\mathbb{E}(|S_{\mathcal{D}(\omega)} f(x)|^2) \right]^{1/2} dx < \infty;$$

here $S_{\mathcal{D}(\omega)}$ is the multi-parameter random dyadic lattice defined above. Moreover, the latter quantity gives an equivalent norm on H^1.

The same result is true with the square function $\widetilde{S}_{\mathcal{D}(\omega)}$ in place of $S_{\mathcal{D}(\omega)}$.

1.3. Some remarks.

REMARK 1.3. It is a simple exercise to show that in both one-parameter and multi-parameter cases

$$\mathbb{E}(|S_{\mathcal{D}(\omega)} f(x)|^2) = \lim_{Q \to \mathbb{R}^N} \frac{1}{|Q|} \int_Q |S_{\mathcal{D}} f(x+u)|^2 \, du,$$

where $\mathcal{D}$ is a fixed dyadic lattice and the limit is taken over cubes (or parallelepipeds) $Q \subset \mathbb{R}^N$ ($Q \subset X$ in the multi-parameter case) centered at 0 whose sidelengths tend to ∞.

REMARK 1.4. Note that for every dyadic lattice $\mathcal{D}$ one has $\|f\|_{H^1} \le C\|S_{\mathcal{D}}f\|_1$; thus by Tonelli's theorem and Hölder's inequality we have for $p \le 2$,

$$\|f\|_{H^1} \le C \int_{\mathbb{R}^N} \left[\mathbb{E}(|S_{\mathcal{D}(\omega)}f(x)|^p)\right]^{1/p} dx \le C \int_{\mathbb{R}^N} \left[\mathbb{E}(|S_{\mathcal{D}(\omega)}f(x)|^2)\right]^{1/2} dx.$$

In the one-parameter case the estimate $\|f\|_{H^1} \le C\|S_{\mathcal{D}}f\|_1$ is trivial and well known. Indeed, H^1-BMO duality ($(H^1)^* = \mathrm{BMO}$, $(H^1_{\mathcal{D}})^* = \mathrm{BMO}_{\mathcal{D}}$; see the definitions in Subsection 4.1) and the trivial inclusion $\mathrm{BMO} \subset \mathrm{BMO}_{\mathcal{D}}$ imply that $H^1_{\mathcal{D}} \subset H^1$ with the desired estimates of the norms.

Since $\mathrm{BMO} \ne \mathrm{BMO}_{\mathcal{D}}$, one can conclude that the inclusion $H^1_{\mathcal{D}} \subset H^1$ is proper.

In the multi-parameter case the estimate $\|f\|_{H^1} \le C\|S_{\mathcal{D}}f\|_1$ is also known to the specialists, but it is less trivial. In fact, the only place that the author is aware of where this is proved is the Ph.D. thesis of J. Pipher; this proof is far from trivial and the calculations are quite tedious. In Subsection 3.3 we present a different, quite simple prove of this fact. This proof is based on the one-parameter (Hilbert-space-valued) H^1-BMO theory.

REMARK 1.5. Applying Hölder's inequality to Theorem 1.1 (to Theorem 1.2 in the multi-parameter case), we see that for $p \le 2$,

$$\int_{\mathbb{R}^N} \left[\mathbb{E}(|S_{\mathcal{D}(\omega)}f(x)|^p)\right]^{1/p} dx \le \int_{\mathbb{R}^N} \left[\mathbb{E}(|S_{\mathcal{D}(\omega)}f(x)|^2)\right]^{1/2} dx \le C\|f\|_{H^1}. \tag{1.5}$$

If $p = 1$, Tonelli's theorem and (1.5) imply that

$$\mathbb{E}\|S_{\mathcal{D}(\omega)}f\|_1 \le C\|f\|_{H^1}. \tag{1.6}$$

This was proved in the one-parameter case in [**2**]. Note that, by duality, (1.6) is equivalent to "BMO from dyadic BMO" statements: if f_ω, $\omega \in \Omega$, is a measurable family of functions, $f_\omega \in \mathrm{BMO}_{\mathcal{D}(\omega)}$, $\|f_\omega\|_{\mathrm{BMO}_{\mathcal{D}(\omega)}} \le 1$, then for f defined by $f(x) = \mathbb{E}_\omega f_\omega(x)$ one has $f \in \mathrm{BMO}$, $\|f\|_{\mathrm{BMO}} \le C$.

This "BMO from dyadic BMO" result was proved directly in [**4**] (the one-parameter case) and in the recent paper [**8**] (the two-parameter case).

2. Proof of Theorem 1.1

2.1. "Vectorization" of the square function. There is a standard way of converting the nonlinear operator $S_{\mathcal{D}}$ into a linear one by treating $S_{\mathcal{D}}f$ as a vector-valued function.

Namely, consider the space $L^1(\ell^2)$ of functions on $\mathbb{Z} \times \mathbb{R}^N$ such that

$$\|f\|_{L^1(\ell^2)} := \int_{\mathbb{R}^N} \left(\sum_{k\in\mathbb{Z}} |f(k,x)|^2 \right)^{1/2} dx < \infty.$$

Define the vector-valued square function $\mathbf{S}_{\mathcal{D}}f$ by

$$\mathbf{S}_{\mathcal{D}}f(k,x) := \Delta_k^{\mathcal{D}}f(x), \qquad k \in \mathbb{Z}, \quad x \in \mathbb{R}^N$$

($\mathbf{S}_{\mathcal{D}}f$ is a function on $\mathbb{Z} \times \mathbb{R}^N$). We will treat this function as a function of the argument $x \in \mathbb{R}^N$ with values in $\ell^2 = \ell^2(\mathbb{Z})$.

Clearly, $\|\mathbf{S}_{\mathcal{D}}f(\,\cdot\,,x)\|_{\ell^2} = S_{\mathcal{D}}f(x)$, so $f \in H^1_{\mathcal{D}}$ if and only if $\mathbf{S}f \in L^1(\ell^2)$. Moreover, $\|\mathbf{S}_{\mathcal{D}}f\|_{L^1(\ell^2)} = \|S_{\mathcal{D}}f\|_1 = \|f\|_{H^1_{\mathcal{D}}}$.

Now let $\mathcal{D}(\omega)$ be a random dyadic lattice, and let $\Omega, \mathbb{P}$ be the corresponding probability space. Consider the space $\mathcal{L} = L^1(\ell^2 \otimes L^2(\Omega, \mathbb{P}))$,

$$\|f\|_{\mathcal{L}} := \int_{\mathbb{R}^N} \left(\int_\Omega \sum_{k \in \mathbb{Z}} |f(k, \omega, x)|^2 \, d\mathbb{P}(\omega) \right)^{1/2} dx.$$

It is an L^1 space with values in the Hilbert space $\ell^2 \otimes L^2(\Omega, \mathbb{P})$; here again f is treated as function of the argument $x \in \mathbb{R}^N$ with values in $\ell^2 \otimes L^2(\Omega, \mathbb{P})$.

Define the vector-valued square function $\mathbf{S}$ with values in $\ell^2 \otimes L^2(\Omega, \mathbb{P})$ by

$$\mathbf{S}f(k, \omega, x) = \mathbf{S}_{\mathcal{D}(\omega)} f(k, x), \qquad x \in \mathbb{R}^N, \ k \in \mathbb{Z}, \ \omega \in \Omega. \tag{2.1}$$

Here and below we will use the notation $\mathbf{S}(\,\cdot\,, \,\cdot\,, x) =: \mathbf{S}f(x) \in \ell^2 \otimes L^2(\Omega, \mathbb{P})$.

Clearly, $\|\mathbf{S}f(x)\|_{\ell^2 \times L^2(\Omega, \mathbb{P})} = \left[\mathbb{E}(|S_{\mathcal{D}(\omega)} f(x)|^2)\right]^{1/2}$, so

$$\|\mathbf{S}f\|_{\mathcal{L}} = \int_{\mathbb{R}^N} \left[\mathbb{E}(|S_{\mathcal{D}(\omega)} f(x)|^2)\right]^{1/2} dx$$

(the norm in $\mathcal{L}$ was constructed so as to ensure this).

Similar "vectorization" can be performed for the square function $\widetilde{S}$. Namely, fix an ordering of the "children" of a dyadic cube (the same one for all cubes), and for each dyadic cube Q define an operator from the functions constant on the "children" of Q to $\mathbb{C}^{2^N}$ by

$$U_Q \mathbf{1}_{Q_k} = |Q_k|^{1/2} e_k, \qquad k = 1, 2, \ldots, 2^{2^N},$$

where Q_k are the "children" of Q and $\{e_k\}_{k=1}^{2^N}$ is the standard basis in $\mathbb{R}^{2^N}$. Now let $\mathcal{L} := L^1(\ell^2 \otimes L^2(\Omega, \mathbb{P}) \otimes \mathbb{C}^{2^N})$ with the norm

$$\|f\|_{\mathcal{L}} := \int_{\mathbb{R}^N} \left(\int_\Omega \sum_{k \in \mathbb{Z}} \|f(k, \omega, x)\|^2_{\mathbb{C}^{2^N}} \, d\mathbb{P}(\omega) \right)^{1/2} dx.$$

Define the square function $\widetilde{\mathbf{S}}$ by

$$\widetilde{\mathbf{S}} f(k, \omega, x) := U_Q \Delta_Q f \in \mathbb{C}^{2^N},$$

where $Q \in \mathcal{D}(\omega)$ is the cube of size 2^k containing x. From the construction it is clear that $\|\widetilde{\mathbf{S}}f\|_{\mathcal{L}} = \int_{\mathbb{R}^N} \left[\mathbb{E}(|\widetilde{S}_{\mathcal{D}(\omega)} f(x)|^2)\right]^{1/2} dx$.

Now the proof of Theorem 1.1 can be outlined in a few sentences. First, it is not hard to show that $\mathbf{S}$ (or $\widetilde{\mathbf{S}}$) is a Calderón–Zygmund operator whose kernel takes values in the Hilbert space $\ell^2 \otimes L^2(\Omega, \mathbb{P})$. It is a well-known fact that such Calderón–Zygmund operators map H^1 to L^1, so

$$\int_{\mathbb{R}^N} \left[\mathbb{E}(|S_{\mathcal{D}(\omega)} f(x)|^2)\right]^{1/2} dx = \|\mathbf{S}f\|_{\mathcal{L}} \le C\|f\|_{H^1}.$$

As we discussed before in Subsection 1.1, the opposite inequality $\|f\|_{H^1} \le C\|\mathbf{S}f\|_{\mathcal{L}}$ is trivial.

Of course, the classical theory of Calderón–Zygmund operators deals with scalar-valued kernels. But, as is well known to the specialists, all the facts that we need are valid in the case of Hilbert-space-valued kernels too.

However, the blind trust is not expected from the reader: all relevant facts will be presented in what follows.

2.2. S as a Calderón–Zygmund operator. We recall that a classical Calderón–Zygmund kernel on $\mathbb{R}^N$ is a function $K(\cdot,\cdot)$ defined on $\mathbb{R}^N \times \mathbb{R}^N \setminus \{(x,x) : x \in \mathbb{R}^N\}$ and satisfying

(1) $|K(x,y)| \le C|x-y|^{-N}$;

(2) there exists $\delta > 0$ such that, whenever $|y - x_0| \ge 2|x - x_0|$,

$$|K(x,y) - K(x_0,y)|,\ |K(y,x) - K(y,x_0)| \le C\frac{|x-x_0|^\delta}{|y-x_0|^{N+\delta}}.$$

One can also consider operator-valued kernels, $K(x,y) \in B(X,Y)$ for arbitrary Banach spaces X and Y. In this case, $|\cdot|$ on the left hand side should be replaced by the norm in $B(\mathcal{X},\mathcal{Y})$.

2.2.1. *Vector-valued Calderón–Zygmund operators.* A Calderón–Zygmund operator with (operator-valued) Calderón–Zygmund kernel K is a bounded operator $T : L^2(\mathbb{R}^N;\mathcal{X}) \to L^2(\mathbb{R}^N;\mathcal{Y})$ such that for all compactly supported $f \in L^2(\mathbb{R}^N;\mathcal{X})$ and $g \in L^2(\mathbb{R}^N;\mathcal{Y}^*)$ with separated supports,

$$\langle Tf, g\rangle = \iint_{\mathbb{R}^N\times\mathbb{R}^N} \langle K(x,y)f(y), g(y)\rangle\, dy\, dx.$$

Such operators with operator-valued kernels were considered, for example, in [**9**], and it was proved there (see Theorem 1.2 in Chapter 3) that if $\mathcal{X}$ and $\mathcal{Y}$ are Hilbert spaces, then the operator acts from H^1 to L^1, $\|Tf\|_{L^1(\mathbb{R}^N,\mathcal{X})} \le C\|f\|_{H^1(\mathbb{R}^N,\mathcal{Y})}$.

In fact, in [**9**] a much more general situation was considered. The kernel was only assumed to satisfy a weaker version of condition (2), and there was no condition (1). Moreover, it was assumed that T was bounded in some L^p, $1 < p < \infty$, not necessarily for $p = 2$.

Note that condition (1) in some form is required for the proof of $T1$ and Tb theorems, but if one assumes that T is bounded in some L^p, $1 < p < \infty$, condition (2) alone is sufficient for the action from H^1 to L^1.

Also, in [**9**] the theorem was proved for the case where $\mathcal{X}$ and $\mathcal{Y}$ are arbitrary Banach spaces, if one defines H^1 via atomic decomposition.

It is well known that for the case of Hilbert-space-valued functions all the definitions of H^1 (via atomic decompositions, via maximal function, via different square functions, or via Riesz transforms) are equivalent,[1] so one can use the result from [**9**] without worrying about what definition of H^1 is used.

In this paper we consider the case where $\mathcal{X} = \mathbb{C}$ and $\mathcal{Y} = \mathcal{H} = \ell^2 \otimes L^2(\Omega,\mathbb{P})$, so we can say that K takes values in the Hilbert space $\mathcal{H}$.

Operators with such kernels also act naturally from $L^2(\mathbb{R}^N;\mathcal{K})$ to $L^2(\mathbb{R}^N; \mathcal{H}\otimes\mathcal{K})$, where $\mathcal{K}$ is a Hilbert space, and we will need this interpretation later in Section 3. Indeed, with each vector $h \in \mathcal{H}$ we can associate an operator $\mathcal{K} \ni f \mapsto f\otimes h \in \mathcal{K}\otimes\mathcal{H}$ (and the norm of this operator is $\|h\|$). So, if K is an $\mathcal{H}$-valued Calderón–Zygmund kernel, then a Calderón–Zygmund operator on $L^2(\mathbb{R}^N;\mathcal{K})$ is a bounded operator $T : L^2(\mathbb{R}^N;\mathcal{K}) \to L^2(\mathbb{R}^N;\mathcal{H}\otimes\mathcal{K})$ such that

$$\langle Tf, g\rangle_{L^2(\mathbb{R}^N;\mathcal{H}\otimes\mathcal{K})} = \iint_{\mathbb{R}^N\times\mathbb{R}^N} \langle K(x,y)\otimes f(y), g(y)\rangle_{\mathcal{H}\otimes\mathcal{K}}\, dy\, dx$$

for all compactly supported f and g with separated supports.

[1] Unfortunately, the author cannot point to a paper where all such equivalences are proved; but following the proofs for the scalar-valued case, one can see that everything works for the Hilbert-space-valued case as well.

2.2.2. *Why* $\mathbf{S}$ *is a Calderón–Zygmund operator*? To find the kernel K of $\mathbf{S}$ we need to compute $\mathbf{S}\delta_y$, where δ_y is the unit mass at $y \in \mathbb{R}^N$:

$$K(x, y) = \left\{\Delta_k^{\mathcal{D}(\omega)} \delta_y(x)\right\}_{k\in\mathbb{Z},\,\omega\in\Omega} \in \ell^2 \otimes L^2(\Omega, \mathbb{P}).$$

To get that expression rigorously, one needs to approximate δ_y by appropriate bump functions; notice that $\Delta_k^{\mathcal{D}_\omega}\delta_y$ is well defined, i.e., does not depend on the choice of an approximating sequence.

Notice that for a dyadic cube Q, $\Delta_Q\delta_y(x) \neq 0$ only if $x \in Q, y \in Q$. Therefore $\Delta_Q\delta_y(x) = 0$ if $\ell(Q) < |x - y|_\infty$, so $\Delta_k^{\mathcal{D}(\omega)}\delta_y(x) = 0$ if $2^k < |x - y|_\infty$.

Noticing that $\|\Delta_k^{\mathcal{D}_\omega}\delta_y\|_\infty \leq C2^{-kN}$ and summing the geometric series we get property (1) of the Calderón–Zygmund kernels.

To show property (2) notice that $\Delta_k^{\mathcal{D}(\omega)}\delta_y(x) = \Delta_k^{\mathcal{D}(\omega)}\delta_y(x_0)$ if all three points x, x_0, y are in the same cube $Q \in \mathcal{D}(\omega)$, $\ell(Q) = 2^k$, and the points x, x_0 are in the same "child" of Q.

The probability that it fails for a given k can be bounded from above by $C|x - x_0|^N 2^{-kN}$. Using the estimate $\|\Delta_k^{\mathcal{D}_\omega}\delta_y\|_\infty \leq C2^{-kN}$ we conclude that

$$\int_\Omega |\Delta_k^{\mathcal{D}(\omega)}\delta_y(x) - \Delta_k^{\mathcal{D}(\omega)}\delta_y(x_0)|^2 d\mathbb{P}(\omega) \leq C2^{-2kN}|x - x_0|^N 2^{-kN} = C|x - x_0|^N 2^{-3kN}.$$

Let k_0 be the maximum $k \in \mathbb{Z}$ such that $2^k < \min\{|x - y|_\infty, |x_0 - y|_\infty\}$. Then $\Delta_k^{\mathcal{D}_\omega}\delta_y(x) = \Delta_k^{\mathcal{D}_\omega}\delta_y(x_0) = 0$ for $k < k_0$, and so

$$\begin{aligned}\|K(x, y) - K(x_0, y)\|^2 \leq C \sum_{k\geq k_0} 2^{-3kN}|x - x_0|^N &\leq C|x - x_0|^N 2^{-3k_0N} \\ &\leq C|x - x_0|^N |x_0 - y|^{-3N}.\end{aligned}$$

Interchanging x and y and repeating the above reasoning we also get that

$$\|K(y, x) - K(y, x_0)\|^2 \leq C|x - x_0|^N |x_0 - y|^{-3N}.$$

This means condition (2) holds with $\delta = N/2$.

The proof for $\widetilde{\mathbf{S}}$ is absolutely the same.

2.3. A remark about conditions $\mathbf{S1} = 0$, $\mathbf{S^*1} = 0$. The material in this subsection is not needed for the proof of the main results. However, it might be of interest for specialists; one can use it to present a different proof of the main results, without employing the result from [**9**] cited in Subsection 2.2.1 about Calderón–Zygmund operators with operator-valued kernels.

Note that the operator $\mathbf{S}$ introduced above satisfies the conditions $\mathbf{S1} = 0$ and $\mathbf{S^*1} = 0$ (more precisely, the second condition should read $\mathbf{S^*1}e = 0$ for all $e \in \mathcal{H}$), which are well known to everybody familiar with the $T1$-theorem.

If one formally plugs 1 into $\mathbf{S}$ or $\mathbf{S}^*$, the result will be 0. Of course, it is only a formal reasoning, since 1 is not in the domain of $\mathbf{S}$, but any reasonable interpretation of $\mathbf{S1}$ gives the same result. For example, it is not hard to show that

$$\mathbf{S1}_Q \to 0, \quad \mathbf{S^*1}_Q e \to 0 \qquad \text{as } \ell(Q) \to \infty \tag{2.2}$$

uniformly on compact subsets, where cubes Q are centered at 0.

It is also easy to see that $\widetilde{\mathbf{S}}\mathbf{1} = 0$, but unfortunately $\widetilde{\mathbf{S}}^*\mathbf{1} \neq 0$. However, it is easy to modify $\widetilde{\mathbf{S}}$ to make $\widetilde{\mathbf{S}}^*\mathbf{1} = 0$.

Namely, let φ be a function on the cube $[0,1)^N$ taking values ± 1 and such that $\int_Q \varphi\, dx = 0$ and let $\varphi_Q(x) = \varphi((x - x_Q)/\ell(Q))$, where x_Q is the base of Q, i.e., the point in Q with smallest coordinates.

Define the square function $\widetilde{\mathbf{S}}$ by

$$\widetilde{\mathbf{S}} f(k, \omega, x) := \varphi_Q(x) U_Q \Delta_Q f \in \mathbb{C}^{2^N},$$

where $Q \in \mathcal{D}(\omega)$ is the cube of size 2^k containing x.

The function φ_Q in the definition of $\widetilde{\mathbf{S}}$ is introduced to ensure that $\widetilde{\mathbf{S}}^* 1 = 0$. Now it is easy to show that $\widetilde{\mathbf{S}}\mathbf{1} = 0$, $\widetilde{\mathbf{S}}^*\mathbf{1} = 0$ (in the sense of (2.2)).

Calderón–Zygmund operators satisfying $T\mathbf{1} = 0$ and $T^*\mathbf{1} = 0$ map H^1 to H^1. To show that one can consider (for instance) the matrix of such an operator in a wavelet basis. It was shown in [**6**] that under rather mild assumption about the wavelet basis, the coefficient space of H^1 in this basis is the Triebel–Lizorkin space $\dot{\mathbf{f}}_1^{0,2}$; see Appendix (Section 4) for the definition. In [**6**] the scalar-valued case was treated, but one can easily see that everything works for the Hilbert-space-valued case, and one just get the vector-valued space $\dot{\mathbf{f}}_1^{0,2}$. Moreover, in [**5**] the H^1 spaces with values in UMD Banach spaces were characterized in terms of coefficient in the wavelet basis; in the Hilbert-space-valued case the result gives exactly $\dot{\mathbf{f}}_1^{0,2}$ with values in the Hibert space.

Using the standard estimates with Calderón–Zygmund kernels one can see that if a Calderón–Zygmund operator T (even with the operator-valued kernel) satisfies $T\mathbf{1} = 0$, $T^*\mathbf{1} = 0$, then its matrix in the wavelet basis is what is called in [**3**] *almost diagonal* for $\dot{\mathbf{f}}_p^{0,q}$, $1 \le p, q < \infty$; cf. Section 4 below for the definition.

It was shown in [**3**] that *almost diagonal* operators are bounded on all $\dot{\mathbf{f}}_p^{0,q}$, $1 \le p, q < \infty$. Since the almost diagonality is a condition on the magnitude of the entries, the result holds for vector-valued Triebel–Lizorkin spaces. Of course, instead of considering a wavelet basis, one can consider a frame decomposition, given by what is called in [**3**] *φ-transform*; all the estimates will be the same.

3. Proof for the multi-parameter case

Proof of the main result for the multi-parameter case (Theorem 1.2) follows the lines of the proof for the one parameter.

Without loss of generality we can assume that the probability space Ω is represented as a product, $(\Omega, \mathbb{P}) = (\Omega_1 \times \Omega_2 \times \cdots \times \Omega_n, \mathbb{P}_1 \times \mathbb{P}_2 \times \cdots \times \mathbb{P}_n)$ and that the random dyadic grid $\mathcal{D}_k(\omega)$ on X_k depends only on ω_k.

For a dyadic lattice $\mathcal{D} = \mathcal{D}_1 \times \mathcal{D}_2 \times \cdots \times \mathcal{D}_n$ on $X_1 \times X_2 \times \cdots \times X_n$ define the vector-valued square function $\mathbf{S}_\mathcal{D}$, with values in $\ell^2(\mathbb{Z}^n)$,

$$\mathbf{S}_\mathcal{D} f(\mathbf{k}, x) := \Delta_\mathbf{k}^\mathcal{D} f(x), \qquad \mathbf{k} \in \mathbb{Z}^n, \ x \in \mathbb{R}^N.$$

Consider the space $\mathcal{L} = L^1(\ell^2(\mathbb{Z}^n) \otimes L^2(\Omega, \mathbb{P}))$,

$$\|f\|_\mathcal{L} := \int_{\mathbb{R}^N} \left(\int_\Omega \sum_{\mathbf{k} \in \mathbb{Z}^n} |f(\mathbf{k}, \omega, x)|^2 d\mathbb{P}(\omega) \right)^{1/2} dx.$$

It is an L^1 space with values in the Hilbert space $\ell^2(\mathbb{Z}^n) \otimes L^2(\Omega, \mathbb{P})$. Note that this Hilbert space can be decomposed as

$$\ell^2(\mathbb{Z}^n) \otimes L^2(\Omega, \mathbb{P}) = (\ell^2 \otimes L^2(\Omega_1 \mathbb{P}_1)) \otimes (\ell^2 \otimes L^2(\Omega_2 \mathbb{P}_2)) \otimes \cdots \otimes (\ell^2 \otimes L^2(\Omega_n \mathbb{P}_n));$$

here $\ell^2 = \ell^2(\mathbb{Z})$.

Define the vector-valued square function with values in the space $\ell^2(\mathbb{Z}^n) \otimes L^2(\Omega, \mathbb{P})$ by

$$\mathbf{S}f(\mathbf{k}, \omega, x) = \mathbf{S}_{\mathcal{D}(\omega)} f(\mathbf{k}, x), \qquad x \in \mathbb{R}^N, \ \mathbf{k} \in \mathbb{Z}^n, \ \omega \in \Omega.$$

Note that $\mathbf{S}$ can be decomposed as a tensor product $\mathbf{S} = \mathbf{S}_1 \otimes \mathbf{S}_2 \otimes \cdots \otimes \mathbf{S}_n$, where $\mathbf{S}_k$ is the one-parameter square function defined by (2.1) in variables $x_k \in X_k = \mathbb{R}^{N_k}$, $\omega_k \in \Omega_k$.

Clearly, as in the one-parameter case, for the multi-parameter square functions we have $\|\mathbf{S}f\|_1 = \int_X \left[\mathbb{E}_\omega(|S_{\mathcal{D}(\omega)} f(x)|^2)\right]^{1/2} dx$.

Similarly, for the square function $\widetilde{S}$ one can construct its vector version $\widetilde{\mathbf{S}}$ with values in $\mathcal{H} = \mathcal{H}_1 \otimes \mathcal{H}_2 \otimes \cdots \otimes \mathcal{H}_n$, $\mathcal{H}_k = \ell^2 \otimes L^2(\Omega_k, \mathbb{P}_k) \otimes \mathbb{C}^{2^{N_k}}$. Again, it is easy to see that $\|\widetilde{\mathbf{S}} f\|_{L^1(X;\mathcal{H})} = \int_X \left[\mathbb{E}_\omega(|\widetilde{S}_{\mathcal{D}(\omega)} f(x)|^2)\right]^{1/2} dx$.

As was already discussed in Subsection 2.2, the operators $\mathbf{S}_k$ are (Hilbert-space-valued) one-parameter Calderón–Zygmund operators, so $\mathbf{S}$ is the tensor product of such operators. And it must be immediately clear to the experts that such operators map $H^1(X_1 \otimes X_2 \otimes \cdots \otimes X_n)$ to L^1.

One way to see that is to notice that $\mathbf{S}$ is a trivial case of multi-parameter Calderón–Zygmund operators, and according to Theorem 2.2 in [**7**] such operators map $H^1(X_1 \otimes X_2 \otimes \cdots \otimes X_n)$ to L^1. Of course, one needs to use a Hilbert-space-valued version of the theorem, but it is clear to the specialists that the proof from [**7**] works in this case. It is also clear that while Theorem 2.2 in [**7**] is stated for $\mathbb{R} \times \mathbb{R} \times \cdots \times \mathbb{R}$, the proof works for $X_1 \otimes X_2 \otimes \cdots \otimes X_n$, $X_k = \mathbb{R}^{N_k}$.

For the reader who is not well familiar with multi-parameter H^1 spaces we present below an alternative proof, which exploits the tensor product structure of $\mathbf{S} = \mathbf{S}_1 \otimes \mathbf{S}_2 \otimes \cdots \otimes \mathbf{S}_n$; it uses only the theory of one parameter H^1-spaces. Of course, we will need the theory of H^1-spaces with values in a Hilbert space, but we need the vector-valued theory in the above multi-parameter reasoning as well.

There is one more reason for presenting the one-parameter proof below: the reasoning above gives the estimate $\int_X \left[\mathbb{E}_\omega(|S_{\mathcal{D}(\omega)} f(x)|^2)\right]^{1/2} dx \le C\|f\|_{H^1(X)}$. The opposite estimate follows from the inequality $\|f\|_{H^1} \le \int_X |S_{\mathcal{D}} f(x)|\, dx$, which is trivial in the one-parameter case. In the multi-parameter case, the same estimate, while true and known to specialists, requires some work to prove it. The one-parameter approach presented below gives a reasonably simple proof of this estimate.

3.1. Multi-parameter H^1-spaces. Recall (cf. [**10**, III.4.4]) that for the one-parameter Hardy space $H^1(\mathbb{R}^N)$ the norm $\|S^{\mathrm{L}} f\|_1$, where S^{L} is the Lusin square function (aka Lusin Area Integral)

$$S^{\mathrm{L}} f(x) = \int_{\Gamma_x} |\nabla f(y,t)|^2 t^{1-n}\, dy\, dt$$

(here $\Gamma_x := \{(y,t) : y \in \mathbb{R}^N, t \ge 0, |y-x| < t\}$, and $f(y,t)$ is the harmonic extension of f from $\mathbb{R}^N$ to $\mathbb{R}^{N+1}_+ := \mathbb{R}^N \times \mathbb{R}_+ = \{(x,t) : x \in \mathbb{R}^N, t \in \mathbb{R}_+\}$) gives an equivalent norm.

One can consider the *vectorization* $\mathbf{S}^{\mathrm{L}}$ of S^{L} as follows. Let $\Gamma = \Gamma_0$ and define

$$\mathbf{S}^{\mathrm{L}} f(x,y,t) = t^{(1-n)/2} \nabla f(x+y,t), \qquad x \in \mathbb{R}^N, \quad (y,t) \in \Gamma.$$

By construction $\mathbf{S}^{\mathrm{L}} f(x,\,\cdot\,,\,\cdot\,) \in L^2(\Gamma) \otimes \mathbb{C}^2$ and $\|\mathbf{S}^{\mathrm{L}} f(x,\,\cdot\,,\,\cdot\,)\|_{L^2(\Gamma)\otimes\mathbb{C}^2} = S^{\mathrm{L}} f(x)$; therefore $\|\mathbf{S}^{\mathrm{L}} f\|_{L^1(L^2(\Gamma))} = \|S^{\mathrm{L}} f\|_1$.

One can define multi-parameter square functions $\vec{S}^{\mathrm{L}}$ and $\vec{\mathbf{S}}^{\mathrm{L}}$ by

$$\vec{S}^{\mathrm{L}} f(x_1, x_2, \ldots, x_n) := \left[\int_{\Gamma_{x_1}\times\cdots\times\Gamma_{x_n}} |\nabla_1 \cdots \nabla_n f(y_1, t_1, y_2, t_2, \ldots, y_n, t_n)|^2 \right. \\ \left. \times\, t_1^{1-N_1} t_2^{(1-N_2)} \cdots t_n^{1-N_n}\, dy_1\, dt_1 \cdots dy_n\, dt_n \right]^{1/2};$$

here $f(y_1, t_1, \ldots, y_n, t_n)$ is harmonic in each variable (y_k, t_k), $y_k \in \mathbb{R}^{N_k}$, $t_k \in \mathbb{R}_+$ and extends f from $\mathbb{R}^{N_1} \times \cdots \times \mathbb{R}^{N_n}$ to $\mathbb{R}_+^{N_1+1} \times \cdots \times \mathbb{R}_+^{N_n+1}$, and ∇_k is the gradient in the variable (y_k, t_k).

Following [**1**] we say that $f \in H^1(X) = H^1(X_1 \otimes \cdots \otimes X_n)$ if $\vec{S}^{\mathrm{L}} f \in L^1(X)$ and $\|\vec{S}^{\mathrm{L}} f\|_1$ defines one of the possible equivalent norms in $H^1(X)$.

Also we define the vector-valued linear square function $\vec{\mathbf{S}}^{\mathrm{L}}$ as $\vec{\mathbf{S}} f(x) = \mathbf{S}_1^{\mathrm{L}} \otimes \mathbf{S}_2^{\mathrm{L}} \otimes \cdots \otimes \mathbf{S}_n^{\mathrm{L}} f(x) \in \mathcal{H} = \mathcal{H}_1 \otimes \mathcal{H}_2 \otimes \cdots \otimes \mathcal{H}_n$, where $\mathbf{S}_k^{\mathrm{L}}$ is the one-parameter square function defined above, taken in the variable x_k, $\mathcal{H}_k = L^2(\Gamma_k) \otimes \mathbb{C}^2$, and Γ_k is the cone in $\mathbb{R}_+^{N_k+1}$ with the vertex at 0.

Again, by construction, $\|\vec{\mathbf{S}}^{\mathrm{L}} f(x)\|_{\mathcal{H}} = |\vec{S}^{\mathrm{L}} f(x)|$, so $\|\vec{\mathbf{S}}^{\mathrm{L}} f\|_{L^1(X;\mathcal{H})}$ gives the norm in $H^1(X)$.

We will use $\vec{\mathbf{S}}^{\mathrm{L}}$ to define multi-parameter H^1 with values in a Hilbert space $\mathcal{K}$; in this case $\vec{\mathbf{S}}^{\mathrm{L}} f(x) \in \mathcal{H} \times \mathcal{K}$. We introduced such spaces only for notational purposes, so while most of the theory of multi-parameter H^1-spaces can be transfered to the Hilbert-space-valued case, we do not need this.

3.2. Proof of the estimate $\int_X \left[\mathbb{E}_\omega(|S_{\mathcal{D}(\omega)} f(x)|^2)\right]^{1/2} dx \le C\|f\|_{H^1(X)}$. Consider a multi-parameter square function $\widetilde{\mathbf{S}}_2 \otimes \cdots \otimes \widetilde{\mathbf{S}}_n$, where each $\widetilde{\mathbf{S}}_k$, $2 \le k \le n$, is either the one-parameter $\mathbf{S}^{\mathrm{L}}$ or the one-parameter "random" square function $\mathbf{S}$, defined in (2.1), taken in the variable x_k.

Assume that the choice of $\widetilde{\mathbf{S}}_k$ is fixed. For a scalar-valued f the function $\widetilde{\mathbf{S}}_2 \otimes \cdots \otimes \widetilde{\mathbf{S}}_n f$ takes values in $\widetilde{\mathcal{H}}^1 := \widetilde{\mathcal{H}}_2 \otimes \cdots \otimes \widetilde{\mathcal{H}}_n$, where each $\widetilde{\mathcal{H}}_k$ is either $L^2(\Gamma_k) \otimes \mathbb{C}^2$ or $\ell^2 \otimes L^2(\Omega_k, \mathbb{P}_k)$, depending on what square function $\widetilde{\mathbf{S}}_k$ is.

Let $\mathbf{S}_k^{\mathrm{L}}$, $\mathbf{S}_k$ be the Lusin and "random" square functions, taken in the variable x_k, and let $\mathcal{H}_k := L^2(\Gamma_k) \otimes \mathbb{C}^2$ and $\mathcal{H}'_k := \ell^2 \otimes L^2(\Omega_k, \mathbb{P}_k)$ be the corresponding target spaces.

LEMMA 3.1.

$$\int_X \|\mathbf{S}_1 \otimes \widetilde{\mathbf{S}}_2 \otimes \cdots \otimes \widetilde{\mathbf{S}}_n f(x)\|_{\mathcal{H}'_1 \otimes \widetilde{\mathcal{H}}^1}\, dx \le C \int_X \|\mathbf{S}_1^{\mathrm{L}} \otimes \widetilde{\mathbf{S}}_2 \otimes \cdots \otimes \widetilde{\mathbf{S}}_n f(x)\|_{\mathcal{H}_1 \otimes \widetilde{\mathcal{H}}^1}\, dx.$$

Since the tensor products of square functions we consider do not depend on the order (the square functions taken in different variables obviously commute), the above lemma tells us that one can replace a factor $\mathbf{S}_k^{\mathrm{L}}$ by $\mathbf{S}_k$ in $\widetilde{\mathbf{S}}_1 \otimes \widetilde{\mathbf{S}}_2 \otimes \cdots \otimes \widetilde{\mathbf{S}}_n f$ and increase the norm by at most the factor C.

Starting with $\mathbf{S}_1^{\mathrm{L}} \otimes \mathbf{S}_2^{\mathrm{L}} \otimes \cdots \otimes \mathbf{S}_n^{\mathrm{L}} f$ and applying Lemma 3.1 successively to each factor, we get

$$\int_X \|\mathbf{S}_1 \otimes \mathbf{S}_2 \otimes \cdots \otimes \mathbf{S}_n f(x)\|_{\mathcal{H}'}\, dx \le C \int_X \|\mathbf{S}_1^{\mathrm{L}} \otimes \mathbf{S}_2^{\mathrm{L}} \otimes \cdots \otimes \mathbf{S}_n^{\mathrm{L}} f(x)\|_{\mathcal{H}}\, dx,$$

which is exactly the desired estimate (here $\mathcal{H}' = \mathcal{H}'_1 \otimes \mathcal{H}'_2 \otimes \cdots \otimes \mathcal{H}'_n$ and $\mathcal{H} = \mathcal{H}_1 \otimes \mathcal{H}_2 \otimes \cdots \otimes \mathcal{H}_n$).

PROOF OF LEMMA 3.1. We introduce notation $x = (x_1, x^1) \in X = X_1 \times X^1$, where $x^1 = (x_2, x_3, \dots, x_n) \in X^1 := X_2 \times \dots \times X_n$.

Consider the vector-valued function

$$\mathbf{S}_1^{\mathrm{L}} \otimes \widetilde{\mathbf{S}}_2 \otimes \dots \otimes \widetilde{\mathbf{S}}_n f = \mathbf{S}_1^{\mathrm{L}} \otimes (\widetilde{\mathbf{S}}_2 \otimes \dots \otimes \widetilde{\mathbf{S}}_n) f.$$

If

$$\int_X \|\mathbf{S}_1^{\mathrm{L}} \otimes (\widetilde{\mathbf{S}}_2 \otimes \dots \otimes \widetilde{\mathbf{S}}_n) f(x)\|_{\mathcal{H}_1' \otimes \widetilde{\mathcal{H}}^1} \, dx < \infty,$$

we conclude that for almost all x^1

$$\widetilde{\mathbf{S}}_2 \otimes \dots \otimes \widetilde{\mathbf{S}}_n f(\,\cdot\,, x^1) \in H^1(X_1; \widetilde{\mathcal{H}}^1), \tag{3.1}$$

and

$$\begin{aligned} \int_X \|\mathbf{S}_1^{\mathrm{L}} \otimes (\widetilde{\mathbf{S}}_2 \otimes \dots \otimes \widetilde{\mathbf{S}}_n) f(x)\|_{\mathcal{H}_1 \otimes \widetilde{\mathcal{H}}^1} \, dx \\ = \int_{X^1} \|\widetilde{\mathbf{S}}_2 \otimes \dots \otimes \widetilde{\mathbf{S}}_n f(\,\cdot\,, x^1)\|_{H^1(X_1; \widetilde{\mathcal{H}}^1)} \, dx^1. \end{aligned} \tag{3.2}$$

Note that in (3.1) we have the usual one-parameter H^1-space (although vector-valued).

As we discussed in Subsection 2.2, $\mathbf{S}_1$ is a vector-valued Calderón–Zygmund operator, so it maps one-parameter H^1 to L^1 (even in the Hilbert-space-valued case). Thus, for almost all x^1 we have

$$\begin{aligned} \int_{X_1} \|\mathbf{S}_1 \otimes (\widetilde{\mathbf{S}}_2 \otimes \dots \otimes \widetilde{\mathbf{S}}_n) f(x_1, x^1)\|_{\mathcal{H}_1' \otimes \widetilde{\mathcal{H}}^1} \, dx_1 \\ \le C \|\widetilde{\mathbf{S}}_2 \otimes \dots \otimes \widetilde{\mathbf{S}}_n f(\,\cdot\,, x^1)\|_{H^1(X_1; \widetilde{\mathcal{H}}^1)} \\ = C \int_{X_1} \|\mathbf{S}_1^{\mathrm{L}} \otimes (\widetilde{\mathbf{S}}_2 \otimes \dots \otimes \widetilde{\mathbf{S}}_n) f(x_1, x^1)\|_{\mathcal{H}_1 \otimes \widetilde{\mathcal{H}}^1} \, dx_1. \end{aligned} \tag{3.3}$$

Integrating over X^1 and taking into account (3.2), we get the conclusion of the lemma. □

3.3. Estimate $\|f\|_{H^1(X)} \le C \int_X |S_{\mathcal{D}} f(x)| \, dx$. Proof of this estimate follows the lines of Section 3.2 almost word-for-word. It is based on the following analogue of Lemma 3.1, which allows us to replace the one-parameter Lusin square functions by the dyadic ones.

Consider again a multi-parameter square function $\widetilde{\mathbf{S}}_2 \otimes \dots \otimes \widetilde{\mathbf{S}}_n$, where now each $\widetilde{\mathbf{S}}_k$, $2 \le k \le n$, is either the one-parameter $\mathbf{S}^{\mathrm{L}}$ or the one-parameter dyadic square function $\mathbf{S}_{\mathcal{D}_k}$, defined in (2.1), taken in the variable x_k. We assume here that in each X_k dyadic lattices $\mathcal{D}_k$ are fixed.

We will use the same notation as in Subsection 3.2, with the only exception that now $\mathcal{H}_k' = \ell^2$.

LEMMA 3.2.

$$\int_X \|\mathbf{S}_1^{\mathrm{L}} \otimes \widetilde{\mathbf{S}}_2 \otimes \dots \otimes \widetilde{\mathbf{S}}_n f(x)\|_{\mathcal{H}_1 \otimes \widetilde{\mathcal{H}}^1} \, dx \le C \int_X \|\mathbf{S}_{\mathcal{D}_1} \otimes \widetilde{\mathbf{S}}_2 \otimes \dots \otimes \widetilde{\mathbf{S}}_n f(x)\|_{\mathcal{H}_1' \otimes \widetilde{\mathcal{H}}^1} \, dx.$$

Applying Lemma 3.1 successively to each variable, as we did in Subsection 3.2, we get

$$\int_X \|\mathbf{S}_1^{\mathrm{L}} \otimes \mathbf{S}_2^{\mathrm{L}} \otimes \dots \otimes \mathbf{S}_n^{\mathrm{L}} f(x)\|_{\mathcal{H}} \, dx \le C \int_X \|\mathbf{S}_{\mathcal{D}_1} \otimes \mathbf{S}_{\mathcal{D}_2} \otimes \dots \otimes \mathbf{S}_{\mathcal{D}_n} f(x)\|_{\mathcal{H}'} \, dx,$$

which is exactly what we need.

PROOF OF LEMMA 3.2. As in (3.2) we get

$$\int_X \|\mathbf{S}_{\mathcal{D}_1} \otimes (\widetilde{\mathbf{S}}_2 \otimes \cdots \otimes \widetilde{\mathbf{S}}_n) f(x)\|_{\mathcal{H}_1 \otimes \widetilde{\mathcal{H}}^1}\, dx = \int_{X^1} \|\widetilde{\mathbf{S}}_2 \otimes \cdots \otimes \widetilde{\mathbf{S}}_n f(\,\cdot\,, x^1)\|_{H^1_{\mathcal{D}_1}(X_1; \widetilde{\mathcal{H}}^1)}\, dx^1. \tag{3.4}$$

We now use the fact that $\|\varphi\|_{H^1} \le C\|\varphi\|_{H^1_{\mathcal{D}}}$ for Hilbert-space-valued functions. Again, as in the scalar-valued case, this follows from H^1-BMO duality ($(H^1)^* =$ BMO, $(H^1_{\mathcal{D}})^* = \mathrm{BMO}_{\mathcal{D}}$) and the trivial inclusion BMO $\subset \mathrm{BMO}_{\mathcal{D}}$, which implies the inclusion $H^1_{\mathcal{D}} \subset H^1$ with the desired estimates of the norms.

Using this inequality we see that for almost all x^1,

$$\|\widetilde{\mathbf{S}}_2 \otimes \cdots \otimes \widetilde{\mathbf{S}}_n f(\,\cdot\,, x^1)\|_{H^1(X_1; \widetilde{\mathcal{H}}^1)} \le C\|\widetilde{\mathbf{S}}_2 \otimes \cdots \otimes \widetilde{\mathbf{S}}_n f(\,\cdot\,, x^1)\|_{H^1_{\mathrm{d}}(X_1; \widetilde{\mathcal{H}}^1)}.$$

Integrating over X^1 and using (3.2), (3.4) we get the conclusion of the lemma. □

4. Appendix: Some facts about H^1 and BMO spaces

4.1. Hilbert-space-valued BMO spaces. Let us recall that a function on $X = \mathbb{R}^N$ with values in a Hilbert space $\mathcal{H}$ belongs to the space BMO $=$ BMO$(X, \mathcal{H})$ if

$$\|f\|_{\mathrm{BMO}} := \sup_Q \int_Q \|f(x) - f_Q\|_{\mathcal{H}}\, dx < \infty; \tag{4.1}$$

here $f_Q := |Q|^{-1} \int_Q f(x)dx$ and the supremum is taken over all cubes $Q \subset \mathbb{R}^N$.

If we fix a dyadic lattice $\mathcal{D}$ and take the supremum in (4.1) only over *dyadic* cubes $Q \in \mathcal{D}$, we get the dyadic space $\mathrm{BMO}_{\mathcal{D}}$ associated with this lattice.

It is well known that $(H^1(\mathbb{R}^N; \mathcal{H}))^* = \mathrm{BMO}(\mathbb{R}^N; \mathcal{H})$ and $(H^1_{\mathcal{D}}(\mathbb{R}^N; \mathcal{H}))^* = \mathrm{BMO}_{\mathcal{D}}(\mathbb{R}^N; \mathcal{H})$; any standard proof of H^1-BMO duality would work for the Hilbert-space-valued functions.

4.2. Triebel–Lizorkin spaces $\dot{\mathbf{f}}_p^{\alpha,q}$ and the equivalence of $\|Sf\|_1$ and $\|\widetilde{S}f\|_1$. In this section we fix a dyadic lattice $\mathcal{D}$, for example take for $\mathcal{D}$ the standard dyadic lattice.

4.2.1. *Spaces* $\dot{\mathbf{f}}_p^{\alpha,q}$. Following [3] define the space $\dot{\mathbf{f}}_p^{\alpha,q}$, $\alpha \in \mathbb{R}$, $1 \le p, p < \infty$, consisting of sequences $s = \{s_Q\}_{Q \in \mathcal{D}}$ such that

$$\|s\|_{\dot{\mathbf{f}}_p^{\alpha,q}} := \left\| \left(\sum_{Q \in \mathcal{D}} (|Q|^{-\alpha/n} |s_Q| \cdot |Q|^{-1/2} \mathbf{1}_Q)^q \right)^{1/q} \right\|_{L^p} < \infty.$$

For $p = \infty$ the norm is defined using BMO-like norm

$$\|s\|_{\dot{\mathbf{f}}_\infty^{\alpha,q}} := \sup_{P \in \mathcal{D}} \left(\frac{1}{|P|} \int_P \sum_{Q \in \mathcal{D},\, Q \subset P} (|Q|^{-\alpha/n} |s_Q| \cdot |Q|^{-1/2} \mathbf{1}_Q)^q \right)^{1/q}.$$

We are interested in the case where the smoothness parameter α is 0; to simplify the notation in this case we will use $\dot{\mathbf{f}}_p^q := \dot{\mathbf{f}}_p^{0,q}$.

We will need the following facts about duality for spaces $\dot{\mathbf{f}}_p^{\alpha,q}$:

$$(\dot{\mathbf{f}}_p^{\alpha,q})^* = \dot{\mathbf{f}}_{p'}^{-\alpha,q'}, \qquad 1 \le p, q < \infty;$$

here $1/p + 1/p' = 1$, $1/q + 1/q' = 1$.

4.2.2. *Almost diagonal operators.* Following [**3**] we say that an operator A with matrix $\{a_{Q,P}\}_{Q,P\in\mathcal{D}}$ is *almost diagonal* (for spaces $\dot{\mathbf{f}}_p^q = \dot{\mathbf{f}}_p^{0,q}$) if there exist $\varepsilon > 0$ and $C < \infty$ such that

$$|a_{Q,P}| \le C\left(1 + \frac{|x_P - x_Q|}{\max\{\ell(P),\ell(Q)\}}\right)^{-N-\varepsilon} \times \min\left\{\left(\frac{\ell(Q)}{\ell(P)}\right)^{(N+\varepsilon)/2}, \left(\frac{\ell(P)}{\ell(Q)}\right)^{(N+\varepsilon)/2}\right\}$$

(the definition is a bit more complicated for $\dot{\mathbf{f}}_p^{s,q}$ with $s \ne 0$).

It was shown in [**3**] that an almost diagonal operator is bounded in all $\dot{\mathbf{f}}_p^q$ spaces, $1 \le q, p < \infty$.

4.2.3. *Equivalence of* $\|Sf\|_1$ *and* $\|\widetilde{S}f\|_1$. From the above result one can easily obtain the equivalence of $\|Sf\|_1$ and $\|\widetilde{S}f\|_1$. First, since $\max_Q |\Delta_Q f|^2 \le 2^N E_Q(|\Delta_Q f|^2)$, we have the pointwise estimate $Sf(x) \le 2^{N/2}\widetilde{S}f(x)$ and so $\|Sf\|_1 \le 2^{N/2}\|\widetilde{S}f\|_1$.

To get the estimate $\|\widetilde{S}f\|_1 \le C\|Sf\|_1$, we express the conditions $Sf \in L^1$, $\widetilde{S}f \in L^1$ in terms of the Triebel–Lizorkin space $\dot{\mathbf{f}}_1^2$. Namely, with each function $f \in L^1_{\mathrm{loc}}$ we associate two sequences $a = \{a_Q\}_{Q\in\mathcal{D}}$ and $b = \{b_Q\}_{Q\in\mathcal{D}}$,

$$a_Q = \left[\left(E_Q|\Delta_Q f|^2\right)(x)\right]^{1/2}, \qquad b_Q = (\Delta_R f)(x),$$

where R is the "parent" of Q and x is an arbitrary point in Q (the result does not depend on x). Then clearly

$$\|\widetilde{S}f\|_1 = \|a\|_{\dot{\mathbf{f}}_1^2}, \qquad \|Sf\|_1 = \|b\|_{\dot{\mathbf{f}}_1^2}.$$

Note that

$$a_R = \left(2^{-N} \sum_{Q \text{ is child of } R} |b_Q|^2\right)^{1/2} \le 2^{-N/2} \sum_{Q \text{ is child of } R} |b_Q| =: T|b|,$$

where $|b| := \{|b_Q|\}_{Q\in\mathcal{D}}$ and, for $s = \{s_Q\}_{Q\in\mathcal{D}}$,

$$(Ts)_R := 2^{-N/2} \sum_{Q \text{ is child of } R} s_Q.$$

The operator T is almost diagonal (it has only finitely many "diagonals"), so

$$\|\widetilde{S}f\|_1 = \|a\|_{\dot{\mathbf{f}}_1^2} \le \|T|b|\|_{\dot{\mathbf{f}}_1^2} \le C\|\,|b|\,\|_{\dot{\mathbf{f}}_1^2} = C\|b\|_{\dot{\mathbf{f}}_1^2} = C\|Sf\|_1.$$

References

[1] S.-Y. A. Chang and R. Fefferman, *Some recent developments in Fourier analysis and H^p-theory on product domains*, Bull. Amer. Math. Soc. (N.S.) **12** (1985), no. 1, 1–43.

[2] B. Davis, *Hardy spaces and rearrangements*, Trans. Amer. Math. Soc. **261** (1980), no. 1, 211–233.

[3] M. Frazier and B. Jawerth, *A discrete transform and decompositions of distribution spaces*, J. Funct. Anal. **93** (1990), no. 1, 34–170.

[4] J. B. Garnett and P. W. Jones, *BMO from dyadic BMO*, Pacific J. Math. **99** (1982), no. 2, 351–371.

[5] T. Hytönen, *Vector-valued wavelets and the Hardy space $H^1(\mathbb{R}^n, X)$*, Studia Math. **172** (2006), no. 2, 125–147.

[6] Y. Meyer, *Wavelets and operators*, Cambridge Studies in Advanced Mathematics, vol. 37, Cambridge University Press, Cambridge, 1992, Translated from the 1990 French original by D. H. Salinger.

[7] J. Pipher, *Journé's covering lemma and its extension to higher dimensions*, Duke Math. J. **53** (1986), no. 3, 683–690.

[8] J. Pipher and L. A. Ward, *BMO from dyadic BMO on the bidisc*, J. Lond. Math. Soc. (2) **77** (2008), no. 2, 524–544.

[9] J. L. Rubio de Francia, F. J. Ruiz, and J. L. Torrea, *Calderón–Zygmund theory for operator-valued kernels*, Adv. in Math. **62** (1986), no. 1, 7–48.

[10] E. M. Stein, *Harmonic analysis: real-variable methods, orthogonality, and oscillatory integrals*, Princeton Mathematical Series, vol. 43, Princeton University Press, Princeton, 1993, With the assistance of Timothy S. Murphy, Monographs in Harmonic Analysis, III.

Department of Mathematics, Brown University, 151 Thayer Street, Box 1917, Providence, RI 02912, USA

E-mail address: treil@math.brown.edu

URL: http://www.math.brown.edu/~treil

Amer. Math. Soc. Transl.
(2) Vol. **226**, 2009

Monge–Ampère Equation and Bellman Optimization of Carleson Embedding Theorems

Vasily Vasyunin and Alexander Volberg

ABSTRACT. The Monge–Ampère equation plays an important part in Analysis. For example, it is instrumental in mass transport problems, where it serves as the Euler–Lagrange equation. On the other hand, the Bellman function technique has recently appeared as a way to view certain Harmonic Analysis problems as problems of Stochastic Optimal Control. This brings us to the Bellman PDE, which, in stochastic setting, is often a Monge–Ampère equation or its close relative. We explore the way of solving the Monge–Ampère equation by a sort of the method of characteristics to find the Bellman function of certain classical Harmonic Analysis problems, and, therefore, to determine the full structure of sharp constants and extremal sequences for those problems.

1. Introduction

Our goal in this article is to show how the Monge–Ampère equation allows us to solve a certain class of Harmonic Analysis problems. Usually an equation of the form

$$\frac{\partial^2 B}{\partial x_1^2}\frac{\partial^2 B}{\partial x_2^2} - \left(\frac{\partial^2 B}{\partial x_1 \partial x_2}\right)^2 = a\frac{\partial^2 B}{\partial x_1^2} + 2b\frac{\partial^2 B}{\partial x_1 \partial x_2} + c\frac{\partial^2 B}{\partial x_2^2} + \varphi$$

is called the Monge–Ampère equation. It appears in many physical and mathematical problems. We need to be able to explicitly solve boundary value problems for the homogeneous Monge–Ampère equation (i.e., with zero on the right-hand side) for some specific domains.

We choose two problems to illustrate the method: the John–Nirenberg inequality (JNI) and the Carleson embedding theorem (CET). In these problems, we determine the Bellman function of the problem (see below). The Bellman function carries all the information about the problem: the sharp constant, construction of extremal functions/extremal sequences; and sometimes it also carries auxiliary information.

2000 *Mathematics Subject Classification.* Primary 28A80; Secondary 28A75, 60D05.

Key words and phrases. Bellman function, Monge–Ampère equation, John–Nirenberg inequality, Carleson embedding theorem.

The first author was supported in part by RFBR grant 08-01-00723a. The second author was supported in part by NSF grant DMS-0501067.

In both JNI and CET, the Bellman functions were found before: in [**SV**] for JNI and in [**M**] for CET. However, the way they were found is rather specific for each problem and not "from basic principles", so to speak.

We propose here a universal method, which fits many problems (we count at least seven of them). Among those we chose JNI and CET because of their classical nature and because the reader can easily compare our approach and our Bellman functions with those in the literature [**SV, M**]. Our approach will differ completely from the previous ones, but the functions will coincide.

We borrowed the Monge–Ampère solution by "the method of characteristics" from the literature, and the most important inspiration came from Slavin and Stokolos [**SlSt**], although the result to which we refer is very old; see the book by Pogorelov [**Pog**], p. 606.

Let us note that our aim in the paper is not only to present a new and unified approach to finding Harmonic Analysis Bellman functions. In fact, for CET we plan to prove some new results, not merely to reprove the old ones from [**M**].

1.1. Bellman functions. The method of Bellman functions in Harmonic Analysis appeared probably in the 1980s in the series of papers by Burkholder devoted to sharp L^p-estimates of martingale transforms. It appeared there under certain disguise (the Bellman function and Bellman PDE were not even mentioned). The method was rediscovered in 1994 in the first preprint version of [**NTV1**]. It was very much developed in [**NT**]. The method turned out to be useful. Sometimes it solves certain important Harmonic Analysis problems that do not have an alternative (classical) approach so far; see e.g. [**PVo**]. There is now a certain amount of literature on the Bellman function technique. We mention an excellent paper [**M**], which serves as our inspiration here, and two small surveys: [**NTV2**] and [**Vo**].

We fix $\varepsilon \in (0,1)$ and an interval I. Subintervals of I will be denoted by J. Here is what is called in [**SV**] the Bellman function for the John–Nirenberg inequality:

$$\begin{aligned}\mathbf{B}_{JNI}(x_1,x_2;\varepsilon) := \sup\big\{\langle e^{\phi}\rangle_I\colon\ &\langle\phi\rangle_I = x_1,\ \langle\phi^2\rangle_I = x_2,\\ &\langle\phi\rangle_J^2 \le \langle\phi^2\rangle_J \le \langle\phi\rangle_J^2+\varepsilon^2 \quad \forall J\subset I\big\}.\end{aligned} \tag{1.1}$$

In what follows, $\langle\phi\rangle_I$ stands for the average of the function ϕ over I, i.e.,

$$\langle\phi\rangle_I = \frac{1}{|I|}\int_I \phi(t)\,dt\,,$$

$|I|$ means Lebesgue measure of I.

The function $\mathbf{B}_{JNI}$ is defined on

$$\Omega_{JNI} := \{x=(x_1,x_2)\colon\ x_1^2 \le x_2 \le x_1^2+\varepsilon^2\}\,. \tag{1.2}$$

Recall that the John–Nirenberg inequality in its integral form is as follows:

$$\langle e^{\phi}\rangle_I \le Ce^{\langle\phi\rangle_I}.$$

It is valid for all BMO-functions ϕ of a sufficiently small BMO-norm. The function $\mathbf{B}_{JNI}$ supplies us with such an estimate with a sharp constant C. The details and the relationship with the classical John–Nirenberg inequality can be found e.g. in [**SV**].

Now, given any I, we consider its dyadic lattice of subintervals $D = D(I)$ (it consists of left and right halves (sons) of I and then all left and right grandsons, et cetera). Let μ denote a nonnegative measure on I without atoms. We write the Bellman function for the Carleson Embedding Theorem (dyadic):

$$\begin{aligned}&\mathbf{B}_{CET}(x_1,x_2,x_3;m,M)\\&\quad:=\sup\Big\{\frac{1}{|I|}\Big(\int_I\phi^2(t)\,d\mu(t)+\sum_{J\in D(I)}\alpha_J\langle\phi\rangle_J^2\Big):\\&\qquad\langle\phi\rangle_I=x_1,\ \langle\phi^2\rangle_I=x_2,\ \frac{1}{|I|}\Big(\mu(I)+\sum_{J\in D(I)}\alpha_J\Big)=x_3,\\&\qquad m|J|\le\mu(J)+\sum_{\ell\in D(J)}\alpha_\ell\le M|J|\ \forall J\in D(I)\Big\}.\end{aligned}\tag{1.3}$$

The function $\mathbf{B}_{CET}$ is defined on

$$\Omega_{CET}:=\{x=(x_1,x_2,x_3):\ x_1^2\le x_2,\ m\le x_3\le M\}.\tag{1.4}$$

We introduce also $\mathbf{B}_{cet}(x_1,x_2,x_3)$. This is what is often called the Bellman function for the Carleson Embedding Theorem; see [**M, NT**]:

$$\begin{aligned}\mathbf{B}_{cet}(x_1,x_2,x_3):=\sup\Big\{&\frac{1}{|I|}\sum_{J\in D(I)}\alpha_J\langle\phi\rangle_J^2:\ \langle\phi\rangle_I=x_1,\ \langle\phi^2\rangle_I=x_2,\\&\frac{1}{|I|}\Big(\sum_{J\in D(I)}\alpha_J\Big)=x_3,\ \sum_{\ell\in D(J)}\alpha_\ell\le|J|\ \forall J\in D(I)\Big\}.\end{aligned}\tag{1.5}$$

Recall that the classical Carleson embedding theorem is equivalent to the following dyadic analog:

$$\sum_{J\in D(I)}\alpha_J\langle\phi\rangle_J^2\le C\|\phi\|^2\qquad\text{for all}\quad f\in L^2(I)$$

$$\text{if and only if}\qquad\sum_{\ell\in D(J)}\alpha_\ell\le M|J|\quad\forall J\subset I.$$

Again, the function $\mathbf{B}_{cet}$ supplies us with an estimate with a sharp constant C.

REMARK. The reader can easily see that the definitions of Bellman functions $\mathbf{B}_{CET}$, $\mathbf{B}_{cet}$ do not depend on the interval I.

The formula for the function $\mathbf{B}_{JNI}(x;\varepsilon)$ was found in [**V**] and [**SV**], the formula for $\mathbf{B}_{cet}(x)$ was found in [**M**], and the general formula for $\mathbf{B}_{CET}(x;m,M)$ below is new. Also we note that the formula for $\mathbf{B}_{cet}(x)$ was found in [**M**] from intricate combinatorial considerations. In the present paper we try to explain a natural way to find this formula from "the basic principles". Namely, all formulas will be obtained here by a unified method of solving the Monge–Ampère equation.

THEOREM 1.

$$\mathbf{B}_{JNI}(x_1,x_2;\varepsilon)=\frac{e^{-\varepsilon}}{1-\varepsilon}\left(1-\sqrt{\varepsilon^2-(x_2-x_1^2)}\right)e^{x_1+\sqrt{\varepsilon^2-(x_2-x_1^2)}}.$$

THEOREM 2. *Consider*

$$B(x_1,x_2,x_3)=\frac{x_3x_2}{[1-2a]^2[1-4a(1-x_3)]},\tag{1.6}$$

where $a = a(x)$ is a unique solution of the cubic equation

$$\frac{x_1^2}{x_2} = \left[\frac{1-2a(1-x_3)}{1-2a}\right]^2 \frac{1-4a}{1-4a(1-x_3)}$$

on the interval $[0, \frac{1}{4}]$. Then

$$\mathbf{B}_{cet}(x) = B(x), \qquad x_1^2 \le x_2, \quad 0 < x_3 \le 1\,.$$

THEOREM 3. *Consider*

$$B(x_1, x_2, x_3; m, M) = \frac{(x_3 - m)x_2}{\left[1-2a(M-m)\right]^2\left[1-4a(M-x_3)\right]} + mx_2\,, \tag{1.7}$$

where $a = a^+(x)$ is a unique solution of the cubic equation

$$\frac{x_1^2}{x_2} = \left[\frac{1-2a(M-x_3)}{1-2a(M-m)}\right]^2 \frac{1-4a(M-m)}{1-4a(M-x_3)} \tag{1.8}$$

on the interval $[0, \frac{1}{4(M-m)}]$. Then

$$\mathbf{B}_{CET}(x; m, M) = \begin{cases} B(x; m, M), & x_1^2 \le x_2, \quad m < x_3 \le M; \\ mx_2, & x_1^2 \le x_2, \quad x_3 = m. \end{cases}$$

REMARK. Actually, sometimes the cubic equation (1.8) has two solutions and sometimes, one solution (for the left hand side in $[0,1)$). One solution (called $a^+(x)$ above) is always in $[0, \frac{1}{4(M-m)}]$. The other solution (if it exists) is negative. We will call it $a^- = a^-(x)$. We will see immediately that it is also responsible for a meaningful and interesting extremal problem.

1.2. Lower Bellman function. If we denote the Bellman function (1.3) by $\mathbf{B}_{\max}$ and introduce another Bellman function

$$\begin{aligned}
&\mathbf{B}_{\min}(x_1, x_2, x_3; m, M) \\
&\quad := \inf\Big\{\frac{1}{|I|}\Big(\int_I \phi^2(t)\,d\mu(t) + \sum_{J\in D(I)} \alpha_J \langle\phi\rangle_J^2\Big) : \\
&\qquad \langle\phi\rangle_I = x_1,\ \langle\phi^2\rangle_I = x_2,\ \frac{1}{|I|}\Big(\mu(I) + \sum_{J\in D(I)} \alpha_J\Big) = x_3, \\
&\qquad m|J| \le \mu(J) + \sum_{\ell\in D(J)} \alpha_\ell \le M|J|\ \forall J \in D(I)\Big\},
\end{aligned} \tag{1.9}$$

then we can show that $a = a^-$ corresponds to this extremal problem.

THEOREM 4. *Consider*

$$B(x_1, x_2, x_3; m, M) = \frac{(x_3 - m)x_2}{\left[1-2a(M-m)\right]^2\left[1-4a(M-x_3)\right]} + mx_2\,, \tag{1.10}$$

where $a = a^-(x)$ is a unique negative solution of the cubic equation

$$\frac{x_1^2}{x_2} = \left[\frac{1-2a(M-x_3)}{1-2a(M-m)}\right]^2 \frac{1-4a(M-m)}{1-4a(M-x_3)}\,.$$

Then

$$\mathbf{B}_{\min}(x;m,M)=\begin{cases}B(x;m,M), & x_1^2\le x_2,\quad M-(M-m)\dfrac{x_1^2}{x_2}<x_3\le M\,;\\[2mm] mx_2, & x_1^2\le x_2,\quad m\le x_3\le M-(M-m)\dfrac{x_1^2}{x_2}\,.\end{cases}$$

Discussion.

- 1. We are going to provide a detailed proof of Theorem 3 only. In this proof we shall mark the places where something has to be changed (mainly concavity should be replaced by convexity) to get the proof of Theorem 4. The reader is invited to fill out all the details of the proof.
- 2. An interesting observation follows from the comparison of (1.6) and (1.7). In fact, we can see that the following rescaling relation holds:

$$\mathbf{B}_{CET}(x_1,x_2,x_3;m,M)=(M-m)\,\mathbf{B}_{cet}\big(x_1,x_2,\frac{x_3-m}{M-m}\big)+m\,x_2\,. \tag{1.11}$$

  In principle, (1.11) should have followed directly from the definitions of $\mathbf{B}_{cet}$, $\mathbf{B}_{CET}$ above. Let us provide a short explanation.

  First, we introduce some notation. We denote the center of a dyadic interval ℓ by a_ℓ and put

$$c_\ell:=a_\ell+i\frac{|\ell|}{2}\,,\qquad \alpha:=\sum_{\ell\in D(I)}\alpha_\ell\,\delta_{c_\ell}\,,$$

  where δ_z stands for the delta measure at z as usual. Actually, (1.11) means that if we are forced to have a uniform estimate from below on how much measure we have in any *closed* Carleson square:

$$(\mu+\alpha)(\overline{Q}_J)\ge m\,|J|\,,\quad \forall J\in D(I)\,,$$

  then to maximize the quantity in the definition of $\mathbf{B}_{CET}(x;m,M)$ over all admissible μ's and α's (that is, to obtain $\mathbf{B}_{CET}(x;m,M)$), one needs to keep "boundary measure" μ to be exactly equal to $m\,dx$. In principle, it is easy to believe that μ should always be grater than or equal to $m\,dx$. But (1.11) claims more, namely, the identity. It means that if we want to maximize the outcome in the definition of $\mathbf{B}_{CET}(x;m,M)$, we need to keep minimal possible mass (namely, $m\,dx$) on the boundary, as a reserve so to speak, and we need to distribute the rest of the mass in the form of measure α somewhere inside the half-plane. Also (1.11) means that this distribution of α mimics and actually equals the rescaled distribution of the extremal measure α in the case where we maximize $\mathbf{B}_{cet}$ and not $\mathbf{B}_{CET}(x;m,M)$.
- 3. Of course, (1.11) shows that the function $\mathbf{B}_{cet}$ is "more equal" among the $\mathbf{B}_{CET}$. It is interesting enough that (1.11) seems to follow not from the definitions of the functions $\mathbf{B}_{cet}$ and $\mathbf{B}_{CET}$ in (1.5) and (1.3) directly, but rather from the proofs of Theorems 2 and 3. In fact, it is not clear *a priori* why the optimal distribution of measure for $\mathbf{B}_{CET}(x;m,M)$ should have these two features: a) it leaves minimal possible "reserve" on the real line; b) it distributes the rest by repeating the best distribution for $\mathbf{B}_{cet}$ (with parameters properly rescaled) as (1.11) shows.

We shall not prove Theorem 1; its proof can be found in [**V**] or [**SV**], and in the latter paper, comparison of the dyadic and nondyadic cases for JNI can be found as well. We use this theorem as a "simple" example to illustrate how to find the Bellman function by using the general method of solving the Monge–Ampère equation. Then we use the same approach in much more involved situations of Theorem 3. This Bellman function is also not quite new; for the case of $m = 0$, $M = 1$, the function was found by Melas in [**M**], where the reader can find a complete proof of the theorem for this (most important) case.

However, in [**M**] there is no explanation how the author was able to guess his quite a nontrivial function. And the main accent of our paper is on the explanation how it is possible to guess right, or more precisely, how to solve the corresponding Bellman equation, which is a Monge–Ampère equation in the situation under consideration. Nevertheless, we will not stop after finding the Bellman function in Theorem 3 and we provide a detailed proof for the sake of completeness and convenience, and because in this more general setting the statement is new. Theorem 4 is new; however, its proof is absolutely similar to the proof of Theorem 3 up to a change of some signs somewhere. We shall point out the places where these changes have to be done in the proof of Theorem 3 without separating them as special statements.

2. How do we proceed?

The consideration of the theorems above can be split into four steps.

- I. In the first step we observe that, by the very definition, the Bellman functions $\mathbf{B}$ satisfy a certain concavity condition in their domain and boundary conditions on (a part of) the boundary of this domain.
- II. In the second step we consider **all** function satisfying these concavity and boundary conditions. We denote this class by $\mathcal{V}$. Next we make the following supposition: if b belongs to $\mathcal{V}$ and is the "best" function of this type, not only has it to satisfy the concavity condition but also this concavity should become degenerate, i.e., the inequality must turn into equality along some vector field on our domain Ω. This brings the Monge–Ampère equation to the picture. One solves it using the boundary conditions mentioned above. The result is a function $B \in \mathcal{V}$. The function B carries interesting geometric information to be used later.

In fact, these two steps are not necessary for the proof of the results; they are needed only for finding a function B, a candidate for the role of the Bellman function $\mathbf{B}$. For example, in the excellent paper [**M**], very complicated Bellman functions appear as *deus ex machina*. As we shall see, the analysis of the Monge–Ampère equation not only supplies us with a candidate, but it also helps at the next two steps, namely, in the proof that the candidate found is really the desired Bellman function.

- III. The third step consists in proving that $B \geq \mathbf{B}$. In convex domains this is usually not difficult. Otherwise it may require a nontrivial proof; see [**SV**] for nonconvex Ω_ε.
- IV. The fourth step consists in proving that $B \leq \mathbf{B}$. This is achieved by presenting extremal functions or extremal sequences of functions. In their turn, such functions are found from the geometric structure of B (mentioned in II).

3. Monge–Ampère equation

This section is here for illustrative purposes. Throughout this section, the function $\mathbf{B}$ is the Bellman function $\mathbf{B}_{JNI}$ for the John–Nirenberg inequality, defined in (1.1), and $\Omega = \Omega_{JNI}$ is its domain, defined in (1.2).

Let us briefly review **Step I** for Theorem 1.

It is easy to see that in the dyadic case (i.e., in the case where only dyadic intervals $J \in D(I)$ are considered in (1.1)) the function $\mathbf{B}$ must satisfy the concavity condition

$$\mathbf{B}(x) - \frac{1}{2}(\mathbf{B}(x^+) + \mathbf{B}(x^-)) \geq 0\,, \tag{3.1}$$

for all triples x, x^+, x^- such that $x = \frac{x^+ + x^-}{2}$ and all three points are in Ω.

In the nondyadic case (i.e., in the case where all subintervals $J \subset I$ are considered in (1.1)) it is not clear *a priori* why the Bellman function $\mathbf{B}$ must satisfy the concavity condition (3.1), and in general, condition (3.1) is not fulfilled. However, it turns out that the Bellman function is locally concave, i.e., concave in any convex subdomain of Ω. We shall not discuss this subtle moment here, especially because we shall use merely the local concavity condition, which is of course weaker than global concavity, and coincides with the latter only for convex domains. The reader can find the discussion of the difference between dyadic and nondyadic Bellman functions for the John–Nirenberg inequality in [**SV**]. Here we only add that for the functions smooth enough, the concavity condition can be rewritten in the differential form:

$$\frac{d^2\mathbf{B}(x)}{dx^2} := \begin{pmatrix} \mathbf{B}_{x_1x_1} & \mathbf{B}_{x_1x_2} \\ \mathbf{B}_{x_2x_1} & \mathbf{B}_{x_2x_2} \end{pmatrix} \leq 0\,, \qquad \forall x \in \Omega\,, \tag{3.2}$$

where by $\mathbf{B}_{x_ix_j}$ we denote the partial derivatives $\frac{\partial^2 \mathbf{B}}{\partial x_i \partial x_j}$.

From the definition, we have one obvious boundary condition on the lower parabola of the boundary $\partial\Omega$:

$$\mathbf{B}(x_1, x_1^2) = e^{x_1}\,, \tag{3.3}$$

since only the constant test functions $\phi = x_1$ correspond to points x with $x_2 = x_1^2$.

We also have a simple homogeneity condition

$$\mathbf{B}(x_1 + \tau, x_2 + 2x_1\tau + \tau^2) = e^{\tau}\mathbf{B}(x_1, x_2)\,. \tag{3.4}$$

This follows from the definition of $\mathbf{B}$ if we take $\phi + \tau$ rather than ϕ as a test function.

Let us move to the **Step II.** Put

$$\mathcal{V} = \{v \in C^2(\Omega) \colon v \text{ satisfies (3.2), (3.3), (3.4)}\}\,.$$

Usually any function from $\mathcal{V}$ (and moreover, any function satisfying only the concavity condition (3.1)) supplies us with some estimate. But if we look for a sharp estimate, we need to choose the minimal possible function from this class, which must be the Bellman function $\mathbf{B}$. This function, “a candidate for the true Bellman function”, will be denoted by the usual letter B. For every point $x \in \Omega$ there exists an extremal function ϕ (or an “almost extremal” function ϕ_n, i.e., a sequence of functions) realizing the supremum in the definition (1.1). The usual procedure of employing the Bellman function consists in consecutive application of (3.1) when splitting the interval I where a test function is defined. For the extremal test function there has to be no loss in such procedure; therefore, for the Bellman function equality must occur at least for one splitting of the point x into a pair $\{x^+, x^-\}$. For a concave function this means that it is linear in some direction. If we have almost extremal functions, i.e., an extremal sequence, then we must have "near equality" in (3.1), at least up to terms of second order. In any case this means that the Hessian matrix (3.2) must degenerate. Thus, as we are looking for the "best" B, on the top of this condition of negativity of the Hessian, we will impose the following **degeneration** condition:

$$\forall x \in \Omega \quad \exists \Theta \in \mathbb{R}^2 \setminus \{0\}:$$

$$\left(\frac{d^2B}{dx^2}\Theta, \Theta\right) = \left(\begin{pmatrix} B_{x_1x_1} & B_{x_1x_2} \\ B_{x_2x_1} & B_{x_2x_2} \end{pmatrix} \begin{pmatrix} \Theta_1 \\ \Theta_2 \end{pmatrix}, \begin{pmatrix} \Theta_1 \\ \Theta_2 \end{pmatrix}\right) = 0\,. \tag{3.5}$$

Since the matrix $\frac{d^2B}{dx^2}$ is negative, we deduce from (3.5) the following degeneration condition for the Hessian:

$$\det\left(\frac{d^2B}{dx^2}\right) = \det\begin{pmatrix} B_{x_1x_1} & B_{x_1x_2} \\ B_{x_2x_1} & B_{x_2x_2} \end{pmatrix} = 0\,, \quad \forall x \in \Omega\,. \tag{3.6}$$

We underline once again that at the first two steps we can admit not too rigorous arguments and various assumptions, because this is not the proof, but rather a heuristic way of finding a candidate for the Bellman function. The rigorous proof that the candidate found is indeed the required Bellman function starts from Step III.

CLAIM. *There is a simple algorithm to find the family of functions indexed by a parameter* δ $(\varepsilon \le \delta \le 1)$,

$$B = B_\delta(x_1, x_2) := \frac{e^{-\delta}}{1-\delta}\left(1 - \sqrt{\delta^2 - (x_2 - x_1^2)}\right) e^{x_1 + \sqrt{\delta^2 - (x_2 - x_1^2)}},$$

that solves the Monge–Ampère equation (3.6) *with boundary condition* (3.3) *and homogeneity condition* (3.4).

To prove this claim we need to explain how to solve the special type of a Monge–Ampère equation we have.

3.1. Monge–Ampère equation and the method of "characteristics". Let $v = v(x_1, \dots, x_n)$ be a smooth function satisfying the following Monge–Ampère equation in some domain Ω:

$$\det\frac{d^2v}{dx^2} = \det\begin{pmatrix} v_{x_1x_1} & \cdots & v_{x_1x_n} \\ \cdots & \cdots & \cdots \\ v_{x_nx_1} & \cdots & v_{x_nx_n} \end{pmatrix} = 0\,, \qquad \forall x = (x_1, \dots, x_n) \in \Omega\,,$$

and suppose that this matrix has rank $n-1$. Then there are functions $t_i(x_1, \dots, x_n)$, $i = 0, 1, \dots, n$, such that

$$v(x) = t_0 + t_1x_1 + t_2x_2 + \cdots + t_nx_n \tag{3.7}$$

and the following $n-1$ linear equations hold:

$$dt_0 + x_1dt_1 + x_2dt_2 + \cdots + x_{n-1}dt_{n-1} + x_ndt_n = 0\,. \tag{3.8}$$

Let us explain why there are $n-1$ equations and why they are linear. One needs to read (3.8) as follows: we think that, say, $t_1, \dots, t_{n-1}$ are $n-1$ independent variables and t_n, t_0 are functions of them. Then (3.8) can be rewritten as follows:

$$\Big(\frac{\partial t_0}{\partial t_1}+x_1+x_n\frac{\partial t_n}{\partial t_1}\Big)dt_1+\cdots+\Big(\frac{\partial t_0}{\partial t_{n-1}}+x_{n-1}+x_n\frac{\partial t_n}{\partial t_{n-1}}\Big)dt_{n-1}=0\,,$$

whence

$$x_i+x_n\frac{\partial t_n}{\partial t_i}+\frac{\partial t_0}{\partial t_i}=0\,,\qquad i=1,\dots,n-1\,. \tag{3.9}$$

So we get $n-1$ equations.

REMARK. In general we can choose **any** $n-1$ variables as independent, of course. Since the order of the variables is arbitrary, sometimes the first $n-1$ are not the most convenient choice.

Now why these are linear equations? Let $t_1,\dots,t_{n-1}$ be fixed. Then (3.9) gives us linear relations in x_1, ..., x_n, so $n-1$ hyperplanes. Therefore, (3.8)–(3.9) give the intersection of $n-1$ hyperplanes, i.e., a line. We can call it $L_{t_1,\dots,t_{n-1}}$. These lines foliate the domain Ω, and (3.7) shows that v is a linear function on each line.

We are going to prove all these propositions for the sake of completeness.

Actually (3.7) and (3.8) (under small disguise) exist in the literature; they are proved in Pogorelov's book [**Pog**], page 606. In fact, it is proved there that the surface $y=v(x)$ is foliated by straight lines, and the normal $(v_{x_1},\dots,v_{x_n},-1)$ to the surface stays the same along each line. So we can take $t_1=v_{x_1},\dots,t_n=v_{x_n}$. Consider $t_0:=v-x_1t_1-\cdots-x_nt_n$. All these functions are constants along each line of Pogorelov's foliation. This is not clear only for t_0. Let us represent the line of foliation as a collection of linear functions of x_n: $x_1=x_1(x_n),\dots,x_{n-1}=x_{n-1}(x_n)$. Now consider

$$G(x_n):=v(x_1(x_n),\dots,x_{n-1}(x_n),x_n)-t_1x_1(x_n)-\cdots-t_{n-1}x_{n-1}(x_n)\,.$$

As v is linear along this line, $G=T_nx_n+T_{n0}$, where T_n, T_{n0} are constants (but they depend on the line chosen). So

$$v=t_1x_1+\cdots+t_{n-1}x_{n-1}+T_nx_n+T_{n0}\,.$$

But there is nothing special in the variable x_n. We can repeat the previous argument with x_1 instead of x_n. Then symmetrically

$$v=T_1x_1+\cdots+t_{n-1}x_{n-1}+t_nx_n+T_{10}\,.$$

Then we get $(t_1-T_1)x_1+(T_n-t_n)x_n+T_{n0}-T_{10}=0$. Varying x_1, x_n along this line (we can always assume without loss of generality that the variables x are not frozen on a line of foliation) we get $t_1=T_1$, $t_n=T_n$ and so $T_{n0}=T_{10}=t_0$. Therefore, t_0 is also constant on the line.

After that we obviously have (3.7). To establish (3.8) it is sufficient to compute the full differential of both sides of (3.7). We do this below.

Even if we were unaware of [**Pog**], it is easy to prove (3.7), (3.8) directly. The matrix $\frac{d^2v}{dx^2}$ annihilates one vector $\Theta(x)$ at every $x=(x_1,\dots,x_n)\in\Omega$. So we get a vector field Θ. Consider its integral curve $x_1=x_1(x_n)$, ..., $x_{n-1}=x_{n-1}(x_n)$.

The vector $\Theta(x)$ is a tangent vector to that curve, i.e.,

$$\Theta = \Theta_n \begin{pmatrix} x_1' \\ x_2' \\ \cdots \\ x_{n-1}' \\ 1 \end{pmatrix}. \tag{3.10}$$

Consider a new function $g(x_n) = v(x_1(x_n), \dots, x_{n-1}(x_n), x_n)$. Due to (3.10), its second derivative is

$$\begin{aligned} g'' &= \left(\frac{d^2 v}{dx^2} \begin{pmatrix} x_1' \\ x_2' \\ \cdots \\ x_{n-1}' \\ 1 \end{pmatrix}, \begin{pmatrix} x_1' \\ x_2' \\ \cdots \\ x_{n-1}' \\ 1 \end{pmatrix} \right) + v_{x_1} x_1'' + \cdots + v_{x_{n-1}} x_{n-1}'' \\ &= v_{x_1} x_1'' + \cdots + v_{x_{n-1}} x_{n-1}''. \end{aligned} \tag{3.11}$$

Now, let us also show that $v_{x_1}, \dots, v_{x_{n-1}}$ are constants on this integral curve. Suppose we are standing on the integral curve. The surface $v_{x_1} = t_1 = \text{const}$ has normal $(v_{x_1x_1}, \dots, v_{x_1x_n})$, that is the first row of the matrix $\frac{d^2v}{dx^2}$, which is orthogonal to the directional vector Θ of the integral curve. Hence Θ is in the tangent hyperplane to the surface $v_{x_1} = t_1$. The same is true for the surfaces $v_{x_i} = t_i = \text{const}$, $i = 2, \dots, n-1$. The intersection of these surfaces gives us our integral curve because Θ is in the intersection of all tangent planes to these surfaces; therefore the curves $L_{t_1,\dots,t_{n-1}}$ indexed by the constants $t_1, \dots, t_n$ are just the integral curves of the tangent bundle Θ. Thus, (3.11) can be rewritten as

$$\frac{d^2}{dx_n^2}\big(g - (t_1\, x_1 + \cdots + t_{n-1}\, x_{n-1})\big) = 0. \tag{3.12}$$

We see that the second derivative of a function (of x_n) in (3.12) is zero. So, the function is linear in x_n, i.e., is equal to $t_n x_n + t_0$, where the constants t_n, t_0 depend only on the curve $L_{t_1,\dots,t_{n-1}}$, that is,

$$t_n = t_n(t_1, \dots, t_{n-1}), \qquad t_0 = t_0(t_1, \dots, t_{n-1}).$$

We obtain on $L_{t_1,\dots,t_{n-1}}$:

$$v(x_1(x_n), \dots, x_{n-1}(x_n), x_n) = t_0 + x_1 t_1 + \cdots + x_{n-1} t_{n-1} + t_n x_n.$$

Since we assumed our vector field to be smooth and its integral curves foliate the whole domain, we get (3.7) by varying the parameters $t_1, \dots, t_n$. To check (3.8), we take the full differential in (3.7). Then

$$[v_{x_1} dx_1 + \cdots + v_{x_n} dx_n] = dv = [t_1 dx_1 + \cdots + t_n dx_n] + dt_0 + x_1 dt_1 + \cdots + x_n dt_n. \tag{3.13}$$

We are on $L_{t_1,\dots,t_n}$ and so $v_{x_i} = t_i$, $i = 1, \dots, n-1$, as we have already established. But it is also easy to see that

$$v_{x_n} = t_n(t_1, \dots, t_{n-1})$$

on $L_{t_1,\dots,t_{n-1}}$. In fact, all v_{x_i} and all t_i are symmetric. We could have chosen to represent the integral curve of Θ not as $x_i = x_i(x_n)$, $i = 1, \dots, n-1$, but as $x_j = x_j(x_1)$, $j = 2, \dots, n$. Now we see that the two expressions in brackets in (3.13) are equal. Then we obtain $x_1 dt_1 + \cdots + x_n dt_n + dt_0 = 0$, which gives the desired $n-1$ linear relations (3.8).

3.2. The proof of the claim. We can use the previous Subsection 3.1. But we prefer to repeat the argument for the simple case of only two variables!

PROOF. Equation (3.6) is clearly solved by a linear function of x_1, x_2, but the linear function cannot satisfy the boundary condition. So we assume that the matrix $\frac{d^2B}{dx^2}$ is negative definite, nonzero, and there exists exactly one vector $\Theta \in \mathbb{R}^2 \setminus \{0\}$ that annihilates it: $\frac{d^2B}{dx^2}\Theta = 0$. This vector may of course depend on (x_1, x_2), and we get a vector field $\Theta(x)$ in Ω. We now show that B_{x_i}, $i = 1, 2$, are constants on the integral curves of this vector field. Locally our integral curves are level sets of a certain function $s = s(x_1, x_2)$. Of course, the function $s(x)$ is not uniquely defined; it is defined up to a change of variables $s \mapsto \varphi \circ s$, where φ is a function of one variable. But in any parametrization, the vector $(-s_{x_2}, s_{x_1})$ is a tangent vector to the integral curve, i.e., it is collinear to our vector field Θ and thus annihilates $\frac{d^2B}{dx^2}$:

$$\begin{pmatrix} B_{x_1x_1} & B_{x_1x_2} \\ B_{x_2x_1} & B_{x_2x_2} \end{pmatrix} \begin{pmatrix} -s_{x_2} \\ s_{x_1} \end{pmatrix} = 0\,. \tag{3.14}$$

Notice that both functions $B_{x_i}(x_1, x_2)$ instead of $s(x_1, x_2)$ satisfy the last equation as well. In fact, the vector $(-(B_{x_1})_{x_2}, (B_{x_1})_{x_1})$ obviously annihilates the first row of $\frac{d^2B}{dx^2}$. Then it annihilates the second row simply by (3.6). The same is true for the vector $(-(B_{x_2})_{x_2}, (B_{x_2})_{x_1})$. Therefore, the integral curves of our vector field are the level sets of the two functions $B_{x_1}(x_1, x_2)$ and $B_{x_2}(x_1, x_2)$. Thus, we can conclude that (at least locally) $B_{x_i} = t_i \circ s$ for some functions t_i of one variable. Moreover, this means that we can take any of B_{x_i} to play the role of s: to define the integral curves of our vector field Θ as its level sets.

We now check that the function $t_0 := B - t_1x_1 - t_2x_2$ is also constant along any integral curve. For this aim, we must check that we can put t_0 instead of s into (3.14), but this is evident, because at each point (x_1, x_2) we have

$$\begin{aligned} -\frac{\partial t_0}{\partial x_2} &= -\frac{\partial B}{\partial x_2} + \frac{\partial t_1}{\partial x_2}x_1 + \frac{\partial t_2}{\partial x_2}x_2 + t_2 = B_{x_1x_2}x_1 + B_{x_2x_2}x_2, \\ \frac{\partial t_0}{\partial x_1} &= \frac{\partial B}{\partial x_1} - \frac{\partial t_1}{\partial x_1}x_1 - \frac{\partial t_2}{\partial x_1}x_2 - t_1 = -B_{x_1x_1}x_1 - B_{x_2x_1}x_2. \end{aligned}$$

So, we have obtained

$$B(x) = t_0 + t_1x_1 + t_2x_2. \tag{3.15}$$

The integral curves of the kernel field of the Hessian of this function are the straight lines given by the equation

$$dt_0 + x_1dt_1 + x_2dt_2 = 0\,. \tag{3.16}$$

Indeed, on the one hand,

$$dB = B_{x_1}dx_1 + B_{x_2}dx_2\,,$$

and on the other hand, the differential of (3.15) is

$$dB = dt_0 + x_1dt_1 + t_1dx_1 + x_2dt_2 + t_2dx_2\,.$$

Comparison of the these two equalities yields (3.16).

If the vertical integral curves of our vector field are nowhere dense (so they could be given by $x_2 = k_1x_1 + k_0$), then they can be regarded as the level sets $t_2 = t$, and the two other parameters are functions of t. Equations (3.15) and (3.16) could be rewritten as

$$B(x) = t_0(t) + t_1(t)x_1 + tx_2\,, \qquad t_0(t)' + t_1(t)'x_1 + x_2 = 0\,. \tag{3.17}$$

If the horizontal integral curves of our vector field are nowhere dense, they can be written as $x_1 = k_1x_2 + k_0$, or as the level sets $t_1 = t$, and the two other parameters are functions of t. Equations (3.15) and (3.16) then take the form

$$B(x) = t_0(t) + tx_1 + t_2(t)x_2\,, \qquad t_0(t)' + x_1 + t_2(t)'x_2 = 0\,. \tag{3.18}$$

Until now we have considered a general equation where the determinant of the Hessian for a function of two variables is zero. Now we return to our specific problem and use (3.4) to determine the structure of the family of integral lines.

Upon differentiating (3.4) with respect to τ and setting $\tau = 0$, we get

$$B_{x_1} + 2x_1B_{x_2} = B$$

or

$$t_1 + 2x_1t_2 = t_0 + t_1x_1 + t_2x_2\,.$$

This is the equation of a straight line,

$$(t_0 - t_1) + x_1(t_1 - 2t_2) + x_2t_2 = 0, \tag{3.19}$$

representing a level set of the functions t_i; therefore it has to be the same straight line that is given by equation (3.16). Hence the following two equations must be satisfied:

$$\frac{dt_0}{t_0 - t_1} = \frac{dt_1}{t_1 - 2t_2} = \frac{dt_2}{t_2}\,.$$

They can be rewritten as

$$d\Big(\frac{t_0}{t_2}\Big) = -t_1\frac{dt_2}{t_2^2}\,, \tag{3.20}$$

$$d\Big(\frac{t_1}{t_2}\Big) = -2\frac{dt_2}{t_2}\,. \tag{3.21}$$

The solution of (3.21) is

$$t_1 = -2t_2\log|t_2| + 2c_1t_2 \tag{3.22}$$

(we have denoted the integration constant by $2c_1$ for future convenience). After plugging this solution in (3.20), we get

$$t_0 = t_2\log^2|t_2| - 2c_1t_2\log|t_2| + c_2t_2\,. \tag{3.23}$$

Now we rewrite equation (3.19) using the expressions obtained for t_i:

$$\Big(\log^2|t_2| + 2(1-c_1)\log|t_2| + (c_2 - 2c_1)\Big) - 2x_1\Big(\log|t_2| - c_1 + 1\Big) + x_2 = 0\,.$$

It is convenient to introduce a new variable

$$a := \log|t_2| - c_1 + 1\,.$$

Then formulas (3.22) and (3.23) can be rewritten as follows:

$$t_1 = 2(1-a)t_2\,, \tag{3.24}$$

$$t_0 = \big((1-a)^2 - c_1^2 + c_2\big)t_2\,, \tag{3.25}$$

and the extremal lines are

$$x_2 - 2ax_1 + a^2 + c_2 - 1 - c_1^2 = 0\,. \tag{3.26}$$

Since

$$c_1^2+1-c_2=x_2-2ax_1+a^2=(x_2-x_1^2)+(x_1-a)^2\geq 0\,,$$

we can introduce a new nonnegative constant δ such that

$$\delta^2=c_1^2+1-c_2$$

and rewrite equation (3.26) in the form

$$x_2-2ax_1+a^2-\delta^2=0\,. \tag{3.27}$$

We observe that these are lines tangent to the parabola $x_2=x_1^2+\delta^2$ at the point $x_1=a$, $x_2=a^2+\delta^2$ and intersecting the lower boundary $x_2=x_1^2$ at the points with $x_1=a\pm\delta$. We want these lines to foliate the domain $\Omega=\{(x_1,x_2): x_1^2\leq x_2\leq x_1^2+\varepsilon^2\}$. Hence,

$$\delta\geq\varepsilon\,.$$

Now, $B(x_1,x_2)$ can almost be written out. The variable a can be found in terms of x_i by solving equation (3.27):

$$a=x_1+\sqrt{\delta^2-(x_2-x_1^2)}\,.$$

We have chosen the solution with the plus sign in front of the square root. The explanation why it corresponds to our problem, and what extremal problem is linked with the opposite sign, can be found in [**V, SV**]. Now, using (3.24), (3.25), and (3.27), we obtain the expression for B:

$$\begin{aligned}B(x_1,x_2)&=t_0+t_1x_1+t_2x_2\\&=\big((1-a)^2-c_1^2+c_2+2(1-a)x_1+x_2\big)t_2\\&=2(1-a+x_1)t_2=\pm2(1-a+x_1)e^{a+c_1-1}\\&=\pm2\Big(1-\sqrt{\delta^2-(x_2-x_1^2)}\Big)e^{x_1+\sqrt{\delta^2-(x_2-x_1^2)}+c_1-1}\,.\end{aligned}$$

Since B must be positive and satisfy the boundary condition (3.3), we choose the appropriate sign and the value of the constant c_1. Finally, we get a family of solutions

$$B(x_1,x_2;\delta)=\frac{1-\sqrt{\delta^2-(x_2-x_1^2)}}{1-\delta}e^{x_1+\sqrt{\delta^2-(x_2-x_1^2)}-\delta}$$

depending on a parameter δ. Acceptable values of the parameter δ are $\varepsilon\leq\delta<1$. Why this is so, and how to choose δ for a given ε (by the way, the choice is different for the dyadic and nondyadic cases), was explained in [**SV**]. □

The above family of functions B was found in [**V**] and [**SV**] using different approaches. After finding this candidate, Steps I, II are completely finished for the John–Nirenberg inequality. The reader is referred to [**SV**], where Steps III and IV are done. The fact that Ω is not convex makes Step III rather delicate.

Now we are going to present all four steps for another more difficult problem of finding the Bellman functions of the Carleson embedding theorem (CET).

4. Step I for the Carleson embedding theorem

From the definition of the function $\mathbf{B} = \mathbf{B}_{CET}$ above, we can immediately see that it satisfies the following **main inequality**:

$$\begin{aligned}(4.1)\qquad &\mathbf{B}(x) - \frac{1}{2}(\mathbf{B}(x^-) + \mathbf{B}(x^+)) \geq x_1^2(x_3 - \frac{1}{2}(x_3^+ + x_3^-))\,,\\ &\forall x, x^\pm \in \Omega\,, \quad x_i = \frac{1}{2}(x_i^+ + x_i^-) \text{ for } i = 1, 2, \quad \text{and } x_3 \geq \frac{1}{2}(x_3^+ + x_3^-)\,.\end{aligned}$$

(For brevity, we usually omit the parameters m and M.) Since the domain $\Omega = \Omega_{CET}$ (see (1.4)) is convex, the **main inequality** is equivalent to the following pair of differential inequalities:

$$(4.2)\qquad \frac{d^2\mathbf{B}}{dx^2}(x) \leq 0\,,$$

which is just concavity, and

$$(4.3)\qquad \frac{\partial \mathbf{B}}{\partial x_3}(x) \geq x_1^2\,.$$

The **main inequality** (4.1), as well as its equivalence to the pair of conditions (4.2) and (4.3), is explained in [**NT**], where the Bellman function $\mathbf{B}_{cet}$ was introduced. But for the reader's convenience (especially because we are working in a slightly more general setting of $\mathbf{B}_{CET}$) we repeat these arguments.

Let a test function ϕ be defined on an interval I. We get a point $x \in \Omega$ with the coordinates

$$x_1 = \langle \phi \rangle_I, \quad x_2 = \langle \phi^2 \rangle_I, \quad x_3 = \frac{1}{|I|}\Big(\mu(I) + \sum_{J \in D(I)} \alpha_J\Big)\,.$$

Splitting I into two halves $I = I^+ \cup I^-$ we get two more points $x^\pm \in \Omega$:

$$x_1^\pm = \langle \phi \rangle_{I^\pm}, \quad x_2^\pm = \langle \phi^2 \rangle_{I^\pm}, \quad x_3^\pm = \frac{1}{|I^\pm|}\Big(\mu(I^\pm) + \sum_{J \in D(I^\pm)} \alpha_J\Big)\,.$$

Then we have

$$x_i = \frac{1}{2}(x_i^+ + x_i^-) \quad \text{for } i = 1, 2, \quad \text{and} \quad x_3 - \frac{1}{2}(x_3^+ + x_3^-) = \frac{\alpha_I}{|I|} \geq 0\,,$$

and

$$\begin{aligned}&\frac{1}{|I|}\Big\{\int_I \phi(t)^2 d\mu(t) + \sum_{J \in D(I)} \alpha_J \langle \phi \rangle_J^2\Big\}\\ &\quad = \frac{1}{2|I^+|}\Big\{\int_{I^+} \phi(t)^2 d\mu(t) + \sum_{J \in D(I^+)} \alpha_J \langle \phi \rangle_J^2\Big\}\\ &\quad\quad + \frac{1}{2|I^-|}\Big\{\int_{I^-} \phi(t)^2 d\mu(t) + \sum_{J \in D(I^-)} \alpha_J \langle \phi \rangle_J^2\Big\} + \frac{\alpha_I}{|I|}\langle \phi \rangle_I^2\,.\end{aligned}$$

Passing in this formula to the supremum over all admissible test functions $\phi|I^\pm$ (i.e., such that $\langle \phi \rangle_{I^\pm} = x_1^\pm$, $\langle \phi^2 \rangle_{I^\pm} = x_2^\pm$) and all admissible measures (i.e., such that $\frac{1}{|I^\pm|}(\mu(I^\pm) + \sum_{J \in D(I^\pm)} \alpha_J) = x_3^\pm$ and $m|J| \leq (\mu(J) + \sum_{\ell \in D(J)} \alpha_\ell) \leq M|J|$), we obtain

$$\mathbf{B}(x) \geq \frac{1}{2}\mathbf{B}(x^-) + \frac{1}{2}\mathbf{B}(x^+) + \frac{\alpha_I}{|I|}x_1^2\,,$$

because the set of admissible test functions ϕ for $\mathbf{B}(x)$ (i.e., such that $\langle\phi\rangle_I = x_1$, $\langle\phi^2\rangle_I = x_2$) is larger. So we come to the necessity of the **main inequality** (4.1).

The concavity of $\mathbf{B}$,

$$\mathbf{B}(x) \geq \frac{1}{2}\big(\mathbf{B}(x^-) + \mathbf{B}(x^+)\big)\,,$$

is an evident consequence of the **main inequality**. Since Ω is a convex domain, it is equivalent (assuming sufficient smoothness of $\mathbf{B}$) to the differential inequality (4.2). The second differential condition emerges if we choose $x_1^\pm = x_1$, $x_2^\pm = x_2$, and $x_3^\pm = x_3 - \Delta$ $(\Delta > 0)$,

$$\frac{\mathbf{B}(x_1,x_2,x_3) - \mathbf{B}(x_1,x_2,x_3-\Delta)}{\Delta} \geq x_1^2,$$

and pass to the limit as $\Delta \to 0$.

Conversely, if (4.2) and (4.3) are fulfilled, then due to the convexity of Ω we have

$$\begin{aligned}
&\mathbf{B}(x) - \frac{1}{2}\big(\mathbf{B}(x^-) + \mathbf{B}(x^+)\big)\\
&\quad= \left(\mathbf{B}(x_1,x_2,x_3) - \mathbf{B}\Big(x_1,x_2,\frac{x_3^+ + x_3^-}{2}\Big)\right)\\
&\qquad+ \left(\mathbf{B}\Big(\frac{x^+ + x^-}{2}\Big) - \frac{1}{2}\Big(\mathbf{B}(x^-) + \mathbf{B}(x^+)\Big)\right)\\
&\quad\geq \frac{\partial\mathbf{B}}{\partial x_3}(x_1,x_2,\zeta)\left(x_3 - \frac{1}{2}(x_3^+ + x_3^-)\right) \geq x_1^2\left(x_3 - \frac{1}{2}(x_3^+ + x_3^-)\right).
\end{aligned}\tag{4.4}$$

In addition to the differential inequalities (4.2) and (4.3) the Bellman function must satisfy the following boundary condition:

$$\mathbf{B}(x_1, x_1^2, x_3) = x_1^2\, x_3\,.\tag{4.5}$$

Indeed, the set of test functions corresponding to a boundary point (x_1, x_1^2, x_3) consists of the constant function $\phi(t) = x_1$ only, which immediately gives (4.5).

As in the preceding problem, our Bellman function has to satisfy a homogeneity condition, namely,

$$\mathbf{B}(\tau x_1, \tau^2 x_2, x_3) = \tau^2\,\mathbf{B}(x_1,x_2,x_3)\,,\quad \tau \in \mathbb{R}\,.\tag{4.6}$$

This follows from the definition of $\mathbf{B}$ if we take a test function $\tau\phi$ in place of ϕ.

Consider all smooth functions in Ω satisfying (4.1) (i.e., (4.2) and (4.3)), the boundary condition (4.5), and the homogeneity property (4.6). We denote this family by $\mathcal{V}$.

As before, we are looking for the "best" $B \in \mathcal{V}$, i.e., such that "near" equality occurs in the **main inequality** at any point x for at least one way of splitting x into a pair $x^\pm$. Exact equality could occur if the supremum in the definition of the Bellman function would be attained. In our problem, there is no extremal test function ϕ and extremal measure (μ,α) realizing the supremum in the definition of $\mathbf{B}$; we can only look for an extremal sequence.

Let us represent the expression from the **main inequality** as a sum of two nonnegative summands, as was done in (4.4):

$$\begin{aligned}(4.7)\qquad &\mathbf{B}(x)-\frac{1}{2}\big(\mathbf{B}(x^-)+\mathbf{B}(x^+)\big)-x_1^2\Big(x_3-\frac{1}{2}(x_3^++x_3^-)\Big)\\ &=\Big(\mathbf{B}\big(\frac{x^++x^-}{2}\big)-\frac{1}{2}\big(\mathbf{B}(x^-)+\mathbf{B}(x^+)\big)\Big)\\ &\quad+\Big(\frac{\partial\mathbf{B}}{\partial x_3}(x_1,x_2,\zeta)-x_1^2\Big)\Big(x_3-\frac{1}{2}(x_3^++x_3^-)\Big)\,.\end{aligned}$$

Since both summands are nonnegative, each of them has to be "almost zero" for some splitting. The positivity of the first summand is guaranteed by the negativity of the Hessian (4.2); therefore for every point x we need to have at least one direction, say Θ, along which the function $\mathbf{B}$ is linear. So, we will impose the following **degeneration** condition:

$$\forall x=(x_1,x_2,x_3)\in\Omega\quad \exists\Theta\in\mathbb{R}^3\setminus\{0\}:$$

$$(4.8)\qquad \Big(\frac{d^2\mathbf{B}}{dx^2}\Theta,\Theta\Big)=\left(\begin{pmatrix}\mathbf{B}_{x_1x_1}&\mathbf{B}_{x_1x_2}&\mathbf{B}_{x_1x_3}\\ \mathbf{B}_{x_2x_1}&\mathbf{B}_{x_2x_2}&\mathbf{B}_{x_2x_3}\\ \mathbf{B}_{x_3x_1}&\mathbf{B}_{x_3x_2}&\mathbf{B}_{x_3x_3}\end{pmatrix}\begin{pmatrix}\Theta_1\\ \Theta_2\\ \Theta_3\end{pmatrix},\begin{pmatrix}\Theta_1\\ \Theta_2\\ \Theta_3\end{pmatrix}\right)=0\,.$$

As the matrix $\frac{d^2\mathbf{B}}{dx^2}$ is negative, we derive from (4.8) the following degeneration condition on the Hessian:

$$(4.9)\qquad \det\frac{d^2\mathbf{B}}{dx^2}=\det\begin{pmatrix}\mathbf{B}_{x_1x_1}&\mathbf{B}_{x_1x_2}&\mathbf{B}_{x_1x_3}\\ \mathbf{B}_{x_2x_1}&\mathbf{B}_{x_2x_2}&\mathbf{B}_{x_2x_3}\\ \mathbf{B}_{x_3x_1}&\mathbf{B}_{x_3x_2}&\mathbf{B}_{x_3x_3}\end{pmatrix}=0\,,\qquad \forall x\in\Omega\,.$$

So, we come again to the Monge–Ampère equation.

To make the second summand in (4.7) "almost zero", we cannot require the equality

$$(4.10)\qquad \frac{\partial\mathbf{B}}{\partial x_3}(x)=x_1^2$$

to be fulfilled everywhere, because the latter equation together with (4.9), with the boundary condition (4.5), and with the homogeneity property (4.6) can be easily solved, and its solution $x_1^2x_3+b(x_1^2-x_2)$ does not satisfy the concavity condition (4.2). Therefore we shall require the equality

$$(4.11)\qquad x_3=\frac{1}{2}(x_3^++x_3^-)\,.$$

However, (4.11) cannot be true for every splitting, because in such a situation the discrete measure α would be zero. Therefore, for some points x we need to split with inequality $x_3>\frac{1}{2}(x_3^++x_3^-)$ and to require (4.10) only at such points.

We must have $x_3>\frac{1}{2}(x_3^++x_3^-)$ on the boundary $x_3=M$. Otherwise one of the points $x^\pm$ would jump out of Ω. Hence, to make the second summand in (4.7) "almost zero", we can and do require the identity (4.10) on the boundary $x_3=M$.

Of course, one can imagine that (4.10) could have happened also at some points inside Ω. But we optimistically disregard this case, and we will win (we will find $\mathbf{B}$) just by imposing (4.10) only on the boundary $x_3=M$, where we showed it to be necessary.

In such a way we get one more boundary condition

$$\frac{\partial \mathbf{B}}{\partial x_3}(x_1, x_2, M; m, M) = x_1^2\,. \tag{4.12}$$

CLAIM. *There is a simple algorithm to find the function $B(x; m, M)$ that solves the Monge–Ampère equation* (4.9) *in the domain* (1.4) *with the boundary conditions* (4.5), (4.12) *and the homogeneity condition* (4.6).

PROOF. Again we consider the vector field Θ such that $\frac{d^2B}{dx^2}(x)\Theta(x) = 0$ for every point $x = (x_1, x_2, x_3) \in \Omega = \Omega_{CET}$. We examine its integral curves. Subsection 3.1 shows that these integral curves are straight lines (segments of straight lines). Subsection 3.1 and the method of characteristics allow us to write the following "parametric" equation for these lines, which we first write in the invariant form (compare with (3.16))

$$x_1 dt_1 + x_2 dt_2 + x_3 dt_3 + dt_0 = 0\,. \tag{4.13}$$

If we choose (t_1, t_2) as the set of independent parameters defining the lines L_{t_1,t_2} that foliate our domain, then (4.13) can be rewritten as follows:

$$\begin{cases} x_1 + \dfrac{\partial t_3}{\partial t_1} \cdot x_3 + \dfrac{\partial t_0}{\partial t_1} = 0\,, \\ x_2 + \dfrac{\partial t_3}{\partial t_2} \cdot x_3 + \dfrac{\partial t_0}{\partial t_2} = 0\,. \end{cases} \tag{4.14}$$

As in the previous section, the solution of the Monge–Ampère equation is given by

$$B(x_1, x_2, x_3) = t_1 \cdot x_1 + t_2 \cdot x_2 + t_3 \cdot x_3 + t_0\,.$$

Here each line is given by fixing two free parameters (t_1, t_2), which are $t_1 = \frac{\partial B}{\partial x_1}$, $t_2 = \frac{\partial B}{\partial x_2}$. The parameters t_3, t_0 are not free; they are unknown functions of t_1, t_2, for which we will find (*a priori* nonlinear) PDE's. They will be simple in our case (and linear). Solving them will bring us the formula for B in the same way as it was in the proof of the claim of Section 3.

Let us use (4.6) now. Differentiating with respect to τ and setting $\tau = 1$, we get

$$t_1 x_1 + 2t_2 x_2 = 2B\,, \tag{4.15}$$

whence

$$t_1 x_1 + 2t_3 x_3 + 2t_0 = 0\,.$$

If we subtract the first equation of (4.14) multiplied by t_1 from (4.15), we get

$$\Big(2t_3 - t_1 \frac{\partial t_3}{\partial t_1}\Big) x_3 + \Big(2t_0 - t_1 \frac{\partial t_0}{\partial t_1}\Big) = 0$$

for every x_3. Hence

$$\begin{cases} 2t_3 - t_1 \dfrac{\partial t_3}{\partial t_1} = 0, \\ 2t_0 - t_1 \dfrac{\partial t_0}{\partial t_1} = 0\,. \end{cases}$$

From these "PDE" we easily deduce that

$$\begin{cases} t_3 = A(t_2) t_1^2, \\ t_0 = D(t_2) t_1^2\,. \end{cases} \tag{4.16}$$

Then the equations of the extremal lines L_{t_1,t_2} can be rewritten in the form

$$\begin{cases} x_1 + 2t_1 A x_3 + 2t_1 D = 0, \\ x_2 + t_1^2 A' x_3 + t_1^2 D' = 0\,. \end{cases} \tag{4.17}$$

We need to work a bit to find the functions A and D. We assume that our line L_{t_1,t_2} intersects the boundary $\partial\Omega$ at a point $\zeta = \zeta(t_1,t_2)$ on the "upper lid" $x_3 = M$ and at a point $\xi = \xi(t_1,t_2)$ on the "side" $x_2 = x_1^2$. Then we have two pairs of equations (4.17) asserting that our points are on the line L_{t_1,t_2},

$$\zeta_1 + 2t_1 AM + 2t_1 D = 0\,, \tag{4.18}$$

$$\zeta_2 + t_1^2 A' M + t_1^2 D' = 0\,, \tag{4.19}$$

$$\xi_1 + 2t_1 A\xi_3 + 2t_1 D = 0\,, \tag{4.20}$$

$$\xi_1^2 + t_1^2 A'\xi_3 + t_1^2 D' = 0\,, \tag{4.21}$$

and two boundary conditions (4.5) and (4.12),

$$\frac{1}{2}\xi_1 t_1 + \xi_1^2 t_2 = \xi_1^2 \xi_3\,, \tag{4.22}$$

$$A t_1^2 = \zeta_1^2\,. \tag{4.23}$$

Thus, we have six equations with six unknown functions: ζ_1, ζ_2, ξ_1, ξ_3, A, and D. Equation (4.19) determines the function ζ_2. If we take ζ_1 from (4.18) and plug it in (4.23), we get

$$A = 4(AM + D)^2\,. \tag{4.24}$$

So, introducing a new function

$$a := 2(AM + D) \tag{4.25}$$

of t_2, we can express the two functions A and D in terms of a: (4.24) yields

$$A = a^2 \tag{4.26}$$

and directly from the definition (4.25) of a we get

$$D = \frac{1}{2}a - Ma^2\,.$$

Note that deducing (4.24) we divided both parts of (4.23) by t_1. This is a lawful operation because the Bellman function $\mathbf{B}$ clearly depends on all variables x_i, and therefore its partial derivatives t_i cannot be identically zero. By the way, since t_3 is not identically zero, neither is a (see (4.16)). Moreover, we can assert that a is not a constant function. Indeed, assuming a to be a constant, we have $\zeta_2 = 0$ from (4.19), whence $\zeta_1 = 0$ due to $\zeta_1^2 \le \zeta_2 = 0$. But (4.18) can be written as $\zeta_1 + at_1 = 0$, and since neither t_1 nor a is zero, we arrive at a contradiction.

As a result we reduce our system to the system of three equations with three unknown functions $a(t_2)$, $\xi_1(t_1,t_2)$, and $\xi_3(t_1,t_2)$. Indeed, equations (4.20), (4.21), and (4.22) can be rewritten as follows:

$$\xi_1 + t_1 a\big(1 - 2a(M - \xi_3)\big) = 0, \tag{4.27}$$

$$\xi_1^2 + t_1^2 a'\Big(\frac{1}{2} - 2a(M - \xi_3)\Big) = 0, \tag{4.28}$$

$$\frac{1}{2}\xi_1 t_1 - \xi_1^2(\xi_3 - t_2) = 0\,. \tag{4.29}$$

To be able to divide (4.29) by ξ_1, we need to check that $\xi_1 = 0$ cannot be an appropriate solution. Assuming that $\xi_1 = 0$ identically, we get from (4.27) that $1 - 2a(M - \xi_3) = 0$, which converts (4.28) into $-\frac{1}{2}t_1^2 a' = 0$. But this is impossible because $t_1 \neq 0$ and a is not a constant function. So, we can rewrite (4.29) as

$$\xi_3 = t_2 + \frac{t_1}{2\xi_1}\,. \tag{4.30}$$

In fact ξ_3 does not depend on t_1. Indeed, if we introduce

$$\eta = 1 - 2a(M - \xi_3)\,, \tag{4.31}$$

equations (4.27) and (4.28) turn into

$$\xi_1 + t_1 a\eta = 0\,, \tag{4.32}$$

$$\xi_1^2 + t_1^2 a'\Big(\eta - \frac{1}{2}\Big) = 0\,, \tag{4.33}$$

whence

$$a'\Big(\eta - \frac{1}{2}\Big) + a^2\eta^2 = 0\,. \tag{4.34}$$

Since a does not depend on t_1, the function η (as a solution of this equation) also depends only on t_2. Soon we shall show that ξ_3 does not depend on t_2 either; it is a constant function!

But now we temporarily remove ξ_3 from the play. From (4.31) we have

$$\xi_3 = M - \frac{1-\eta}{2a}\,, \tag{4.35}$$

and, together with (4.32), formula (4.30) yields

$$\xi_3 = t_2 - \frac{1}{2a\eta}\,. \tag{4.36}$$

The resulting equation is

$$t_2 - \frac{1}{2a\eta} = M - \frac{1-\eta}{2a}\,,$$

or

$$t_2 - M = \frac{\eta^2 - \eta + 1}{2a\eta}\,. \tag{4.37}$$

As a result, we have two equations (4.34) and (4.37) for two unknown functions a and η of the variable t_2. Of course, now we would be able to solve the quadratic equation (4.34) with respect to η and plug the solution in (4.37) trying to solve the resulting differential equation with respect to a. This is possible, but leads to many complications. It is much easier to differentiate (4.37),

$$1 = \frac{a\eta'(\eta^2 - 1) - a'(\eta^3 - \eta^2 + \eta)}{2a^2\eta^2}\,,$$

and then replace the denominator by using (4.34). Then

$$-a'(2\eta - 1) = a\eta'(\eta^2 - 1) - a'(\eta^3 - \eta^2 + \eta)\,,$$

or

$$(\eta^2 - 1)(\eta - 1)a' = a(\eta^2 - 1)\eta'\,. \tag{4.38}$$

First we consider the possible constant solutions $\eta(t_2) = \pm 1$. The solution $\eta = 1$ is not suitable. Indeed, if $\eta = 1$, then from (4.37) we get

$$a = \frac{1}{2(t_2 - M)},$$

whence $a' = -2a^2$ and

$$A' = 2aa' = -4a^3, \qquad D' = \Big(\frac{1}{2} - 2aM\Big)a' = -a^2(1 - 4aM).$$

Therefore, in this case, the extremal line L_{t_1,t_2} defined by (4.17) has the form

$$\begin{cases} x_1 = -at_1\big(1 - 2a(M - x_3)\big), \\ x_2 = a^2t_1^2\big(1 - 4a(M - x_3)\big). \end{cases}$$

But a point of this line belongs to our domain Ω only if $x_3 = M$, because for other points we have

$$x_2 - x_1^2 = -4a^4t_1^2(M - x_3)^2 < 0.$$

Now, we check that the solution $\eta = -1$ is also impossible. If $\eta = -1$, then (4.37) yields

$$t_2 = M - \frac{3}{2a},$$

whence $a' = \frac{2}{3}a^2$ and

$$A' = 2aa' = \frac{4}{3}a^3, \qquad D' = \Big(\frac{1}{2} - 2aM\Big)a' = \frac{1}{3}a^2(1 - 4aM).$$

In this case the extremal line L_{t_1,t_2} has the form

$$\begin{cases} x_1 = at_1\big(2a(M - x_3) - 1\big), \\ x_2 = \dfrac{1}{3}a^2t_1^2\big(4a(M - x_3) - 1\big). \end{cases}$$

If we introduce $\omega = 2a(M - x_3) - 1$, then we get

$$\frac{x_1^2}{x_2} = \frac{3\omega^2}{2\omega + 1} \in [0, 1],$$

i.e., $-\frac{1}{3} \le \omega \le 1$, and ω is a function of x_1 and x_2 only; it does not depend on x_3. Therefore (see (4.16))

$$t_3 = a^2t_1^2 = \frac{3x_2}{2\omega + 1}$$

does not depend on x_3, and

$$B_{x_3x_3} = \frac{\partial t_3}{\partial x_3} = 0.$$

Since

$$t_2 = M - \frac{3}{2a} = M - 3\frac{M - x_3}{\omega + 1},$$

we have

$$B_{x_2x_3} = \frac{\partial t_2}{\partial x_3} = \frac{3}{\omega + 1},$$

and therefore

$$\det\begin{pmatrix} B_{x_2x_2} & B_{x_2x_3} \\ B_{x_3x_2} & B_{x_3x_3} \end{pmatrix} = -\frac{9}{(\omega + 1)^2} < 0, \qquad \forall x \in \Omega.$$

Hence the concavity condition (4.2) is not fulfilled and $\eta = -1$ is not an appropriate solution either.

So, we can return to (4.38) and cancel the common factor $(\eta^2 - 1)$. We arrive at

$$(\eta - 1)a' = a\eta',$$

which implies $a = C \cdot (\eta - 1)$. Then (4.35) yields

$$\xi_3(t_2) = M + \frac{1}{2C} =: c\,.$$

Recall that $\xi_3(t_1, t_2) = c$ is the level at which the line L_{t_1,t_2} intersects the boundary $x_2 = x_1^2$, $0 \le x_3 \le 1$, i.e., the third coordinate of the intersection point. So, $m \le c \le M$. By our assumptions, these lines intersect the boundary $x_3 = M$; therefore, to foliate the whole domain Ω (at least its interior) we need to have $c = m$. So, from now on we assume that $\xi_3 = c = m$. Of course, we could consider any $c > m$ and try to find an additional solution of the Monge–Ampére equation in the subdomain $m \le x_3 \le c$, glue it to our solution, and hope that this will work. But we shall not do this. First of all, we do not need to look for another solution. The better way to proceed is to try now to check that the function found above is precisely the Bellman function of the problem. Should we have some obstructions in proving, we could continue our search. Another argument explaining why we do not need "to glue" our candidate from two parts is the fact that there is no concave solution in the subdomain $m \le x_3 \le c$ which is a continuous extension of the solution found above.

Recall that we have the following expression for B (see (4.15)):

$$B = \frac{1}{2} t_1 x_1 + t_2 x_2\,. \tag{4.39}$$

Thus, to find B we need to find t_1 and t_2 as functions of a point x running over our domain Ω.

From (4.36) we have

$$t_2 = m + \frac{1}{2a[1 - 2a(M - m)]}\,. \tag{4.40}$$

This gives us the desired a as a function of t_2, but it is clear that a is a more convenient parameter than t_2. We shall express all other functions of t_2 in terms of a and look for a as a function of $x \in \Omega$. For this, we return to the equations of the extremal lines (4.17) rewriting them in terms of a. From (4.40) we have

$$\frac{dt_2}{da} = -\frac{1 - 4a(M - m)}{2a^2[1 - 2a(M - m)]^2}\,;$$

therefore

$$a' = -\frac{2a^2[1 - 2a(M - m)]^2}{1 - 4a(M - m)}$$

and

$$A' = 2aa' = -\frac{4a^3[1 - 2a(M - m)]^2}{1 - 4a(M - m)}\,,$$

$$D' = \frac{1}{2}a'(1 - 4aM) = -\frac{a^2(1 - 4aM)[1 - 2a(M - m)]^2}{1 - 4a(M - m)}\,.$$

Now equation (4.17) of the line L_{t_1,t_2} becomes

$$\begin{cases} x_1 = -t_1 a\,[1-2a(M-x_3)]\,, \\ x_2 = t_1^2 a^2 \dfrac{[1-2a(M-m)]^2[1-4a(M-x_3)]}{1-4a(M-m)}\,. \end{cases} \tag{4.41}$$

As we have known, the side surface $\{x_2 = x_1^2\}$ is intersected by L_{t_1,t_2} at the points of the "bottom lid" $\{x_3 = m\}$,

$$\xi = (\xi_1, \xi_1^2, m)\,, \qquad \xi_1 = -t_1 a\,[1-2a(M-m)]\,. \tag{4.42}$$

If we take ξ_1 for parameterizing the extremal lines, then their equations will be more symmetric. Indeed, (4.42) implies

$$t_1 = -\frac{\xi_1}{a[1-2a(M-m)]}\,,$$

and therefore (4.41) turns into

$$\begin{cases} x_1 = \dfrac{1-2a(M-x_3)}{1-2a(M-m)}\,\xi_1\,, \\ x_2 = \dfrac{1-4a(M-x_3)}{1-4a(M-m)}\,\xi_1^2\,. \end{cases} \tag{4.43}$$

These equations immediately supply us with an expression for a; namely, $a = a(x)$ is a root of the following cubic equation:

$$s := \frac{x_1^2}{x_2} = \left[\frac{1-2a(M-x_3)}{1-2a(M-m)}\right]^2 \frac{1-4a(M-m)}{1-4a(M-x_3)}\,. \tag{4.44}$$

To determine which of the three possible roots of the equation (4.44) gives us the desired value of a, we investigate the above function $s = s(a)$ defined by (4.44) as a function of the parameter a, assuming all other parameters to be fixed.

First, we note that the extremal line (4.43) intersects the plane $\{x_3 = M\}$ at the point

$$\zeta = (\zeta_1, \zeta_2, M)\,, \qquad \zeta_1 = \frac{\xi_1}{1-2a(M-m)}\,, \qquad \zeta_2 = \frac{\xi_1^2}{1-4a(M-m)}\,. \tag{4.45}$$

Since $\zeta_2 \geq 0$, we have the first restriction for a:

$$a < \frac{1}{4(M-m)}\,. \tag{4.46}$$

The behavior of $s(a)$ on the semiaxis (4.46) is the same for all values of other parameters: on the negative half-line it increases monotonically from $\frac{M-x_3}{M-m}$ to 1, and on the interval $[0, \frac{1}{4(M-m)}]$ it decreases monotonically from 1 to 0. Indeed, taking the logarithmic derivative of (4.44), we get

$$\begin{aligned} \frac{1}{s}\frac{\partial s}{\partial a} &= 2\left(\frac{2(M-m)}{1-2a(M-m)} - \frac{2(M-x_3)}{1-2a(M-x_3)}\right) \\ &\quad - \left(\frac{4(M-m)}{1-4a(M-m)} - \frac{4(M-x_3)}{1-4a(M-x_3)}\right) \\ &= \frac{4(x_3-m)}{[1-2a(M-m)][1-2a(M-x_3)]} - \frac{4(x_3-m)}{[1-4a(M-m)][1-4a(M-x_3)]}\,, \end{aligned}$$

whence we immediately conclude that

$$\operatorname{sgn}\frac{\partial s}{\partial a} = -\operatorname{sgn} a \tag{4.47}$$

if (4.46) is fulfilled.

Therefore, for a given s, $0 \le s \le 1$, we have one or two solutions of (4.44) satisfying (4.46): for all s there exists a positive solution $a = a^+$, and for $s > \frac{M-x_3}{M-m}$ there exists a negative solution $a = a^-$ as well. We shall see that our solution is $a = a^+$, but the solution $a = a^-$ is not meaningless: unexpectedly we get a solution of another extremal problem, namely (1.9).

To complete Step II of constructing a Bellman function candidate we need to write down formula (1.7) or (1.10). Everything is ready to do this: we have the expression (4.39) for B, where we need to substitute (4.40) for t_2 and to take t_1 from the first line of (4.41). So, we get

$$B(x;m,M) = -\frac{x_1^2}{2a\,[1-2a(M-x_3)]} + \left(m + \frac{1}{2a[1-2a(M-m)]}\right)x_2 \tag{4.48}$$

$$= \frac{(x_3-m)x_2}{\big[1-2a(M-m)\big]^2\big[1-4a(M-x_3)\big]} + mx_2\,, \tag{4.49}$$

where $a = a^+$ is the positive root ($4(M-m)a^+ \le 1$) of the cubic equation (4.44) for Theorem 3, and $a = a^-$ is the negative root for Theorem 4. □

5. Foliation according to B

We have already seen that the lines L_{t_1,t_2} are given by equations (4.43), and in the parametrization of the family of the extremal lines it is more convenient to use the parameter a instead of t_2 and ξ_1 instead of t_1:

$$\ell_{a,\xi_1} := L_{t_1,t_2}\,.$$

For a given ξ_1, we have a "fan" of extremal lines starting at an arbitrary point $\xi = (\xi_1, \xi_1^2, m)$ on the "edge" $\{x\colon x_2 = x_1^2,\ x_3 = m\}$:

$$F^+_{\xi_1} = \Big\{\ell_{a,\xi_1} : 0 \le a < \frac{1}{4(M-m)}\Big\}\,,$$
$$F^-_{\xi_1} = \big\{\ell_{a,\xi_1} :\ -\infty < a \le 0\big\}\,,$$

where $F^+_{\xi_1}$ is the fan of extremal lines of $B_{\max}$, and $F^-_{\xi_1}$ is the fan of extremal lines of $B_{\min}$.

Let us find the trace of our fan on the boundary $x_3 = M$. We recall that the intersection point ζ is given by (4.45):

$$\begin{cases} \zeta_1 = \dfrac{\xi_1}{1-2a(M-m)}\,, \\[2mm] \zeta_2 = \dfrac{\xi_1^2}{1-4a(M-m)}\,. \end{cases}$$

This is a parametric equations of the hyperbola C_{ξ_1},

$$\zeta_2 = \frac{\zeta_1\xi_1^2}{2\xi_1 - \zeta_1}\,,$$

tangent to the parabola $\zeta_2 = \zeta_1^2$ at the point (ξ_1, ξ_1^2).

The intersection of the fan $F^+_{\xi_1}$ with the plane $x_3 = M$ is the piece $C^+_{\xi_1}$ of this hyperbola between the points ξ_1 and $2\xi_1$, and for the fan $F^-_{\xi_1}$ it is the piece $C^-_{\xi_1}$ of this hyperbola between ξ_1 and zero. For $\xi_1 = 0$ it is the axis $C^+_0 = \{(0, \zeta_2)\}$ or the singleton $C^-_0 = \{(0,0)\}$.

Everything is on the "upper lid" $\{x_3 = M,\ x_1^2 \le x_2\}$ of Ω, and the curves $C^+_{\xi_1}$, as well as the curves $C^-_{\xi_1}$, foliate this "upper lid". The lines $\ell_{a,\xi_1} \in F^\pm_{\xi_1}$ connect the points of $C^\pm_{\xi_1}$ to the point (ξ_1, ξ_1^2, m) on the boundary of Ω, and each of the two sets of lines $\{F^\pm_{\xi_1}\colon -\infty < \xi_1 < +\infty\}$ foliates a subdomain of Ω. The lines of the fan $F^+_{\xi_1}$ foliate the whole Ω, whereas those of $F^-_{\xi_1}$ foliate the subdomain

$$\Big\{x = \{x_1, x_2, x_3\}\colon\ \frac{M - x_3}{M - m} < \frac{x_1^2}{x_2} \le 1,\ \ x_3 \le M\Big\}, \tag{5.1}$$

where the function $B_{\min}$ is defined. Indeed, we have a solution $a = a^+$ of (4.44) for an arbitrary point of Ω, while a solution $a = a^-$ exists if and only if the point x is from (5.1).

6. Lower boundary of Ω

Notice that we have found our function B only in $\{x_1^2 \le x_2,\ m < x_3 \le M\}$ for $B_{\max}$ and in $\{x_1^2 \le x_2,\ M - (M-m)\frac{x_1^2}{x_2} < x_3 \le M\}$ for $B_{\min}$. However, the two functions are continuous in the closed domains. For $B_{\min}$ clearly the limit is mx_2 as $x_3 \to M - (M-m)\frac{x_1^2}{x_2}$, which corresponds to $a \to -\infty$. For $B_{\max}$, formula (4.49) has indeterminacy $\frac{0}{0}$ as $x_3 \to m$, which corresponds to $a \to \frac{1}{4(M-m)}$. But we can easily pass to the limit in (4.48) and obtain the values of the function $B_{\max}(x_1, x_2, m; m, M)$ on the "lower lid" by continuity. Namely,

$$B_{\max}(x_1, x_2, m; m, M) = 4(M-m)(x_2 - x_1^2) + mx_2\,. \tag{6.1}$$

The norm of embedding becomes apparent. In fact, the best constant in the inequality

$$B_{\max} \le C\, x_2$$

can be read from (6.1), and it is at least $4M - 3m$. So for the case of $m = 0$ it is at least $4M$. Actually $4M$ **is** the best constant. This can easily be proved now, but notice that it is not even explicit in the formula for $B_{\max}(x; 0, M)$! So the "nonphysical" boundary $x_3 = 0$ can be useful too, and the limit values of B on it can give interesting information.

7. Reducing parameters M and m

Before we start proving Theorems 3 and 4, we would like to show that it is sufficient to prove them for $m = 0$ and $M = 1$. On the one hand, we have

$$\mathbf{B}_{CET}(x_1, x_2, x_3; m, M) = (M-m)\mathbf{B}_{CET}\Big(x_1, x_2, \frac{x_3 - m}{M - m}; 0, 1\Big) + mx_2. \tag{7.1}$$

This is a direct consequence of the definition. Indeed, while the homogeneity condition (4.6) follows from the definition if we compare the set of test functions $\{\phi\colon \langle\phi\rangle_I = x_1,\ \langle\phi^2\rangle_I = x_2\}$ and the set $\{\tilde\phi = t\phi\}$ with the same measure μ and the same set of point masses α_J, relation (7.1) follows from considering the same set of test functions ϕ but comparing with a "renormalized" measure $\tilde{\mu}(J) = \frac{\mu(J)-m|J|}{M-m}$ and point masses $\tilde{\alpha}_J = \frac{\alpha_J}{M-m}$. This yields $\tilde{m} = 0$, $\tilde{M} = 1$, and $\tilde{x}_3 = \frac{x_3-m}{M-m}$.

Notice that the definition of the function $\mathbf{B}_{cet}(x)$ in (1.5) differs from the definition of $\mathbf{B}_{CET}(x;0,1)$. The first allows no measure μ on the boundary. However, these functions have turned out to be equal, and the extremal μ must be zero. It is not clear how to see this immediately from the definitions (1.5) and (1.3) of $\mathbf{B}_{cet}(x)$ and $\mathbf{B}_{CET}(x;0,1)$. However, since the supremum in the definition of $\mathbf{B}_{cet}(x)$ is taken over the smaller set of test measures ($\mu = 0$), we have the inequality $\mathbf{B}_{cet}(x) \le \mathbf{B}_{CET}(x;0,1)$. So, the proof of Theorems 2 and 3 will consist in proving two inequalities: $\mathbf{B}_{CET}(x;0,1) \le B(x)$ and $\mathbf{B}_{cet}(x) \ge B(x)$, where $B(x)$ is the function defined by (1.6), or $B(x) = B(x;0,1)$ where $B(x;m,M)$ is the function defined by (1.7).

8. $\mathbf{B}_{CET}(x;0,1) \le B(x)$. Concavity

In what follows the symbol $\mathbf{B}(x)$ always stands for $\mathbf{B}_{CET}(x;0,1)$, $\Omega = \Omega_{CET}$ from (1.4) with $m = 0$ and $M = 1$, and a is the root of (4.44) (with $m = 0$ and $M = 1$) satisfying (4.46) ($a < \frac{1}{4}$ in our case).

LEMMA 5. *The function B is concave in the domain Ω.*

PROOF. We need to check that the matrix $\{B_{x_ix_j}\}$ is nonpositive, i.e., that all leading principal minors of $-\{B_{x_ix_j}\}$ are nonnegative. Since we already know that it has zero determinant (we have found B from this condition), we need to check two inequalities:

$$B_{x_1x_1} \le 0 \qquad \text{and} \qquad B_{x_1x_1}B_{x_2x_2} - B_{x_1x_2}B_{x_2x_1} \ge 0\,. \tag{8.1}$$

By the way, the reader could check by direct calculation that the function B satisfies the equation $\det\{B_{x_ix_j}\} = 0$. Probably, the simplest way to do this is not to calculate the determinant itself, but to verify that the directional vector Θ of the line L_{t_1,t_2},

$$\Theta(x) = \Big(\frac{2ax_1}{1-2a(1-x_3)}, \frac{4a(1-2a)^2x_1^2}{(1-4a)[1-2a(1-x_3)]^2}, 1\Big)^T, \tag{8.2}$$

is in the kernel of the Hessian matrix $\{B_{x_ix_j}\}$.

Let us return to the proof of inequalities (8.1). From the first line of (4.41) we deduce that

$$t_1 = -\frac{x_1}{a[1-2a(1-x_3)]}\,, \tag{8.3}$$

and

$$B_{x_1x_1} = \frac{\partial t_1}{\partial x_1} = -\frac{1}{a[1-2a(1-x_3)]} + \frac{1-4a(1-x_3)}{a^2[1-2a(1-x_3)]^2} \cdot \frac{2x_1^2}{x_2} \cdot \frac{\partial a}{\partial s}\,.$$

The sign of both summands here is the opposite of the sign of a (we use (4.47)). So, we have proved the first inequality of (8.1) with $a = a^+$, but simultaneously we get the inequality $(B_{\min})_{x_1x_1} \ge 0$ with $a = a^-$, which we need for proving the convexity of $B_{\min}$.

Using the expression (4.40) for t_2 (with $m = 0$, $M = 1$), we easily calculate other second derivatives we need:

$$B_{x_1x_2} = B_{x_2x_1} = \frac{\partial t_2}{\partial x_1} = -\frac{1-4a}{a^2(1-2a)^2} \cdot \frac{x_1}{x_2} \cdot \frac{\partial a}{\partial s},$$

$$B_{x_2x_2} = \frac{\partial t_2}{\partial x_2} = \frac{1-4a}{2a^2(1-2a)^2} \cdot \frac{x_1^2}{x_2^2} \cdot \frac{\partial a}{\partial s},$$

whence

$$\begin{aligned} B_{x_1x_1}B_{x_2x_2} - B_{x_1x_2}B_{x_2x_1} &= -\frac{B_{x_2x_2}}{a[1-2a(1-x_3)]} \\ &= \frac{1-4a}{2a^3(1-2a)^2[1-2a(1-x_3)]} \cdot \frac{x_1^2}{x_2^2} \cdot \frac{\partial a}{\partial s}, \end{aligned}$$

which is always nonnegative due to (4.47).

Thus, the concavity of $B = B_{\max}$ is fully proved, as well as the convexity of $B = B_{\min}$. □

LEMMA 6. *In* Ω *we have* $\frac{\partial B}{\partial x_3}(x) \geq x_1^2$.

PROOF. The assertion immediately follows from the first line of (4.16) together with (4.26) and (8.3):

$$\frac{\partial B}{\partial x_3} = t_3 = \frac{x_1^2}{[1-2a(1-x_3)]^2} \geq x_1^2$$

for $a = a^+$. It is clear that for $a = a^-$ (i.e., for $B_{\min}$) we have the opposite inequality. □

LEMMA 7. *In* Ω *we have* $\mathbf{B}_{CET}(x;0,1) \leq B(x)$.

PROOF. In the beginning of Section 4 it was already explained that the statements of Lemma 5 and Lemma 6 are equivalent to the **main inequality** (4.1).

Now the **main inequality** (4.1) and **the convexity of the domain** Ω will allow us to finish the proof of the theorem. We need a simple lemma, which is an analog of Green's formula for functions on graphs.

We consider the dyadic tree $\mathcal{T}$ whose vertices are denoted by σ, where σ is the word formed by $\pm$, the empty word σ_0 is the root of the tree, and $|\sigma|$ is the length of the word. The following linear operator on the functions on $\mathcal{T}$ is called the discrete Laplacian:

$$\Delta f(\sigma) = \frac{1}{2}(f(\sigma+) + f(\sigma-)) - f(\sigma).$$

We can associate the boundary of the tree with the interval $[0,1]$ in the following sense. For Lebesgue almost every point $t \in [0,1]$ we have a unique branch $(\sigma_0, \sigma_1(t), \dots)$ of $\mathcal{T}$ associated with it: just consider the dyadic form of t and let its nth digit encode whether to branch to the $+$ or $-$ side on the nth stage. Note that $\sigma_n(t)$ is the same for all t from a dyadic interval of nth generation, and so $f(\sigma_n(t))$ is a step function on $[0,1]$ constant on any dyadic interval of nth generation for every function f on $\mathcal{T}$.

Here is a lemma, which deserves to be called **Green's formula for the dyadic tree**. We identify the "boundary" of the tree with $[0,1]$ as was explained above.

LEMMA 8. *Let f be a positive function on $\mathcal{T}$ satisfying the following conditions:*
1) $f(\sigma_n(t)) \le g(t)\ \forall n,\ g \in L^1$;
2) *almost everywhere on* $[0,1]$ *there exists a limit* $\lim_{n\to\infty} f(\sigma_n(t)) := F(t)$.
Then

$$f(\sigma_0) = \int_0^1 F(t)\,dt - \sum_{\sigma\in\mathcal{T}} 2^{-|\sigma|}\Delta f(\sigma)\,.$$

PROOF. Multiplying the equalities

$$\Delta f(\sigma) = \frac{1}{2}\big(f(\sigma+) + f(\sigma-)\big) - f(\sigma)$$

by $2^{-|\sigma|}$ and summing over all σ of length k, we get

$$\sum_{|\sigma|=k} 2^{-|\sigma|}\Delta f(\sigma) = \sum_{|\sigma|=k+1} 2^{-|\sigma|}f(\sigma) - \sum_{|\sigma|=k} 2^{-|\sigma|}f(\sigma)\,.$$

If we now take the sum of this equalities over all $k < n$, we obtain

$$\sum_{|\sigma|<n} 2^{-|\sigma|}\Delta f(\sigma) = \sum_{|\sigma|=n} 2^{-|\sigma|}f(\sigma) - f(\sigma_0)\,.$$

The sum on the right hand side is the integral of the step function $f(\sigma_n(t))$ over $[0,1]$, i.e.,

$$f(\sigma_0) = \int_0^1 f(\sigma_n(t))\,dt - \sum_{|\sigma|<n} 2^{-|\sigma|}\Delta f(\sigma)\,.$$

To obtain the stated result we need only to pass to the limit in this equality. This is possible by the Lebesgue dominated convergence theorem. □

Now fix $x = (x_1, x_2, x_3) \in \Omega$ and fix any bounded function ϕ on the interval $I = [0,1]$ (we recall that nothing depends on the original interval) with

$$\langle\phi\rangle_I = x_1\,, \quad \langle\phi^2\rangle_I = x_2\,,$$

and a nonnegative measure μ on I such that $0 \le \mu(J) \le |J|$ for all dyadic subintervals J of I, and also an arbitrary collection of nonnegative numbers $\{\alpha_J\}_{J\in D(I)}$ such that

$$\frac{\mu(I)}{|I|} + \frac{1}{|I|}\sum_{J\in D(I)}\alpha_J = x_3 \in (0,1], \qquad \frac{\mu(\ell)}{|\ell|} + \frac{1}{|\ell|}\sum_{J\in D(\ell)}\alpha_J \in [0,1] \quad \forall \ell \in D(I)\,.$$

We see immediately that $d\mu = w\,dt$, where $0 \le w \le 1$.

The intervals of $D(I)$ and the dyadic tree $\mathcal{T}$ are in natural one-to-one correspondence. We call I_σ the interval corresponding to the vertex σ; I corresponds to σ_0. Consider the following function. Take B from (4.48) (with $M = 1, m = 0$) and put

$$M(\sigma) := \frac{1}{|I_\sigma|}\Big(\mu(I_\sigma) + \sum_{J\in D(I_\sigma)}\alpha_J\Big)$$

and

$$f(\sigma) := B\big(\langle\phi\rangle_{I_\sigma}, \langle\phi^2\rangle_{I_\sigma}, M(\sigma)\big)\,. \tag{8.4}$$

Then of course

$$f(\sigma_0) = B(x_1, x_2, x_3)\,. \tag{8.5}$$

A function on $\mathcal{T}$ is called subharmonic if its discrete Laplacian is nonnegative, i.e.,

$$\frac{1}{2}(f(\sigma+)+f(\sigma-))-f(\sigma)\geq 0 \qquad \forall\sigma\in\mathcal{T};$$

it is called superharmonic if its discrete Laplacian is nonpositive, i.e.,

$$\frac{1}{2}(f(\sigma+)+f(\sigma-))-f(\sigma)\leq 0 \qquad \forall\sigma\in\mathcal{T}.$$

By the main inequality (4.1), f is a superharmonic function on the tree $\mathcal{T}$. (Recall that for $B_{\min}$ we have the opposite main inequality, so in that case f is a subharmonic function.) Furthermore, the same (4.1) gives us an estimate of the discrete Laplacian:

$$-\Delta f(\sigma)\geq -\langle\phi\rangle_{I_\sigma}^2\,\Delta M(\sigma)$$

(and the opposite inequality for $B_{\min}$). It readily follows that

$$\Delta M(\sigma)=-\frac{\alpha_{I_\sigma}}{|I_\sigma|}=-2^{|\sigma|}\alpha_{I_\sigma}\,.$$

Therefore, combining the last two inequalities we get

$$-\Delta f(\sigma)\geq 2^{|\sigma|}\langle\phi\rangle_{I_\sigma}^2\,\alpha_{I_\sigma} \tag{8.6}$$

(and the opposite inequality for $B_{\min}$).

Now we want to apply Lemma 8. To do this we need to check conditions 1) and 2) of the lemma. Recall that now we consider bounded functions ϕ only; therefore, there is no problem with condition 1), because for a bounded function the averages over all subintervals are uniformly bounded and the argument of the continuous function B in (8.4) runs over a compact set, whence the functions $f(\sigma_n(t))$ are uniformly bounded as well. So we need to check that the limit of $f(\sigma_n(t))$ exists and to find it. To do this we use the following simple lemma.

LEMMA 9. *Let w be the density of the measure μ with respect to Lebesgue measure. Then for almost every $t\in[0,1]$ there exists a limit*

$$\lim_{n\to\infty} M(\sigma_n(t))=w(t)\,.$$

Let us take this lemma for granted and finish the proof first.

Given one of these "good" t that are also Lebesgue points of ϕ and ϕ^2, we have $\langle\phi\rangle_{I_{\sigma_n(t)}}\to\phi(t)$, $\langle\phi^2\rangle_{I_{\sigma_n(t)}}\to\phi(t)^2$. Looking at the definition (8.4) of $f(\sigma)$ and using Lemma 9, we conclude that

$$f(\sigma_n(t))\to B(\phi(t),\phi(t)^2,w(t))\,,$$

and since $B(x_1,x_1^2,x_3)=x_1^2x_3$, we get

$$f(\sigma_n(t))\to\phi(t)^2w(t)\,.$$

Substituting this into Lemma 8 and taking (8.5) into account, we obtain

$$B(x)=\int_0^1\phi^2d\mu-\sum_{\sigma\in\mathcal{T}}2^{-|\sigma|}\Delta f(\sigma)\,.$$

Estimate (8.6) yields the inequality

$$B(x)\geq\int_0^1\phi^2d\mu+\sum_{\sigma\in\mathcal{T}}\langle\phi\rangle_{I_\sigma}^2\alpha_{I_\sigma}\,.$$

Passing to the supremum over all the admissible test functions ϕ, we come to the desired estimate

$$B(x) \geq \mathbf{B}(x).$$

Note that for $B = B_{\min}$ we have the opposite inequality in (8.6). Using it and taking the infimum over all admissible test function ϕ, we arrive at the opposite estimate

$$B_{\min}(x) \leq \mathbf{B}_{\min}(x).$$

To complete the proof of Lemma 7, we have to return our debt — the proof of Lemma 9.

PROOF OF LEMMA 9. Since

$$M(\sigma_n(t)) = 2^n \Big(\mu(I_{\sigma_n(t)}) + \sum_{\ell \in D(I_{\sigma_n(t)})} \alpha_\ell\Big)$$

and

$$2^n \mu(I_{\sigma_n(t)}) = \frac{\mu(I_{\sigma_n(t)})}{|I_{\sigma_n(t)}|} \xrightarrow[n\to\infty]{} w(t)\,,$$

the only thing we need to prove is that for almost every $t \in [0,1]$ we have

$$\frac{1}{|J|} \sum_{\ell \in D(J)} \alpha_\ell \to 0\,,$$

as the dyadic intervals J containing t shrink to t.

Let

$$E_n = \Big\{ t \in [0,1] \colon \limsup_{\substack{|J|\to 0\\ J\in D(I),\, t\in J}} \frac{1}{|J|} \sum_{\ell \in D(J)} \alpha_\ell > \frac{1}{n} \Big\}.$$

Then for an arbitrary η, $\eta > 0$, and for every point t, $t \in E_n$, we can take a dyadic interval $J = J(\eta, t, n)$ such that

$$\sum_{\ell \in D(J)} \alpha_\ell \geq \frac{1}{n} |J|\,.$$

Since two dyadic intervals are either disjoint or one of them contains the other, we can choose a disjoint covering $J_1, J_2, \dots$ of E from the set of all these J. Then

$$|E_n| \leq \sum_k |J_k| \leq n \sum_k \sum_{\ell \in D(J_k)} \alpha_\ell \leq n \sum_{|\ell|<\eta} \alpha_\ell \xrightarrow[\eta\to 0]{} 0\,.$$

Therefore $|E_n| = 0$ for all n and the desired convergence occurs on the set of full measure $[0,1] \setminus \bigcup_n E_n$. Lemma 9 is proved. □

The proof of Lemma 7 is complete. □

9. $B(x) \leq \mathbf{B}_{cet}(x)$. Extremal sequences

LEMMA 10. *In Ω we have $B(x) \leq \mathbf{B}_{cet}(x)$.*

PROOF. Below $I_0 = [0,1]$. Consider first only the points $x = \zeta = (\zeta_1, \zeta_2, 1)$ on the "upper lid" of Ω. We want to fix a large integer n and to construct a function ϕ_n and a sequence $\{\alpha_\ell^n\}_{\ell \in D(I_0)}$ in such a way that the sum in the definition of $\mathbf{B}_{cet}$ approximates the value of $B(\zeta)$:

$$\sum_{\ell \in D(I_0)} \langle \phi_n \rangle_\ell^2 \alpha_\ell^n = B(\zeta_1, \zeta_2, 1) + o(1)\,,$$

where

$$\langle \phi_n \rangle_{I_0} = \zeta_1 \tag{9.1}$$

and

$$\langle \phi_n^2 \rangle_{I_0} = \zeta_2 \,. \tag{9.2}$$

Fix a large integer n and split $I_0 = [0,1]$ into the union of $n+1$ intervals $I_k = [2^{-k}, 2^{-k+1}]$, $1 \le k \le n$, and $J_0 = [0, 2^{-n}]$. The functions $m_k(x) := 2^k x - 1$ map I_k, $1 \le k \le n$, onto I_0. Let I_{ks} be the preimages of I_s under m_k. Then $m_s \circ m_k$ maps I_{ks} onto I_0. We also consider J_k, $1 \le k \le n$, the preimages of J_0 under m_k. We iterate this procedure obtaining $I_{k_1 k_2 \ldots k_m}$, $J_{k_1 k_2 \ldots k_{m-1}}$. Put $\alpha^n_{I_{k_1 k_2 \ldots k_m}} = |J_{k_1 k_2 \ldots k_m}|$ and $\alpha_\ell = 0$ for all other dyadic intervals ℓ. We call J_0 the 0th generation, and all $J_{k_1 k_2 \ldots k_m}$ the mth generation. It is easy to see that all J's are disjoint (except, perhaps, the endpoints), and moreover, $\bigcup J_{k_1 k_2 \ldots k_m} = I_0$ up to a set of measure zero. Therefore

$$\sum_{\ell \in D(I_0),\, \ell \subset I} \alpha_\ell = |I| \qquad \forall I \in D(I_0)\,,$$

and in particular the total length of all generations of J's is equal to 1. Now we are going to define ϕ_n:

$$\phi_n(t) = \begin{cases} c_n \zeta_1\,, & t \in J_0\,, \\ \phi_n(m_k(t))\,, & t \in I_k\,, \quad k = 2, \ldots, n\,, \\ d_n \phi_n(m_1(t))\,, & t \in I_1\,. \end{cases} \tag{9.3}$$

Here c_n and d_n are constants, which will be defined immediately. First, notice that this recursive definition of ϕ_n really defines it a.e. (as long as c_n, d_n are prescribed) merely because we have it already defined on the 0th generation of J's (that is on J_0). But then it is defined on the m_k-preimages of J_0 (and so, on the first generation of J's), but then it is defined on the preimages of preimages (the second generation of J's), *et cetera*. The disjointness of the J's guarantees that the ϕ_n are defined consistently, and since the union of all generations of J's gives I_0 up to a set of zero Lebesgue measure, the ϕ_n are defined almost everywhere.

We define c_n, d_n from the requirements (9.1), (9.2):

$$\frac{1}{2^n} c_n \zeta_1 + \Big(1 - \frac{1}{2^n} - \frac{1}{2}\Big)\zeta_1 + \frac{1}{2} d_n \zeta_1 = \zeta_1\,,$$
$$\frac{1}{2^n} c_n^2 \zeta_1^2 + \Big(1 - \frac{1}{2^n} - \frac{1}{2}\Big)\zeta_2 + \frac{1}{2} d_n^2 \zeta_2 = \zeta_2\,.$$

Then we get (using our notation $s = \zeta_1^2/\zeta_2 = (1-4a)(1-2a)^{-2}$) two possible solutions

(9.4)
$$c_n^\pm = \frac{1 + 2^{-n+1} \mp \sqrt{(1+2^{-n+1})(1-s)}}{s + 2^{-n+1}} \to \frac{1 \mp \sqrt{1-s}}{s} = \frac{1}{1 \pm \sqrt{1-s}} = 1 - 2a^\pm\,.$$

LEMMA 11. *We have*

$$\phi_n(t) = c_n \langle \phi_n \rangle_{I_{k_1 \ldots k_m}}, \quad \forall t \in J_{k_1 \ldots k_m}\,, \tag{9.5}$$

where the empty sequence $k_1 \ldots k_m$ corresponds to I_0 and J_0. In particular,

$$\langle \phi_n \rangle^2_{I_{k_1 \ldots k_m}} = \frac{1}{c_n^2} \langle \phi_n^2 \rangle_{J_{k_1 \ldots k_m}}. \tag{9.6}$$

PROOF. Let us start with the empty sequence. Then we should check that

$$\phi_n(t) = c_n\langle\phi_n\rangle_{I_0}, \quad \forall t \in J_0 .$$

But the average over I_0 is ζ_1, and $\phi_n(t) = c_n\zeta_1$ on J_0 by definition. The rest of the lemma follows by induction from the self-similarity of ϕ_n; in fact, it is either the same function ϕ_n "preshrunk" to a smaller interval, or it is a fixed multiple of this. So we are done. □

Now, using (9.6), (9.4), and the definition of α_ℓ^n we get

$$\begin{aligned}\sum_{\ell\in D(I_0)} \langle\phi_n\rangle_\ell^2\,\alpha_\ell^n &= \sum_{\mathbf{k}=k_1\dots k_m} \langle\phi_n\rangle_{I_\mathbf{k}}^2\,\alpha_{I_\mathbf{k}}^n = \frac{1}{c_n^2}\sum_{\mathbf{k}=k_1\dots k_m} \langle\phi_n^2\rangle_{J_\mathbf{k}}\,|J_\mathbf{k}| \\ &= \frac{1}{c_n^2}\int_0^1 \phi_n^2(t)dt = \frac{\zeta_2}{c_n^2} = \frac{\zeta_2}{(1-2a)^2} + o(1)\,.\end{aligned}$$

This completes the proof because, by (1.6) with $x_3 = 1$,

$$B(\zeta_1,\zeta_2,1) = \frac{\zeta_2}{(1-2a)^2}\,.$$

So, choosing $a = a^+$ we get the statement of the theorem, but simultaneously with $a = a^-$ we have proved the assertion $\mathbf{B}_{\min} \geq B(\zeta)$.

Now, using ϕ_n from (9.3) we can construct an extremal sequence of test functions for an arbitrary point x. For this aim we consider the extremal line $L = L_{t_1,t_2}$ passing through x. Let ζ be the point of intersection of L with the upper lid (i.e., $\zeta = (\zeta_1,\zeta_2,1)$), and ξ the point of L on the side boundary, or, what is the same in our case ($m = 0$), on the bottom lid (i.e., $\xi = (\xi_1,\xi_1^2,0)$). The expressions for the coordinates of ζ and ξ in terms of x_i can be deduced from (4.43):

$$\begin{aligned}\zeta &= \Big(\frac{x_1}{1-2a(1-x_3)}, \frac{x_2}{1-4a(1-x_3)}, 1\Big)\,, \\ \xi &= \Big(\frac{(1-2a)x_1}{1-2a(1-x_3)}, \frac{(1-4a)x_2}{1-4a(1-x_3)}, 0\Big)\,,\end{aligned}$$

where, as usual, a is the solution of (4.44) with $M = 1$ and $m = 0$.

Consider the binary representation of the number x_3:

$$x_3 = \sum_{l=1}^{\infty}\beta_l 2^{-l}\,, \qquad \beta_l \in \{0,1\}\,,$$

i.e., the β_l are binary digits of x_3. Then we define

$$\psi_n = \sum_{l=1}^{\infty}\big(\beta_l\,\phi_n\circ m_l + (1-\beta_l)\xi_1\big)\chi_{(2^{-l},2^{-l+1})}\,. \tag{9.7}$$

The same definition can be formulated as follows: if $\beta_l = 1$, then on the interval $(2^{-l}, 2^{-l+1})$ we put the properly rescaled function ϕ_n from (9.3) corresponding to the point ζ; otherwise we put the constant function equal to ξ_1. The masses α_I are defined by the same rescaling that was used for the functions ϕ_n. Let us explain in more detail what this means. First of all, we introduce a system of intervals $I_\mathbf{k}^l$, $J_\mathbf{k}^l$ as the preimages of $I_\mathbf{k}$, $J_\mathbf{k}$ (defined above) under the mapping m_l, $l = 1, 2, \dots$. Then we put $\alpha_{I_\mathbf{k}^l}^n = |J_\mathbf{k}^l|$ if $\beta_l = 1$, and $\alpha_\ell^n = 0$ for all other $\ell \in D(I_0)$.

Let us check that ψ_n and α^n constitute extremal sequences of test functions and masses. Using (9.1) we get:

$$\langle\psi_n\rangle_{[0,1]} = \int_0^1 \psi_n(t)\,dt = \sum_{l=1}^{\infty} 2^{-l}\Big(\beta_l \int_0^1 \phi_n(t)\,dt + (1-\beta_l)\xi_1\Big)$$
$$= x_3\zeta_1 + (1-x_3)\xi_1 = \frac{x_3x_1}{1-2a(1-x_3)} + \frac{(1-x_3)(1-2a)x_1}{1-2a(1-x_3)} = x_1\,.$$

Now we use (9.2) and take into account that $\beta_l^2 = \beta_l$:

$$\langle\psi_n^2\rangle_{[0,1]} = \int_0^1 \psi_n^2(t)\,dt = \sum_{l=1}^{\infty} 2^{-l}\Big(\beta_l \int_0^1 \phi_n^2(t)\,dt + (1-\beta_l)\xi_1^2\Big)$$
$$= x_3\zeta_2 + (1-x_3)\xi_2 = \frac{x_3x_2}{1-4a(1-x_3)} + \frac{(1-4a)x_2}{1-4a(1-x_3)} = x_2\,.$$

Finally, we need to check that $\sum_{\ell\in D(I_0)}\langle\psi_n\rangle_\ell^2\alpha_\ell^n$ tends to $B(x)$:

$$\sum_{\ell\in D(I_0)} \langle\psi_n\rangle_\ell^2\,\alpha_\ell^n = \sum_{l,\mathbf{k}=k_1\dots k_m} \langle\psi_n\rangle_{I_\mathbf{k}^l}^2\,\alpha_{I_\mathbf{k}^l}^n = \sum_l \beta_l \sum_\mathbf{k} \langle\phi_n\rangle_{I_\mathbf{k}}^2\,|J_\mathbf{k}^l|$$
$$= \sum_l \beta_l \sum_\mathbf{k} \frac{1}{c_n^2}\langle\phi_n\rangle_{J_\mathbf{k}}^2\,2^{-l}|J_\mathbf{k}| = \frac{x_3}{c_n^2}\int_0^1 \phi_n^2(t)dt$$
$$= \frac{x_3\zeta_2}{c_n^2} \to \frac{x_3\zeta_2}{(1-2a)^2} = B(x)\,.$$

□

9.1. Geometry of the foliation: explanation of the choice of ϕ_n in (9.3). This subsection contains a heuristic explanation why the function ϕ_n was chosen in this form in (9.3).

The main point here is a certain geometric observation concerning the foliation of the domain Ω by the lines L_{t_1,t_2}. Consider the foliation of the upper lid $\{x_3 = 1\} \cap \Omega$ by the parabolas $P_s := \{x\colon sx_2 = x_1^2,\ x_3 = 1\}$, $0 \le s \le 1$. Then we make the following easy geometric observation.

LEMMA 12. *Let L_{t_1,t_2} intersects the upper lid at the point $\zeta = (\zeta_1, \zeta_2, 1)$ and let $\zeta \in P_s$. Then the projection of the line L_{t_1,t_2} to the upper lid is tangent to P_s at ζ.*

PROOF. The line $L_{t_1,t_2} = \ell_{a,\xi_1}$ passes through the points

$$\xi = (\xi_1, \xi_1^2, 0) \qquad \text{and} \qquad \zeta = \Big(\frac{\xi_1}{1-2a}, \frac{\xi_1^2}{1-4a}, 1\Big)$$

(see Section 5), and therefore, the slope of its projection to the boundary $x_3 = 1$ is

$$\frac{\dfrac{\xi_1^2}{1-4a} - \xi_1^2}{\dfrac{\xi_1}{1-2a} - \xi_1} = 2\xi_1\frac{1-2a}{1-4a}\,.$$

But the slope of the tangent line to P_s at ζ is

$$\frac{2\zeta_1}{s} = \frac{2\xi_1}{1-2a}\cdot\frac{(1-2a)^2}{1-4a} = 2\xi_1\frac{1-2a}{1-4a}\,.$$

(We have used the expression (4.44) for s with $m = 0$ and $x_3 = M = 1$.) The coincidence of these two quantities proves the lemma. □

Let us see what influence this lemma has on the choice of ϕ_n.

We are given $L = L_{t_1,t_2}$ intersecting the upper lid at $\zeta = (\zeta_1, \zeta_2, 1)$ and the bottom lid at $\xi = (\xi_1, \xi_1^2, 0)$. To construct ϕ corresponding to ζ we make infinitesimally small jumps $\zeta + \varepsilon\ell$, $\zeta - \varepsilon\ell$ along L (here ℓ is a unit vector parallel to L and with positive third coordinate). Should we know how to build the extremal functions $\phi_\pm$ for $\zeta \pm \varepsilon\ell$, we would just put the one for $\zeta + \varepsilon\ell$ to the right half $I^+ = I_1$ of $I = I_0$ and the one for $\zeta - \varepsilon\ell$ to the left half I^- of I_0. The one for $\zeta + \varepsilon\ell$ is definitely unknown. But the one for $\zeta - \varepsilon\ell$ is "known" in the following sense: we can make the next splitting $I^- = I^{--} \cup I^{-+}$ by jumping from $\zeta - \varepsilon\ell$ to $\zeta - 2\varepsilon\ell = (\zeta - \varepsilon\ell) - \varepsilon\ell$ and $\zeta = (\zeta - \varepsilon\ell) + \varepsilon\ell$. Returning to the point ζ for $I^{-+} = I_2$ we have to put there the initial function ϕ_n properly rescaled. Then we can restart the same procedure again for I^{--} and the corresponding point $\zeta - 2\varepsilon\ell$, but now we make a jump of the length 2ε to return to the initial point ζ. This is an explanation for the self-similar structure of ϕ_n and for the second line of (9.3).

If we take ε equal to the $2^{-(n-1)}$th part of the distance between the points ζ and ξ, then after n jumps along L we will find ourselves at the endpoint $\zeta - 2^n\varepsilon\ell = \xi$ of L. At this moment (i.e., for the interval J_0) we know that the test function is just the constant ξ_1. This is an explanation for the first line in the definition of ϕ_n.

To explain d_n, we recall that we do not know yet how to build the extremal function on I_1 for the point $\zeta + \varepsilon\ell$. Moreover, we cannot do that. This point is outside of Ω. However, when we are on the upper lid we are not forced to split the point x into the pair $x^\pm$ for which x is the midpoint of the interval $[x^-, x^+]$; moreover we must take $x_3^- + x_3^+ < 2x_3$ to get a nonzero measure α. We can do this just on the upper lid due to the boundary condition (4.12). After fixing $x^- = \zeta - \varepsilon\ell$, we have to take $x_1^+ = \zeta_1 + \varepsilon\ell_1$, $x_2^+ = \zeta_2 + \varepsilon\ell_2$, but the third coordinate can be chosen arbitrarily; it only has to be close enough to $\zeta_3 = 1$ not to disturb too much equality that must approximately occur in the **main inequality**. For any choice we get some extremal sequence. For simplicity we can take x^+ on the upper lid as well, i.e. choose $x_3^+ = 1$. Then the point x^+ will be the projection p_ε of $\zeta + \varepsilon\ell$ to the upper lid. We know from Lemma 12 that p_ε lies on the tangent line to the parabola P_s: $sx_2 = x_1^2$ at the point ζ. We do not know an extremal function for this point, but we "know" extremal functions (sequences) for all points of the parabola P_s, because they are multiples of our original function ϕ_n. Indeed, if ϕ is an extremal function for $\mathbf{B}(x_1, x_2, x_3)$, then $\tau\phi$ is an extremal function for $\mathbf{B}(\tau x_1, \tau^2 x_2, x_3)$, i.e., the extremal functions corresponding to the different points of the parabola $sx_2 = x_1^2$ differ by a constant factor. So, we make one more perturbation and instead of p_ε take x^+ on P_s and put the properly rescaled function ϕ_n times a constant on I_1. This is the explanation for the third line of (9.3).

Let us note that moving from the tangent line to the parabola P_s we disturb first two coordinates of the point. Therefore, instead of splitting along the extremal line, we split along some line close enough to L, whose projection to the upper lid is not a tangent but a secant, which tends to the true extremal line in the limit $n \to \infty$ ($\varepsilon \to 0$). Since the extremal sequence of test functions, of course, is not unique, we can allow such perturbation for simplicity of the construction of an extremal sequence.

Finally, we need to explain how the extremal sequence of the masses α^n was chosen. Nonzero mass appears only when $x_3^- + x_3^+ > 2x_3$. Moreover, we know the value of this mass, it is $\big(x_3 - \frac{1}{2}(x_3^- + x_3^+)\big)|I|$, where I is the interval we split. For example, for $I = I_0$ we have $|I_0| = 1$, $x_3 = x_3^+ = 1$, and therefore $\alpha^n_{I_0} = \frac{1}{2}(1 - x_3^-)$, i.e., the half of the projection of the vector $\varepsilon\ell$ to the plane $\{x_1, x_2\}$. Recall that ε is equal to the $2^{-(n-1)}$th part of the distance between the points ζ and ξ. But the projection of the segment between the points ζ and ξ to the plane $\{x_1, x_2\}$ is just one ($\zeta_3 - \xi_3 = 1$); thus we have to put $\alpha^n_{I_0} = 2^{-n} = |J_0|$. By the recursive procedure of our definition we get $\alpha_{I_{\mathbf{k}}} = |J_{\mathbf{k}}|$ for all intervals $I_{\mathbf{k}}$. Only the intervals $I_{\mathbf{k}}$ with some multiindex $\mathbf{k}$ correspond to points on the upper lid, and for the points corresponding to all other intervals we have $x_3 < 1$, i.e., we split them with equality $x_3^- + x_3^+ = 2x_3$, and therefore we have to put $\alpha^n_\ell = 0$ for all these intervals.

9.2. Geometry of the foliation: explanation of the choice of ψ_n in (9.7). Now we try to explain why the function ψ_n was chosen in this form in (9.7). Again, let L be the extremal line passing through the point x and let $\zeta = (\zeta_1, \zeta_2, 1)$, $\xi = (\xi_1, \xi_1^2, 0)$ be its endpoints on the boundary of Ω. We know that, to keep equality in the main inequality when splitting our interval, we have to take the points $x^\pm$ on L. If $\beta_1 = 1$, i.e., $x_3 \geq \frac{1}{2}$ and the distance between x and ζ is not greater than the distance from x to ξ, then we can take $x^+ = \zeta$, and therefore put to $I^+ = I_1$ the corresponding extremal function and masses already known to us. If $\beta_1 = 0$, i.e., $x_3 < \frac{1}{2}$, the point ξ is closer to x than ζ, and then we take $x^+ = \xi$, where the situation is simple: we have put to I^+ the only possible test function identically equal to ξ_1 and no masses to any subintervals of I^+.

After this step, everything is defined on $I^+ = I_1$, and we continue in the same way to consider the point x^- by splitting the interval $I^- = (0, 1/2)$ into the union of $I^{--} = (0, 1/4)$ and $I^{-+} = (1/4, 1/2) = I_2$. In a similar way, we put on I_2 either the rescaled ϕ_n if ζ is closer to x^- (i.e., $\beta_2 = 1$) or the constant function ξ_1 if the point ξ is closer. Continuing in this way, we get the representation (9.7) for ψ_n.

9.3. Extremal functions for $B_{\min}$. The inequality $\mathbf{B}_{\min}(x) \leq B_{\min}(x)$ is proved only for the points of the domain of $B_{\min}$ when there exists a negative solution of equation (4.44), i.e., for $x_2(1 - x_3) < x_1^2$. To complete the proof of Theorem 4 we need to check that $\mathbf{B}_{\min}(x) = 0$ for all $x = (x_1, x_2, x_3)$ such that $x_2(1 - x_3) \geq x_1^2$.

In this situation there exists an extremal function ϕ and extremal weights α such that

$$\sum_{\ell \in D(I_0)} \langle \phi \rangle_\ell^2 \alpha_\ell^n = 0\,. \tag{9.8}$$

Indeed, if β_l are the binary digits of $\frac{x_1^2}{x_2}$, i.e.,

$$\frac{x_1^2}{x_2} = \sum_{l=1}^{\infty} \beta_l\, 2^{-l}\,,$$

we take

$$\phi = \frac{x_2}{x_1} \sum_{l=1}^{\infty} \beta_l\, \chi_{I_l} \tag{9.9}$$

and put $\alpha_\ell = 0$ for all dyadic ℓ except the interval $\ell = I_l$ such that $\beta_l = 0$. For the latter intervals we put

$$\alpha_{I_l} = \frac{x_2 x_3}{x_2 - x_1^2} \cdot |I_l|. \tag{9.10}$$

Recall that $I_l = (2^{-l}, 2^{-l+1})$, i.e., $\alpha_{I_l} = 2^{-l}$.

For such a pair ϕ, α formula (9.8) is clearly true, because nonzero masses α_ℓ correspond only to intervals where the function ϕ is identically zero. Therefore, we need only to check that the pair ϕ, α is admissible. The condition

$$\sum_{\ell\in D(J)} \alpha_\ell \le |J| \qquad \forall J \in D([0,1])$$

is evidently fulfilled, because the sum above contains at most one interval I_l (except $\ell = [0,1]$), and since $x_2x_3 \le x_2 - x_1^2$, we have

$$\sum_{\ell\in D(J)} \alpha_\ell \le \alpha_{I_l} \le |I_l| \le |J|\,.$$

We have the required averages of ϕ and ϕ^2:

$$\langle\phi\rangle_{[0,1]} = \int_0^1 \phi(t)\,dt = \frac{x_2}{x_1}\sum_{l=1}^{\infty}\beta_l\,2^{-l} = \frac{x_2}{x_1}\cdot\frac{x_1^2}{x_2} = x_1\,,$$

$$\langle\phi^2\rangle_{[0,1]} = \int_0^1 \phi^2(t)\,dt = \frac{x_2^2}{x_1^2}\sum_{l=1}^{\infty}\beta_l\,2^{-l} = \frac{x_2^2}{x_1^2}\cdot\frac{x_1^2}{x_2} = x_2\,,$$

and the total mass is

$$\sum_{\ell\in D([0,1])} \alpha_\ell = \frac{x_2x_3}{x_2-x_1^2}\sum_{l:\,\beta_l=0}|I_l| = \frac{x_2x_3}{x_2-x_1^2}\sum_{l\ge1}(1-\beta_l)\,2^{-l}$$

$$= \frac{x_2x_3}{x_2-x_1^2}\Big(1-\frac{x_1^2}{x_2}\Big) = x_3\,.$$

This completes the proof of Theorem 4.

Now we add a few words to explain the above construction. First of all we note that the subdomain $x_2(1-x_3) \le x_1^2$ of Ω is convex and any straight line in this subdomain is extremal. This means that we can take an extremal line L passing through the given point x and connecting two points of the boundary, where we know the corresponding extremal functions and masses. These are the points

$$\xi = \Big(\frac{x_2}{x_1}, \frac{x_2^2}{x_1^2}, 0\Big) \qquad \text{and} \qquad \zeta = \Big(0, 0, \frac{x_2x_3}{x_2-x_1^2}\Big)\,.$$

The constant function $\frac{x_2}{x_1}$ and zero mass correspond to the point ξ, and the zero function and the mass $\frac{x_2x_3}{x_2-x_1^2}$ correspond to the point ζ. Should we have a nondyadic problem, we could choose $x^- = \xi$ and $x^+ = \zeta$ with $\alpha_{I^-} = 0$ and $\alpha_{I^+} = \frac{x_2x_3}{x_2-x_1^2}$. Since we have a dyadic problem, we have to make a dyadic rearrangement: on each step we choose the nearest point among ξ and ζ, to put the corresponding function and mass, say, to I^+ and to continue splitting I^- following the same rule. This brings us to formulas (9.9) and (9.10).

10. Step I for Carleson embedding theorems in L^p with $p > 1$

Now, we briefly consider how to find the Bellman function for the Carleson embedding operator acting on arbitrary L^p. We shall follow the same method and we start with an evident definition of the Bellman function. Namely,

$$\begin{aligned}
&\mathbf{B}_{\max}(x_1, x_2, x_3; m, M)\\
&\quad:= \sup\Big\{ \int_I |\phi(t)|^p \, d\mu(t) + \sum_{J\in D(I)} \alpha_J |\langle\phi\rangle_J|^p :\\
&\qquad\qquad \langle\phi\rangle_I = x_1,\ \langle|\phi|^p\rangle_I = x_2,\ \frac{1}{|I|}\Big(\mu(I) + \sum_{J\in D(I)} \alpha_J\Big) = x_3,\\
&\qquad\qquad\qquad m|J| \le \mu(J) + \sum_{\ell\in D(J)} \alpha_\ell \le M|J|\ \forall J \in D(I)\Big\},
\end{aligned}$$

$$\begin{aligned}
&\mathbf{B}_{\min}(x_1, x_2, x_3; m, M)\\
&\quad:= \inf\Big\{ \int_I |\phi(t)|^p \, d\mu(t) + \sum_{J\in D(I)} \alpha_J |\langle\phi\rangle_J|^p :\\
&\qquad\qquad \langle\phi\rangle_I = x_1,\ \langle|\phi|^p\rangle_I = x_2,\ \frac{1}{|I|}\Big(\mu(I) + \sum_{J\in D(I)} \alpha_J\Big) = x_3,\\
&\qquad\qquad\qquad m|J| \le \mu(J) + \sum_{\ell\in D(J)} \alpha_\ell \le M|J| \quad \forall J \in D(I)\Big\}.
\end{aligned}$$

The functions $\mathbf{B}$ are defined in

$$\Omega := \{x = (x_1, x_2, x_3) \colon\ |x_1|^p \le x_2,\ m \le x_3 \le M\}. \tag{10.1}$$

As before, the Bellman functions $\mathbf{B}$ do not depend on the interval I.

From the definition of the functions $\mathbf{B}$ above we can immediately see that they satisfy the main inequality

$$\begin{aligned}
&\mathbf{B}_{\max}(x) - \frac{1}{2}(\mathbf{B}_{\max}(x^-) + \mathbf{B}_{\max}(x^+)) \ge |x_1|^p\Big(x_3 - \frac{1}{2}(x_3^+ + x_3^-)\Big),\\
&\mathbf{B}_{\min}(x) - \frac{1}{2}(\mathbf{B}_{\min}(x^-) + \mathbf{B}_{\min}(x^+)) \le |x_1|^p\Big(x_3 - \frac{1}{2}(x_3^+ + x_3^-)\Big),\\
&\forall x,\ x^\pm \in \Omega,\quad x_i = \frac{1}{2}(x_i^+ + x_i^-) \text{ for } i = 1, 2, \quad \text{and } x_3 \ge \frac{1}{2}(x_3^+ + x_3^-).
\end{aligned}$$

Since the domain Ω, defined by (10.1), is convex, the main inequality is equivalent to a pair of differential inequalities. The first inequality of this pair is just concavity for $\mathbf{B}_{\max}$ and convexity for $\mathbf{B}_{\min}$:

$$\frac{d^2\mathbf{B}_{\max}}{dx^2}(x) \le 0, \qquad \frac{d^2\mathbf{B}_{\min}}{dx^2}(x) \ge 0. \tag{10.2}$$

The second means that the function is not only monotonic in x_3, but has an estimate for the corresponding partial derivative:

$$\frac{\partial\mathbf{B}_{\max}}{\partial x_3}(x) \ge |x_1|^p, \qquad \frac{\partial\mathbf{B}_{\min}}{\partial x_3}(x) \le |x_1|^p. \tag{10.3}$$

One boundary condition follows from the definition:

$$\mathbf{B}(x_1, |x_1|^p, x_3) = |x_1|^p\, x_3, \tag{10.4}$$

as well as the homogeneity condition:

$$\mathbf{B}(\tau x_1, |\tau|^p x_2, x_3) = |\tau|^p\, \mathbf{B}(x_1, x_2, x_3). \tag{10.5}$$

As before, we are looking for a "best" function B satisfying (10.2)–(10.5). This means that we subject our function B to two more **sharpness conditions**. These are given by the following degeneration condition:

$$\det \frac{d^2B}{dx^2} = \det \begin{pmatrix} B_{x_1x_1} & B_{x_1x_2} & B_{x_1x_3} \\ B_{x_2x_1} & B_{x_2x_2} & B_{x_2x_3} \\ B_{x_3x_1} & B_{x_3x_2} & B_{x_3x_3} \end{pmatrix} = 0 \qquad \forall x \in \Omega \tag{10.6}$$

(the Monge–Ampère equation), and the second boundary condition:

$$\frac{\partial B}{\partial x_3}(x_1, x_2, M; m, M) = |x_1|^p\,. \tag{10.7}$$

CLAIM. *There is a simple algorithm to find the functions* $B_{\max}(x; m, M)$ *and* $B_{\min}(x; m, M)$ *that satisfy conditions* (10.2)–(10.7).

PROOF. Again we look for a solution B of the form

$$B(x) = t_1 \cdot x_1 + t_2 \cdot x_2 + t_3 \cdot x_3 + t_0\,,$$

where $t_i = \frac{\partial B}{\partial x_i}$, $i = 1, 2, 3$. The integral curves of the vector field Θ such that $\frac{d^2B}{dx^2}(x)\Theta(x) = 0$ are again intervals of straight lines

$$x_1dt_1 + x_2dt_2 + x_3dt_3 + dt_0 = 0\,. \tag{10.8}$$

Each line is given by fixing two free parameters, for example t_1 and t_2. Then we can rewrite equation (10.8) defining lines L_{t_1,t_2} that foliate our domain in the following form:

$$\begin{cases} x_1 + \dfrac{\partial t_3}{\partial t_1} \cdot x_3 + \dfrac{\partial t_0}{\partial t_1} = 0\,, \\ x_2 + \dfrac{\partial t_3}{\partial t_2} \cdot x_3 + \dfrac{\partial t_0}{\partial t_2} = 0\,. \end{cases} \tag{10.9}$$

Differentiating (10.5) with respect to τ at the point $\tau = 1$, we get

$$x_1t_1 + px_2t_2 = pB\,, \tag{10.10}$$

whence

$$(p-1)t_1x_1 + pt_3x_3 + pt_0 = 0\,. \tag{10.11}$$

If we subtract the first equation of (10.9) multiplied by $(p-1)t_1$ from (10.11), we get

$$\Big(pt_3 - (p-1)t_1\frac{\partial t_3}{\partial t_1}\Big)x_3 + \Big(pt_0 - (p-1)t_1\frac{\partial t_0}{\partial t_1}\Big) = 0$$

for every x_3. If we introduce the dual exponent $q = \frac{p}{p-1}$, then the latter equation can be rewritten as follows:

$$\Big(qt_3 - t_1\frac{\partial t_3}{\partial t_1}\Big)x_3 + \Big(qt_0 - t_1\frac{\partial t_0}{\partial t_1}\Big) = 0\,.$$

Hence

$$\begin{cases} qt_3 - t_1\dfrac{\partial t_3}{\partial t_1} = 0\,, \\ qt_0 - t_1\dfrac{\partial t_0}{\partial t_1} = 0\,. \end{cases}$$

From these "PDE" we easily see that

$$\begin{cases} t_3 = A(t_2)|t_1|^q\,, \\ t_0 = D(t_2)|t_1|^q\,. \end{cases}$$

Then the equations (10.9) of the extremal lines L_{t_1,t_2} can be rewritten in the form

$$\begin{cases} x_1 + q|t_1|^{q-2}t_1 A x_3 + q|t_1|^{q-2}t_1 D = 0\,, \\ x_2 + |t_1|^q A' x_3 + |t_1|^q D' = 0\,. \end{cases} \tag{10.12}$$

As before, we assume that our extremal line L_{t_1,t_2} intersects the boundary $\partial\Omega$ at a point $\zeta = \zeta(t_1,t_2)$ on the "upper lid" $x_3 = M$ and at a point $\xi = \xi(t_1,t_2)$ on the "side" $x_2 = |x_1|^p$. Then we have two pairs of equations (10.12) asserting that our points are on the line L_{t_1,t_2}:

$$\zeta_1 + q|t_1|^{q-2}t_1 AM + q|t_1|^{q-2}t_1 D = 0\,, \tag{10.13}$$

$$\zeta_2 + |t_1|^q A'M + |t_1|^q D' = 0\,, \tag{10.14}$$

$$\xi_1 + q|t_1|^{q-2}t_1 A\xi_3 + q|t_1|^{q-2}t_1 D = 0\,, \tag{10.15}$$

$$|\xi_1|^p + |t_1|^q A'\xi_3 + |t_1|^q D' = 0\,, \tag{10.16}$$

and two boundary conditions (10.4) and (10.7):

$$\frac{1}{p}\xi_1 t_1 + |\xi_1|^p t_2 = |\xi_1|^p \xi_3\,, \tag{10.17}$$

$$A|t_1|^q = |\zeta_1|^p\,. \tag{10.18}$$

If we take ζ_1 from (10.13) and plug it in (10.18), we get

$$A = q^p|AM + D|^p\,. \tag{10.19}$$

So, introducing a new function

$$a = q(AM + D) \tag{10.20}$$

of t_2, we can express the two functions A and D in terms of a: (10.19) yields

$$A = |a|^p,$$

and directly from the definition (10.20) of a we get

$$D = \frac{1}{q}a - M|a|^p\,.$$

As a result, equations (10.15), (10.16), and (10.17) can be rewritten as follows:

$$\xi_1 + |t_1|^{q-2}t_1 a\Big(1 - qa|a|^{p-2}(M - \xi_3)\Big) = 0\,, \tag{10.21}$$

$$|\xi_1|^p + \frac{p}{q}|t_1|^q a'\Big(\frac{1}{p} - qa|a|^{p-2}(M - \xi_3)\Big) = 0\,, \tag{10.22}$$

$$\frac{1}{p}\xi_1 t_1 - |\xi_1|^p(\xi_3 - t_2) = 0\,. \tag{10.23}$$

As before, we introduce the following new function:

$$\eta = 1 - qa|a|^{p-2}(M - \xi_3)\,, \tag{10.24}$$

and then equations (10.21) and (10.22) turn into

$$\xi_1 + |t_1|^{q-2}t_1 a\eta = 0\,, \tag{10.25}$$

$$|\xi_1|^p + \frac{p}{q}|t_1|^q a'\Big(\eta - \frac{1}{q}\Big) = 0\,, \tag{10.26}$$

whence

$$a'\Big(\eta - \frac{1}{q}\Big) + \frac{q}{p}|a|^p|\eta|^p = 0\,. \tag{10.27}$$

Since a does not depend on t_1, the function η (as a solution of this equation) also depends only on t_2.

On the one hand, from (10.24) we have

$$\xi_3 = M - \frac{1-\eta}{qa|a|^{p-2}}\,; \tag{10.28}$$

on the other hand, (10.23) together with (10.25) yield

$$\xi_3 = t_2 + \frac{\xi_1 t_1}{p|\xi_1|^p} = t_2 - \frac{a\eta}{p|a|^p|\eta|^p}\,. \tag{10.29}$$

So, excluding x_3 we get

$$t_2 - \frac{a\eta}{p|a|^p|\eta|^p} = M - \frac{1-\eta}{qa|a|^{p-2}}\,.$$

Differentiating the last equation with respect to t_2 we obtain

$$1 + \frac{a'\eta + \eta' a}{q|a|^p|\eta|^p} = \frac{\eta'}{qa|a|^{p-2}} + \frac{p-1}{q}\cdot\frac{(1-\eta)a'}{|a|^p}$$

and then we multiply this formula by the common denominator and replace it using (10.27) by $-pa'(\eta - \frac{1}{q})$:

$$-pa'\Big(\eta - \frac{1}{q}\Big) + a'\eta + \eta' a = \eta' a|\eta|^p + (p-1)(1-\eta)a'|\eta|^p\,,$$

or

$$(p-1)(|\eta|^p - 1)(\eta - 1)a' = a(|\eta|^p - 1)\eta'\,.$$

Thus, we arrive at the following equation:

$$(p-1)(\eta - 1)a' = a\eta'\,,$$

which implies $\eta = 1 + C|a|^{p-1}$. Then (10.28) yields

$$\xi_3(t_2) = M + \frac{C|a|^{p-1}}{qa|a|^{p-2}} = M + \frac{C}{q}\operatorname{sgn} a =: c\,.$$

The last expression is a constant, because $\operatorname{sgn} a$ can take only one value due to $\xi_3 \le M$. As before, the natural choice for the constant c is $c = m$.

Recall that we have the following expression for B (see (10.10)):

$$B = \frac{1}{p}t_1 x_1 + t_2 x_2\,.$$

Thus, to find an expression for B, we need to find t_1 and t_2 as functions of a point x running over our domain Ω.

From (10.24) we have

$$\eta = 1 - qa|a|^{p-2}(M - m)\,,$$

hence (10.29) yields

$$t_2 = m + \frac{a\eta}{p|a\eta|^p} = m + \frac{a - q(M-m)|a|^p}{p\big|a - q(M-m)|a|^p\big|^p}\,. \tag{10.30}$$

This gives us the desired a as a function of t_2, but it is clear that a is a more convenient parameter than t_2. We shall express all other functions of t_2 in terms of a and look for a as a function of $x \in \Omega$. For this, we return to the equations (10.12) of the extremal lines rewriting them in terms of a. From (10.30) we have

$$\frac{dt_2}{da} = -\frac{1}{q|a\eta|^p}\cdot\frac{d}{da}(a\eta) = -\frac{1 - pq(M-m)|a|^{p-2}a}{q\big|a - q(M-m)|a|^p\big|^p}\,;$$

therefore

$$a' = -\frac{qa\big|a - q(M-m)|a|^p\big|^p}{a - pq(M-m)|a|^p}$$

and

$$A' = p|a|^{p-2}aa' = -pq|a|^p\frac{\big|a - q(M-m)|a|^p\big|^p}{a - pq(M-m)|a|^p}\,,$$

$$D' = \frac{1}{q}\big(1 - pqM|a|^{p-2}a\big)a' = -\frac{\big(a - pqM|a|^p\big)\big|a - q(M-m)|a|^p\big|^p}{a - pq(M-m)|a|^p}\,.$$

Now, equation (10.12) of the line L_{t_1,t_2} becomes

$$\begin{cases} x_1 = -|t_1|^{q-2}t_1\left(a - q(M-x_3)|a|^p\right), \\[2mm] x_2 = |t_1|^q\dfrac{\big|a - q(M-m)|a|^p\big|^p\big(a - pq(M-x_3)|a|^p\big)}{a - pq(M-m)|a|^p}\,. \end{cases} \tag{10.31}$$

To get a more symmetric expression, as before, we take a more geometrical parameter ξ_1 instead of t_1, where ξ_1 is the first coordinate of $\xi = (\xi_1, |\xi_1|^p, m)$, the point where the side surface $\{x_2 = |x_1|^p\}$ is intersected by our extremal line. In terms of a and ξ_1, equations (10.31) turn into

$$\begin{cases} x_1 = \dfrac{a - q(M-x_3)|a|^p}{a - q(M-m)|a|^p}\,\xi_1\,, \\[3mm] x_2 = \dfrac{a - pq(M-x_3)|a|^p}{a - pq(M-m)|a|^p}\,|\xi_1|^p\,. \end{cases} \tag{10.32}$$

These equations immediately supply us with an expression for a, namely, $a = a(x)$ is a root of the following equation:

$$s := \frac{|x_1|^p}{x_2} = \left|\frac{a - q(M-x_3)|a|^p}{a - q(M-m)|a|^p}\right|^p\frac{a - pq(M-m)|a|^p}{a - pq(M-x_3)|a|^p}\,. \tag{10.33}$$

To determine which of the possible roots of equation (10.33) gives us the desired value of a, we investigate the above function $s = s(a)$ defined by (10.33) as a function of the parameter a, assuming all other parameters to be fixed.

First, we note that the extremal line (10.32) intersects the plane $\{x_3 = M\}$ at the point

$$\zeta = (\zeta_1, \zeta_2, M)\,, \quad \zeta_1 = \frac{a\xi_1}{a - pq(M-x_3)|a|^p}\,, \quad \zeta_2 = \frac{a|\xi_1|^p}{a - pq(M-x_3)|a|^p}\,.$$

Since $\zeta_2 \ge 0$, we have the first restriction for a:

$$a < \Big(\frac{1}{pq(M-m)}\Big)^{q-1}. \tag{10.34}$$

The behavior of $s(a)$ on the semiaxis (10.34) is the same for all values of other parameters: on the negative half-line it monotonically increases from $\big(\frac{M-x_3}{M-m}\big)^{p-1}$ to 1, and on the interval $[0, \big(\frac{1}{pq(M-m)}\big)^{q-1}]$ it monotonically decreases from 1 to 0. Therefore, for a given s, $0 \le s \le 1$, we have one or two solutions of (10.33) satisfying (10.34): for all s there exists a positive solution $a = a^+$, and for $s > \big(\frac{M-x_3}{M-m}\big)^{p-1}$ there exists also a negative solution $a = a^-$. We already know that the solution $a = a^+$ corresponds to the function $\mathbf{B}_{\max}$, and the solution $a = a^-$ corresponds to the function $\mathbf{B}_{\min}$.

To write down the expression for B, we use (10.10), then we express x_1 using the first equation of (10.31), and by means of the second equation of (10.31) we replace the expression for $|t_1|^q$. For t_2 we use (10.30). As a result we obtain

$$\begin{aligned}
B &= \frac{1}{p}x_1t_1 + x_2t_2 = -\frac{1}{p}|t_1|^q\big(a - q(M-x_3)|a|^p\big) + x_2t_2 \\
&= -\frac{\big(a - q(M-x_3)|a|^p\big)\big(a - pq(M-m)|a|^p\big)}{p\big|a - q(M-m)|a|^p\big|^p\big(a - pq(M-x_3)|a|^p\big)}x_2 \\
&\quad + x_2\Big(m + \frac{a - q(M-m)|a|^p}{p\big|a - q(M-m)|a|^p\big|^p}\Big) \\
&= \frac{a|a|^p(x_3-m)x_2}{\big|a - q(M-m)|a|^p\big|^p\big(a - pq(M-x_3)|a|^p\big)} + mx_2 \\
&= \frac{(x_3-m)x_2}{\big|1 - q(M-m)a|a|^{p-2}\big|^p\big(1 - pq(M-x_3)a|a|^{p-2}\big)} + mx_2\,.
\end{aligned}$$

As has already been mentioned, here we choose the solution $a = a^+$ of (10.33) for $B_{\max}$ and $a = a^-$ for $B_{\min}$. □

So, we can now state the following theorem.

THEOREM 13. *Consider*

$$B(x_1, x_2, x_3; m, M) = \frac{(x_3-m)x_2}{\big|1 - q(M-m)a|a|^{p-2}\big|^p\big(1 - pq(M-x_3)a|a|^{p-2}\big)} + mx_2\,.$$

We shall denote this function by $B_{\max}$ if $a = a^+(x)$ is a unique positive solution of the equation

$$\frac{|x_1|^p}{x_2} = \Big|\frac{a - q(M-x_3)|a|^p}{a - q(M-m)|a|^p}\Big|^p \frac{a - pq(M-m)|a|^p}{a - pq(M-x_3)|a|^p}$$

from the interval $[0, \big(\frac{1}{pq(M-m)}\big)^{q-1}]$, and this function will be denoted by $B_{\min}$ if $a = a^-(x)$ is a unique negative solution of the same equation. The domain of $B_{\max}$ is

$$\mathrm{Dom}(B_{\max}) = \{x\colon\ x_2 \ge |x_1|^p,\ m < x_3 \le M\};$$

the domain of $B_{\min}$ is

$$\mathrm{Dom}(B_{\min}) = \Big\{x\colon\ |x_1|^p \le x_2 < \Big(\frac{M-m}{M-x_3}\Big)^{p-1}|x_1|^p,\ m < x_3 \le M\Big\}.$$

Then

$$\mathbf{B}_{\max}(x;m,M)=\begin{cases} B_{\max}(x;m,M), & m<x_3\le M\,,\\ mx_2, & x_3=m\,. \end{cases}$$

and

$$\mathbf{B}_{\min}(x;m,M)=\begin{cases} B_{\min}(x;m,M), & M-(M-m)\dfrac{|x_1|^q}{x_2^{q-1}}<x_3\le M\,,\\ mx_2, & m\le x_3\le M-(M-m)\dfrac{|x_1|^q}{x_2^{q-1}}\,. \end{cases}$$

The proof of this theorem follows almost *verbatim* the proof of our Theorems 3 and 4 above. So, we are not going to prove it here. Instead, we wish to show that our results encompass a slightly more general situation than it could have been thought.

11. Trees

As in [M], we let (T,μ_0) be a nonatomic probability space. Two measurable subsets A, B of T will be called almost disjoint if $\mu_0(A\cap B)=0$. Then we give the following

DEFINITION. A set $\mathcal{T}$ is called a tree if the following conditions are satisfied:

- $T\in\mathcal{T}$ and for every $I\in\mathcal{T}$ we have $\mu_0(I)>0$.
- For every $I\in\mathcal{T}$, there exists a finite or countable subset $C(I)\subset\mathcal{T}$ containing at least two elements such that:
  - a) the elements of $C(I)$ are pairwise almost disjoint;
  - b) $I=\bigcup_{J\in C(I)}J$.
- $\mathcal{T}=\bigcup_{m\ge0}\mathcal{T}_m$, where $\mathcal{T}_0=\{T\}$, $\mathcal{T}_{m+1}=\bigcup_{I\in\mathcal{T}_m}C(I)$.
- We have $\lim_{m\to\infty}\sup_{I\in\mathcal{T}_m}\mu_0(I)=0$.

For any tree $\mathcal{T}$ we define its exceptional set $E=E(\mathcal{T})$ as

$$E=\bigcup_{I\in\mathcal{T}}\ \bigcup_{J_1,J_2\in C(I),J_1\ne J_2} J_1\cap J_2\,.$$

It is clear that E has μ_0-measure zero.

An easy induction shows that each $\mathcal{T}_m$ consists of pairwise almost disjoint sets whose union is T. Moreover, if $t\in T\setminus E$, then for each m there exists exactly one I_m in $\mathcal{T}_m$ containing t. For every $m>0$ there exists a $J\in\mathcal{T}_{m-1}$ such that $I_m\in C(J)$. Then of course $J=I_{m-1}$. Hence the set $A(t)=\{I\in\mathcal{T}: t\in I\}$ forms a chain of nested elements of $\mathcal{T}_m$, $m=0,1,\dots$. Inclusions are all strict. From this remark it is clear that if $I,J\in\mathcal{T}$ and $I\cap J\cap(T\setminus E)$ is nonempty, then $I\subset J$ or $J\subset I$. In particular, for any I, $J\in\mathcal{T}$, either $\mu_0(I\cap J)=0$ or one of I and J is contained in the other. We denote by $\mathcal{T}(J)$ all elements of $\mathcal{T}$ that are subsets of J. Also for any $I\in\mathcal{T}$, put

$$\langle\phi\rangle_I:=\frac{1}{\mu_0(I)}\int_I\phi\,d\mu_0\,.$$

Given any tree $\mathcal{T}$, consider the collection $\alpha = \{\alpha_I\}_{I\in\mathcal{T}}$ of nonnegative numbers. Let μ be another measure on T. We can introduce the tree Bellman function:

$$\begin{aligned}&\mathbf{B}^{\mathcal{T}}_{CET}(x_1,x_2,x_3;m,M)\\ &\quad:= \sup\Big\{\int_T \phi^2(t)\,d\mu(t) + \sum_{J\in\mathcal{T}}\alpha_J\langle\phi\rangle_J^2 :\\ &\qquad\qquad \langle\phi\rangle_T = x_1,\ \langle\phi^2\rangle_T = x_2,\ \Big(\mu(T)+\sum_{J\in\mathcal{T}}\alpha_J\Big) = x_3,\\ &\qquad\qquad\qquad m\mu_0(J)\le \mu(J) + \sum_{\ell\in\mathcal{T}(J)}\alpha_\ell \le M\mu_0(J)\ \forall J\in\mathcal{T}\Big\}.\end{aligned}$$

Independently of the tree, the same Theorem 3 holds:

THEOREM 14. *Consider*

$$B(x_1,x_2,x_3;m,M) = \frac{(x_3-m)x_2}{\big[1-2a(M-m)\big]^2\big[1-4a(M-x_3)\big]} + mx_2\,,$$

where $a=a(x)$ is a unique solution of the cubic equation

$$\frac{x_1^2}{x_2} = \left[\frac{1-2a(M-x_3)}{1-2a(M-m)}\right]^2 \frac{1-4a(M-m)}{1-4a(M-x_3)} \tag{11.1}$$

on the interval $[0,\frac{1}{4(M-m)}]$. Then

$$\mathbf{B}^{\mathcal{T}}_{CET}(x;m,M) = \begin{cases} B(x;m,M), & x_1^2\le x_2,\quad m< x_3\le M,\\ mx_2, & x_1^2\le x_2,\quad x_3=m.\end{cases}$$

We can introduce the tree Bellman function also for an arbitrary p, $1<p<\infty$. Again the result for the trees will not depend on the tree; it coincides with the dyadic result. The proof of $\mathbf{B}^{\mathcal{T}}_{CET}(x;m,M)\le B(x;m,M)$ is exactly the same as before; the proof of $\mathbf{B}^{\mathcal{T}}_{CET}(x;m,M)\ge B(x;m,M)$ is not too difficult either if one uses the following lemma from [**M**].

LEMMA 15. *For every $I\in\mathcal{T}$ and every α such that $0<\alpha<1$, there exists a subfamily $F(I)\subset\mathcal{T}$ consisting of pairwise almost disjoint subsets of I such that*

$$\mu_0\Big(\bigcup_{J\in F(I)} J\Big) = (1-\alpha)\mu(I)\,.$$

Finally, it should be noted that an arbitrary tree can be isomorphically reduced to a dyadic one. To do this, we can replace every vertex I with more than two outgoing branches (i.e., $\#C(I)>2$) by a vertex with two branches. Namely, take the new successor $C'(I)$ of I consisting of an element J of maximal measure ($J\in C(I)$: $\mu_0(J)\ge\mu_0(\ell)\ \forall\ell\in C(J)$) and the union of the rest, i.e., a new element J' with the set of successors $\bigcup_{\ell\in C(I),\ell\ne J}\ell$. Repeating this procedure by induction we arrive at a dyadic tree.

Furthermore, without loss of generality we can take T to be an interval of length $\mu_0(T)$, and every $I\in\mathcal{T}$ will be represented as a subinterval of length $\mu_0(I)$. Another equivalent isomorphic representation of the underlying space is an arbitrary interval (say $[0,1]$) with the dyadic (with respect to Lebesgue measure) subset of intervals, but the measure on this space will be naturally induced by the measure μ_0 from T. We see that the Bellman function is the same independently of this underlying measure (or the size of the splitting coefficients in the first representation). We would like to underline that a similar feature does not always occur (see, e.g., [**SV**]). Perhaps the main reason why this is the case for the Carleson embedding theorem, but not for the John–Nirenberg inequality, is the convexity of the domain for the corresponding Bellman function.

References

[M] A. Melas, *The Bellman functions of dyadic-like maximal operators and related inequalities*, Adv. in Math. **192** (2005), no. 2, pp. 310–340.

[NT] F. Nazarov and S. Treil, *The hunt for Bellman function: applications to estimates of singular integral operators and to other classical problems in harmonic analysis*, Algebra i Analiz **8** (1997), no. 5, 32–162; English transl. in St. Petersburg Math. J. **8** (1998), no. 5.

[NTV1] F. Nazarov, S. Treil, and A. Volberg, *The Bellman functions and two-weight inequalities for Haar multipliers*, J. Amer. Math. Soc. **12** (1999) no. 4, 909–928.

[NTV2] F. Nazarov, S. Treil, and A. Volberg, *Bellman function in Stochastic Optimal Control and Harmonic Analysis* (*how our Bellman function got its name*), Oper. Theory: Advances and Appl. **129** (2001), 393–424.

[PVo] S. Petermichl and A. Volberg, *Heating of the Ahlfors–Beurling operator: Weakly quasiregular maps on the plane are quasiregular*, Duke Math. J. **112** (2002), no. 2, 281–305.

[Pog] A. V. Pogorelov, *Extrinsic geometry of convex surfaces*, Translations of Mathematical Monographs, Amer. Math. Soc., vol. 35, Providence, RI, 1973.

[SlSt] L. Slavin and A. Stokolos, *The Bellman PDE for the dyadic maximal function and its solution*, Preprint, 2006.

[SV] L. Slavin and V. Vasyunin, *Sharp results in the integral-form John–Nirenberg inequality*, Preprint, 2007; `http://arxiv.org/abs/0709.4332`.

[V] V. Vasyunin, *The sharp constant in the John–Nirenberg inequality*, Preprint PDMI no. 20, 2003; `http://www.pdmi.ras.ru/preprint/2003/index.html`.

[Vo] A. Volberg, *Bellman approach to some problems in Harmonic Analysis*, Seminaire des Equations aux derivées partielles, Ecole Polytechnique, 2002, exposé XX, 1–14.

V. A. Steklov Institute of Mathematics, St. Petersburg, Russia
E-mail address: `vasyunin@pdmi.ras.ru`

Department of Mathematics, Michigan State University, USA, and the University of Edinburgh, United Kingdom
E-mail address: `volberg@math.msu.edu`
E-mail address: `a.volberg@ed.ac.uk`

Amer. Math. Soc. Transl.
(2) Vol. **226**, 2009

Remarks on Nehari's Problem, Matrix A_2 Condition, and Weighted Bounded Mean Oscillation

A. Volberg and P. Yuditskii

ABSTRACT. We consider Nehari's problem in the case of nonuniqueness of a solution. The solution set is then parameterized by the unit ball of H^∞ by means of so-called *regular generators* — bounded holomorphic functions ϕ. The definition of *regularity* is given below, but we mention now that 1) the following assumption on the modulus of ϕ is sufficient for *regularity*: $\frac{1}{1-|\phi|^2} \in L^1(\mathbb{T})$; 2) there is no necessary and sufficient condition of *regularity* for bounded holomorphic ϕ in terms of $|\phi|$ on $\mathbb{T}$, [**12**]. This makes reasonable an attempt to find a weaker sufficient condition on $|\phi|$ than the condition in 1). This is done in this paper. Also we discuss certain new necessary and sufficient conditions of *regularity* in terms of bounded mean (weighted) oscillations of ϕ. They involve the matrix A_2 condition from [**21**].

1. Introduction

Recent developments in the inverse scattering/spectral theory [**23, 22, 9, 15, 17, 18, 19**] stimulated our interest to an old question on the description of the Nehari problem solution set (a precise formulation is given in Problem 1.1 below). The Nehari problem is strongly related to nonlinear Fourier analysis [**22**] or, which is basically the same, to the inverse scattering problem for CMV matrices [**14**].

Here we consider L^p spaces of functions on the unit circle $\mathbb{T}$ and their Hardy subspaces H^p. Recall that the famous Nehari theorem describes the projections of functions of the unit ball of L^∞ onto the Hardy space H^2_-; see e.g. the textbooks [**16, 10**]. Let P_- be the Riesz projector $P_- : L^2 \to H^2_-$. The function $F_- \in H^2_-$ possesses the representation

$$F_- = P_- f, \quad \|f\|_\infty \le 1, \tag{1.1}$$

if and only if the corresponding Hankel operator

$$\Gamma x := P_-(F_- x), \quad x \in H^2, \tag{1.2}$$

2000 *Mathematics Subject Classification*. Primary 47A68, 30D50, 34L25, 78A46.

Key words and phrases. Nehari's problem, scattering theory, A_2-weights, bounded mean oscillation.

Partially supported by NSF grant DMS-0501067 and the Austrian Founds FWF, project number P20413–N18.

has norm less than or equal to one, $\|\Gamma\| \le 1$ (the operator is naturally defined, say, on polynomials and then extended by continuity) .

Let

$$\mathcal{N}(F_-) = \{f \in L^\infty : F_- = P_- f, \quad \|f\|_\infty \le 1\}. \tag{1.3}$$

The Nehari *problem* deals with a description of $\mathcal{N}(F_-)$ for a given F_-. Thus the Nehari theorem is the solvability condition for this problem. The problem was solved by Adamyan, Arov, and Krein [**1, 2, 3**]. In the case of nonuniqueness, the set of solutions is parameterized by the unit ball of the class H^∞. Precisely, there exists $\phi = \phi_{F_-} \in H^\infty$ with the following three properties:

$$\|\phi\|_\infty \le 1, \quad \int_{\mathbb{T}} \log(1 - |\phi|)\, dm > -\infty, \quad \phi(0) = 0. \tag{1.4}$$

This function is accompanied by the outer function ψ,

$$\psi(\zeta) = e^{\frac{1}{2}\int_{\mathbb{T}} \frac{t+\zeta}{t-\zeta} \log(1-|\phi(t)|^2)\, dm(t)}, \tag{1.5}$$

and the function

$$f_0 = -\frac{\bar{\phi}\psi}{\bar{\psi}}. \tag{1.6}$$

The set $\mathcal{N}(F_-)$ is of the form

$$\mathcal{N}(F_-) = \left\{ f = f_{\mathcal{E}} = f_0 + \frac{\psi^2 \mathcal{E}}{1 - \phi\mathcal{E}} : \mathcal{E} \in H^\infty, \ \|\mathcal{E}\|_\infty \le 1 \right\}. \tag{1.7}$$

This is the hard

PROBLEM 1.1. Specify analytic properties of those holomorphic ϕ's of (1.4) that generate the description (1.7). Following Arov [**4**], we call such ϕ's regular.

REMARK 1.2. It is convenient to associate with a function ϕ of the form (1.4) the unitary-valued matrix function

$$\mathcal{S} = \mathcal{S}_\phi = \begin{bmatrix} \phi & \psi \\ \psi & f_0 \end{bmatrix} \tag{1.8}$$

with the entries given by (1.5), (1.6). Then the relation (1.7) between f and $\mathcal{E}$ can be rewritten in the vector form

$$\begin{bmatrix} A \\ f \end{bmatrix} = \mathcal{S} \begin{bmatrix} A\mathcal{E} \\ 1 \end{bmatrix} = \begin{bmatrix} \phi & \psi \\ \psi & f_0 \end{bmatrix} \begin{bmatrix} A\mathcal{E} \\ 1 \end{bmatrix},$$

where A is defined by this relation in a unique way, $A = \frac{\psi}{1-\phi\mathcal{E}}$. The fact that $\mathcal{S}$ is unitary implies that

$$|f|^2 + |A|^2 = 1 + |A|^2|\mathcal{E}|^2,$$

i.e.,

$$1 - |f|^2 = |A|^2(1 - |\mathcal{E}|^2) \ge 0.$$

Let us give an example of nonregular ϕ from (1.4). Choose any inner function Δ, $\Delta(0) > 0$. The point is that a *holomorphic* matrix function,

$$\begin{bmatrix} \frac{\Delta - \Delta(0)}{1+\Delta(0)} & \sqrt{\Delta(0)}\frac{1+\Delta}{1+\Delta(0)} \\ \sqrt{\Delta(0)}\frac{1+\Delta}{1+\Delta(0)} & \frac{\Delta - \Delta(0)}{1+\Delta(0)} \end{bmatrix}, \tag{1.9}$$

is unitary on $\mathbb{T}$. That is, for such $\phi = \frac{\Delta - \Delta(0)}{1+\Delta(0)}$, the corresponding $f_0 = \phi$ belongs to H^∞, and thus the class

$$\left\{ f = f_0 + \frac{\psi^2 \mathcal{E}}{1 - \phi \mathcal{E}} : \mathcal{E} \in H^\infty, \ \|\mathcal{E}\|_\infty \le 1 \right\} \tag{1.10}$$

describes a *proper* subclass of the set $\mathcal{N}(0)$ = unit ball of H^∞ (since $P_- f_0 = 0$ in this case).

We show that, for instance, $f = 0$ cannot be represented in this way. First let us note that $\frac{\psi}{1-\phi\mathcal{E}} \in H^2$. In fact, since $\frac{1+\phi\mathcal{E}}{1-\phi\mathcal{E}}$ is a function in the unit disc with positive real part, we have (in the sense of boundary values on the unit circle)

$$\left| \frac{\psi}{1 - \phi\mathcal{E}} \right|^2 \le \frac{1 - |\phi\mathcal{E}|^2}{|1 - \phi\mathcal{E}|^2} = \operatorname{Re} \frac{1 + \phi\mathcal{E}}{1 - \phi\mathcal{E}} \in L^1, \tag{1.11}$$

and in addition $\frac{\psi}{1-\phi\mathcal{E}}$ is a function of the Smirnov class (the denominator is an outer function). So if 0 is in the set, we get

$$\frac{f_0}{\psi} = -\frac{\psi\mathcal{E}}{1 - \phi\mathcal{E}} \in H^2.$$

On the other hand, this function belongs to H^2_- due to the representation

$$\frac{f_0}{\psi} = -\frac{\bar{\phi}}{\bar{\psi}}$$

(ϕ/ψ is also of the Smirnov class). Thus $\phi = 0$ and $\psi = 1$.

Note that actually this is a general obstruction: according to Arov's theorem, one can always "factor out" (in a certain sense) a holomorphic $\mathcal{S}$-matrix from the given one, so that the remaining part, indeed, will generate the description of a Nehari problem in the form (1.7) (the so-called singular-regular factorization [**4**]).

On a certain stage the answers to Problem 1.1 and to a comparably long list of similar problems (see for instance [**7, 8**], where a similar question with respect to the Hamburger moment problem is discussed) were formulated in terms of the density of a certain class of holomorphic functions in a Hilbert space associated with the data.

We need to recall the Nagy–Foiaş functional model space [**20**]. It can be associated with an arbitrary function ϕ of the unit ball of H^∞ (the Schur class)

$$K_\phi := H^2 \oplus \overline{\Delta L^2} \ominus \{\phi \oplus \Delta\} H^2,$$

where $\Delta := \sqrt{1 - |\phi|^2}$, and

$$\overline{\Delta L^2} = \operatorname{clos}_{L^2}\{f = \Delta g : g \in L^2\}.$$

In our specific case of $\log(1 - |\phi|^2) \in L^1$ we can choose an "analytic" square root instead of "arithmetic" one, i.e., use ψ instead of Δ, and of course $\overline{\Delta L^2} = L^2$. So, the function space is of the form

$$K_\phi := \begin{bmatrix} H^2 \\ L^2 \end{bmatrix} \ominus \begin{bmatrix} \phi \\ \psi \end{bmatrix} H^2 = \begin{bmatrix} 0 \\ H^2_- \end{bmatrix} \oplus \hat{H},$$

where

$$\hat{H} := \hat{H}_\phi = H^2(\mathbb{C}^2) \ominus \begin{bmatrix} \phi \\ \psi \end{bmatrix} H^2. \tag{1.12}$$

That is, we have $\left[\begin{smallmatrix} x_+ \\ g \end{smallmatrix}\right] \in K_\phi$ if and only if $x_+ \in H^2$, $g \in L^2$, and

$$x_- := \bar{\phi} x_+ + \bar{\psi} g \in H^2_-.$$

Alternatively, we can characterize K_ϕ as pairs $\left[\begin{smallmatrix} x_+ \\ x_- \end{smallmatrix}\right]$ such that

$$x_\pm \in H^2_\pm \quad \text{and} \quad g := \frac{x_- - \bar{\phi} x_+}{\bar{\psi}} \in L^2.$$

It looks natural to hope that the pairs

$$\begin{bmatrix} x_+ \\ x_- \end{bmatrix} = \begin{bmatrix} \psi y_+ \\ \bar{\psi} y_- \end{bmatrix}, \quad y_\pm \in H^2_\pm, \tag{1.13}$$

form a dense set in K_ϕ (recall that ψ is an outer function). The corresponding g,

$$g = y_- - \frac{\bar{\phi}\psi}{\bar{\psi}} y_+ = y_- + f_0 y_+,$$

surely belongs to L^2, and the element of K_ϕ is of the form

$$\begin{bmatrix} \psi y_+ \\ g \end{bmatrix} = \begin{bmatrix} 0 \\ y_- + P_- f_0 y_+ \end{bmatrix} \oplus \begin{bmatrix} \psi y_+ \\ P_+ f_0 y_+ \end{bmatrix} \in \begin{bmatrix} 0 \\ H^2_- \end{bmatrix} \oplus \hat{H}. \tag{1.14}$$

However, in fact,

THEOREM 1.3. *A function ϕ is regular if and only if the vectors of the form* (1.14) *constitute a dense set in K_ϕ or, which is the same,*

$$\check{H}_\phi := \operatorname{clos}_{H^2(\mathbb{C}^2)} \left\{ \begin{bmatrix} \psi y_+ \\ P_+ f_0 y_+ \end{bmatrix} : y_+ \in H^2 \right\} = \hat{H}_\phi. \tag{1.15}$$

Moreover, (1.15) *holds as soon as*

$$P_+ \bar{t} \begin{bmatrix} \phi(t) \\ \psi(t) \end{bmatrix} = \begin{bmatrix} \frac{\phi(t)}{t} \\ \frac{\psi(t) - \psi(0)}{t} \end{bmatrix} \in \check{H}_\phi. \tag{1.16}$$

For a proof see e.g. [**13**].

A trivial consequence is the following

PROPOSITION 1.4. *Let*

$$\frac{1}{1 - |\phi|^2} \in L^1. \tag{1.17}$$

Then ϕ is regular.

Indeed, we put $x(t) = \frac{\phi(t)}{t\psi(t)} \in H^2$ and we get

$$\begin{bmatrix} \psi x \\ P_+ f_0 x \end{bmatrix} = \begin{bmatrix} \frac{\phi(t)}{t} \\ P_+ \frac{1}{t}(\psi - 1/\bar{\psi}) \end{bmatrix} = \begin{bmatrix} \frac{\phi(t)}{t} \\ \frac{\psi(t) - \psi(0)}{t} \end{bmatrix},$$

since $\frac{1}{t} 1/\bar{\psi} \in H^2_-$.

One of the main goals of this note is to discuss the following question: Is it possible to give a sufficient condition better than (1.17) in terms of the absolute value of ϕ?

We point out a nice result, which was obtained in [**12**]; see also [**11, 13**]. It was shown that there is no necessary and sufficient condition for the regularity of ϕ in terms of the absolute value $|\phi|$.

THEOREM 1.5. *Let $\phi \in H^\infty$ satisfies (1.4). Then there exists an inner function Φ such that $\phi\Phi$ is regular.*

2. Condition on the modulus $|\psi|$ which ensures regularity of ϕ but is weaker than $1/\psi \in H^2$.

We want to see some nontrivial conditions on $|\phi|$ that guarantee that ϕ is regular. By nontrivial we understand any condition different from

$$\frac{1}{|\psi|^2} = \frac{1}{1-|\phi|^2} \in L^1. \tag{2.1}$$

We denote by h the outer function with the modulus

$$|h|^2 := \frac{1}{1-|\phi|^2}\,.$$

In other words,

$$h = \frac{1}{\psi}\,.$$

An outer h always exists by the assumption (1.4) on ϕ. It is *not* in H^2 throughout this section because we are looking for "nontriviality".

THEOREM 2.1. *Suppose $|h|^2 := \frac{1}{1-|\phi|^2} \notin L^1(\mathbb{T})$. Suppose also that*

$$\liminf_{N\to\infty}\Big(\int_{|h|\le N} |h|^4\,dm\Big)\Big(\int_{|h|>N} (\log|h|)^4\,dm\Big) = 0\,. \tag{2.2}$$

Then ϕ is regular.

COROLLARY 2.2. *Suppose $|h|^2 := \frac{1}{1-|\phi|^2} \notin L^1(\mathbb{T})$. Suppose also that*

$$\liminf_{N\to\infty} N^4 \int_{|h|>N} (\log|h|)^4\,dm = 0\,. \tag{2.3}$$

Then ϕ is regular.

Let us briefly explain the assumptions (2.2), (2.3). It is easy to satisfy them if there exists a sequence $N_n \to \infty$ such that

$$|\{\zeta\in\mathbb{T} : |h| > N_n\}|\,(N_n \log N_{n+1})^4 \le \frac{1}{n^2}\,. \tag{2.4}$$

On the other hand, this is easily reconcilable with the following condition which guarantees nontriviality:

$$|\{\zeta\in\mathbb{T} : |h| = N_{n+1}\}|\,(N_{n+1})^2 \ge (n+1)^2\,. \tag{2.5}$$

In fact, to have both (2.5) and (2.4) one can define $|h|$ to be a step function with values $N_n, n = 1, 2, \ldots$ on sets of measures $\frac{n^2}{N_n^2}$ (this gives (2.5)), and choose N_n going to infinity extremely fast to have firstly $\sum_{k=n+1}^{\infty} \frac{k^2}{N_k^2} < \frac{2(n+1)^2}{N_{n+1}^2}$ and secondly

$$\frac{(n+1)^2}{N_{n+1}^2}(N_n \log N_{n+1})^4 \le \frac{1}{2n^2}\,.$$

Then (2.4) follows.

REMARK. Conditions (2.2), (2.3) are of course conditions only on $|\phi|$ or, which is the same, on $|\psi|$.

Now we prove Theorem 2.1. We are grateful to A. Alexandrov whose idea is used in the proof.

PROOF. We need to use (2.2) to prove the existence of H^2 functions v_n such that

$$\psi v_n \to \frac{\phi(\zeta)}{\zeta} \text{ in } H^2 \text{ and}$$

$$-P_+\left(\frac{\bar{\phi}\psi}{\bar{\psi}} v_n\right) \to \frac{\psi(\zeta)-\psi(0)}{\zeta} \text{ in } H^2 .$$

Notice that if we had $\frac{1}{|\psi|^2} = |h|^2 \in L^1$, we could have taken

$$v_n = \frac{\phi(\zeta)}{\zeta\psi(\zeta)},$$

which would be functions in H^2 in this case. But we have exactly the opposite case: $\frac{1}{|\psi|^2} = |h|^2 \notin L^1$.

Also, notice that to satisfy the above relationships it is enough to construct $g_n \in H^2$ such that

$$g_n h \in H^2 \ \forall n, \tag{2.6}$$

and such that

$$g_n \to 1 \text{ in } H^2 \text{ and } -P_+\left(\frac{|\phi|^2\bar{\zeta}}{\bar{\psi}} g_n\right) \to \frac{\psi(\zeta)-\psi(0)}{\zeta} \text{ in } H^2 . \tag{2.7}$$

In fact, having g_n like that, we put $v_n = \frac{\phi(\zeta)}{\zeta} h g_n$. Then $v_n \in H^2$ by (2.6). And these v_n satisfy the two conditions mentioned above because of (2.7).

Now let us write

$$-P_+\left(\frac{|\phi|^2\bar{\zeta}}{\bar{\psi}} g_n\right) = -P_+\left(\bar{\zeta}\frac{g_n}{\bar{\psi}}\right) + P_+(\bar{\zeta}\psi g_n) =: I_n + II_n .$$

We have

$$II_n = \frac{\psi g_n - (\psi g_n)(0)}{\zeta} \to \frac{\psi - \psi(0)}{\zeta}$$

in H^2 because $g_n \to 1$ in H^2, ψ is from H^∞, and the backward shift operator is bounded in H^2.

Thus, the only thing we need now is to construct g_n such that (2.6) holds, $g_n \to 1$ in H^2, and

$$-I_n = P_+(\bar{\zeta}\bar{h} g_n) \to 0 \tag{2.8}$$

in H^2. To have all this it is enough to have

$$P_+(\bar{\zeta}\bar{h} g_n) \to 0 \text{ in } H^2;\ g_n h \in H^2;\ g_n \to 1 \text{ in } H^2 . \tag{2.9}$$

Let us fix a sequence $N_n \to \infty$ and put

$$\phi_n = \begin{cases} \log|h|, & |h| > N_n, \\ 0, & |h| \le N_n. \end{cases}$$

Here are our g_n's:

$$g_n := e^{-(\phi_n + i\widetilde{\phi_n})}.$$

In fact, obviously,

$$\|P_+(\bar{\zeta}\bar{h} g_n)\|_2^2 \le \|\bar{h} e^{-(\phi_n + i\widetilde{\phi_n})} - \bar{h} e^{-(\phi_n - i\widetilde{\phi_n})}\|_2^2 . \tag{2.10}$$

And, of course, $|g_n||h| \le N_n$.

Now we have $\|\bar{h}e^{-(\phi_n+i\widetilde{\phi_n})}-\bar{h}e^{-(\phi_n-i\widetilde{\phi_n})}\|_2^2=4\|he^{-\phi_n}\sin\widetilde{\phi_n}\|_2^2$. In conjunction with (2.10) this gives (we also use that $|\sin x|\le|x|$ and the definition of ϕ_n)

$$\frac{1}{4}\|P_+(\bar{h}g_n)\|_2^2\le\|he^{-\phi_n}\widetilde{\phi_n}\|_2^2\le\int_{|h|\le N_n}|h|^2|\widetilde{\phi_n}|^2+\int_{\mathbb{T}}|\widetilde{\phi_n}|^2=:J_1+J_2. \tag{2.11}$$

To estimate J_1 we write (using the boundedness of the harmonic conjugation operator $\widetilde{\cdot}$ in $L^4(\mathbb{T})$)

$$\begin{aligned} J_1&\le\int_{|h|\le N_n}|h|^2|\widetilde{\phi_n}|^2\le\left(\int_{|h|\le N_n}|h|^4\,dm\right)^{\frac{1}{2}}\left(\int_{\mathbb{T}}|\widetilde{\phi_n}|^4\,dm\right)^{\frac{1}{2}}\\ &\le C\left(\int_{|h|\le N_n}|h|^4\,dm\right)^{\frac{1}{2}}\left(\int_{\mathbb{T}}|\phi_n|^4\,dm\right)^{\frac{1}{2}}\\ &=C\left(\int_{|h|\le N_n}|h|^4\,dm\right)^{\frac{1}{2}}\left(\int_{|h|>N_n}|\log|h||^4\,dm\right)^{\frac{1}{2}}. \end{aligned} \tag{2.12}$$

In particular, if (2.2) holds, there exists a sequence of numbers $N_n\to\infty$ such that the last expression tends to zero. Thus $J_1\to 0$.

Now let us estimate J_2:

$$\begin{aligned} J_2&=\int_{\mathbb{T}}|\widetilde{\phi_n}|^2\le\left(\int_{\mathbb{T}}|\widetilde{\phi_n}|^4\,dm\right)^{\frac{1}{2}}\\ &\le C\left(\int_{\mathbb{T}}|\phi_n|^4\,dm\right)^{\frac{1}{2}}=C\left(\int_{|h|>N_n}|\log|h||^4\,dm\right)^{\frac{1}{2}}\\ &\le C\left(\int_{|h|\le N_n}|h|^4\,dm\right)^{\frac{1}{2}}\left(\int_{|h|>N_n}|\log|h||^4\,dm\right)^{\frac{1}{2}}. \end{aligned} \tag{2.13}$$

The last inequality holds for large N_n. In fact, for large N_n, the integral $\int_{|h|\le N_n}|h|^4\,dm$ is as large as we wish because we assumed that $\int_{\mathbb{T}}|h|^2\,dm=\infty$.

Therefore, if (2.2) holds, then $J_2\to 0$. Going back to (2.10) we see that we have proved Theorem 2.1. □

3. Strong regularity

The strong regularity means that ϕ is regular and in addition $\|\Gamma\|<1$, where $\Gamma=\Gamma_{f_0}$. In other words, ϕ is strongly regular if and only if ϕ is regular and the operator $(I-\Gamma^*\Gamma)$ is invertible.

The next theorem is a combination of the Helson–Szegő, Hunt–Muckenhoupt–Wheeden, and Adamyan–Arov–Krein theorems (AAK); for a matrix generalization see e.g. [**5, 6**].

THEOREM 3.1. *A function ϕ is strongly regular if and only if*

$$\frac{1-|\phi|^2}{|1-\phi|^2}=w\in A_2. \tag{3.1}$$

Note that the left hand side in (3.1), being a positive harmonic function in the unit disc ($=\operatorname{Re}\frac{1+\phi}{1-\phi}$), is equal to the harmonic extension of an A_2 weight on the circle.

PROOF. Let ϕ be regular and $\|\Gamma_{f_0}\| < 1$. Consider the symbol f_1 that corresponds to the choice $\mathcal{E} = 1$ in (1.7) (recall that f_0 corresponds to $\mathcal{E} = 0$). It is of the form

$$f_1 = \frac{\bar{g}}{g}, \quad g := \frac{1-\phi}{\psi},$$

and we have $\Gamma = \Gamma_{f_1}$. By the Adamyan–Arov–Krein theorem (AAK) (see Remark 3.2),

$$\frac{1}{\psi\psi(0)} = (I - \Gamma^*\Gamma)^{-1}1 \in H^2. \tag{3.2}$$

Thus $g \in H^2$ and $\|\Gamma_{\frac{\bar{g}}{g}}\| < 1$. By the Helson–Szegő and Hunt–Muckenhoupt–Wheeden theorems (see e.g. [**16**]),

$$|g|^2 = \frac{|1-\phi|^2}{1-|\phi|^2} \in A_2.$$

Conversely, from (3.1) we conclude that

$$\frac{1+\phi}{1-\phi} = w + i\widetilde{w}\,, \tag{3.3}$$

where, as before, $\widetilde{w}$ stands for the harmonic conjugate of w (the Hilbert transform on the circle). Then of course

$$\frac{1-\phi}{1+\phi} = \frac{1}{w+i\widetilde{w}}\,,$$

and so

$$\frac{1-|\phi|^2}{|1+\phi|^2} = \operatorname{Re}\frac{1}{w+i\widetilde{w}} = \frac{w}{w^2+\widetilde{w}^2}\,. \tag{3.4}$$

Let us derive that

$$\frac{|1+\phi|^2}{1-|\phi|^2} \in L^1 \tag{3.5}$$

on the unit circle. Indeed,

$$\int_{\mathbb{T}} \frac{|1+\phi|^2}{1-|\phi|^2} = \int_{\mathbb{T}} w + \int_{\mathbb{T}} \frac{\widetilde{w}^2}{w} \le \|w\|_1 + Q_{1/w}\|w\|_1 < \infty\,,$$

where $Q_{1/w}$ stands for the norm of the Hilbert transform from $L^2_{1/w}$ to itself, which is finite as $w \in A_2$.

Combine (3.5) with a simple remark that (3.1) implies $\frac{|1-\phi|^2}{1-|\phi|^2} \in L^1(\mathbb{T})$. Add these two relations and obtain

$$\frac{1}{1-|\phi|^2} \in L^1(\mathbb{T})\,. \tag{3.6}$$

We know that this is sufficient for being regular.

Finally, by the converse statement in the Helson–Szegő and Hunt–Muckenhoupt–Wheeden theorems, we have $\|\Gamma\| = \|\Gamma_{f_1}\| = \|\Gamma_{\frac{\bar{g}}{g}}\| < 1$ if $|g|^2 \in A_2$. The latter is exactly (3.1). □

REMARK 3.2. Let us discuss (3.2) from the point of view of regularity. We still assume that $\|\Gamma\| < 1$, that is, $(I-\Gamma^*\Gamma)^{-1}1$ has the direct meaning. Then it is easy to check that the vector

$$\check{k} := \begin{bmatrix} \psi(I-\Gamma^*\Gamma)^{-1}1 \\ P_+f_0(I-\Gamma^*\Gamma)^{-1}1 \end{bmatrix} \psi(0) \tag{3.7}$$

is the reproducing kernel in $\check{H}^2_\phi$; see (1.15). Indeed,

$$\left\langle \begin{bmatrix} \psi y_+ \\ P_+f_0y_+ \end{bmatrix}, \check{k} \right\rangle = \langle (I-\Gamma^*\Gamma)y_+, (I-\Gamma^*\Gamma)^{-1}1\rangle \psi(0) = y_+(0)\psi(0).$$

Also, it is evident that the reproducing kernel of $\hat{H}^2_\phi$ is $\hat{k} = \left[\begin{smallmatrix} 1 \\ 0 \end{smallmatrix}\right]$, i.e.,

$$\left\langle \begin{bmatrix} x_1 \\ x_2 \end{bmatrix}, \hat{k} \right\rangle = x_1(0), \quad \forall \begin{bmatrix} x_1 \\ x_2 \end{bmatrix} \in \hat{H}^2_\phi.$$

Thus, $\hat{H}_\phi = \check{H}_\phi$ implies $\hat{k} = \check{k}$, and the equality of the first components is precisely (3.2).

Generally, in the AAK theory, for a regular ϕ the following formula holds

$$\frac{1}{\psi(\zeta)\psi(0)} = \lim_{r\downarrow 1}((rI-\Gamma^*\Gamma)^{-1}1)(\zeta),$$

so the last function is not necessarily in H^2.

3.1. An alternative proof. The previous proof exploited a great deal of the AAK theory in the part that proves (3.1) from the strong regularity, and we wish to give a more direct proof for the reader who is not so familiar with this subtle material.

First we need an AAK lemma, which can be found in [**16**] or extracted from AAK papers in our reference list (however, we provide the proof for the sake of completeness).

LEMMA 3.3. *Let $F \in L^\infty$ and $d(F, H^\infty) < 1$. Then the coset $F + H^\infty$ contains a function $\frac{\bar{h}}{h}$, where $|h|^2 \in A_2$. In particular, this coset contains a unimodular function v such that the Toeplitz operator T_v is invertible.*

PROOF. Let $d(f, H^\infty)$ denote the distance between a function $f \in L^\infty$ and the subspace H^∞ of bounded holomorphic functions in the unit disc. Let H^∞_0 denote the set of bounded holomorphic functions in the unit disc that vanish at zero. Given such a function F consider $d(\bar{z}F, H^\infty) = d(F, H^\infty_0)$. Two cases may occur. Suppose first that $d(\bar{z}F, H^\infty) = 1$. Then the Hankel operator $H_{\bar{z}F} := P_-(\bar{z}F\cdot) : H^2 \to H^2_-$ attains its norm. In fact, $\|H_{\bar{z}F}\|_{\mathrm{ess}} \le d(\bar{z}F, \bar{z}H^\infty) = d(F, H^\infty) < 1 = d(\bar{z}F, \bar{z}H^\infty) = \|H_{\bar{z}F}\|$. This is just Nehari's classical theorem (see [**16**]), and $\|\cdot\|_{\mathrm{ess}}$ means the norm modulo compact operators (essential norm). If the essential norm of the operator A in the Hilbert space is strictly less than its norm, then A attains its norm. See [**16**, Ch. VII] again, or just notice that we can reduce our statement to self-adjoint operators by considering A^*A (and polar decomposition $A = U(A^*A)^{1/2}$). But if the essential norm of a self-adjoint operator is strictly smaller than its norm, it means that its norm is a maximal eigenvalue of finite multiplicity (the spectral theorem), and thus the operator attains its norm. Now let our Hankel operator attain its norm 1 at a vector $H \in H^2$, $\|H\|_2 = 1$. Denote $\bar{G}_0 = H_{\bar{z}F}H$. Denote by u a function in the coset $\bar{z}F + H^\infty$ of $\|u\|_\infty = 1$. It always exists by an obvious compactness argument.

Then

$$1 = \|\bar{G}_0\|_2 = \|H_{\bar{z}F}H\| = \|H_u H\|_2 = \|P_-(uH)\| \leq \|uH\|_2 \leq \|H\|_2 = 1\,.$$

This means of course that $|u| = 1$ almost everywhere on the circle and that uH is antianalytic (that is, $P_+(uH) = 0$).

Therefore,

$$uH = P_-(uH) = \bar{G}_0\,.$$

We conclude that the two H^2-functions H and G_0 have the same modulus a.e. on the unit circle. Write $H = S_1 h$, $G_0 = zS_2 h$ for their inner-outer factorizations (h is an outer function here). Then we obtain

$$\frac{\bar{h}}{S_0 h} = u \in \bar{z}F + H^\infty\,,$$

where $S_0 = zS_1S_2$, $h \in H^2$. Consider $S = zS_1S_2$ and

$$v := zu = \frac{\bar{h}}{Sh} \in F + H^\infty\,.$$

We want to prove now that the Toeplitz operator T_v is invertible. From the fact that $d(v, H^\infty) = d(F, H^\infty) < 1$ and from Nehari's theorem we know that $\|H_v\| < 1$. But $|v| = 1$ a.e. on the unit circle and then $T_v^* T_v = I - H_v^* H_v$. Therefore $\|H_v\| < 1$ means that T_v is bounded from below (that is, it is left-invertible). To prove that it is invertible it is sufficient to prove that its adjoint has trivial kernel. Let $R \in \operatorname{Ker} T_v^* = T_{\bar{v}}$. Then

$$T_{\bar{v}}R = 0 \Rightarrow P_+\Big(\frac{h}{\bar{S}\bar{h}}R\Big) = 0 \Rightarrow \frac{h}{\bar{S}\bar{h}}R = \bar{z}\bar{r}\,,$$

where $r \in H^2$. Then $hR = \bar{z}\bar{S}\bar{h}\bar{r}$, the left hand side being from H^1, and the right hand side being from H^1_-. The intersection being zero we conclude that $R = 0$.

So we get v, $v := zu = \frac{\bar{h}}{Sh} \in F + H^\infty$, such that T_v is invertible. By the Helson–Szegő theorem (see [**16**, Ch. VII]), we conclude that $v = \frac{\bar{h}}{h}$, where $|h|^2 \in A_2$.

Now we consider the second case: $d(\bar{z}F, H^\infty) < 1$. We denote $f := \bar{z}F$ and consider the function $\tau(c) := d(f + c\bar{z}, H^\infty)$. We know that $\tau(0) < 1, \tau(\infty) = \infty$, and τ is obviously continuous. So we can find c_0 such that for $f + c_0\bar{z} =: \bar{z}\Phi$

$$\begin{aligned} d(\bar{z}\Phi, \bar{z}H^\infty) &= d(f + c_0\bar{z}, \bar{z}H^\infty) = d(f, H^\infty) < 1,\\ d(\bar{z}\Phi, H^\infty) &= d(f + c_0\bar{z}, H^\infty) = 1. \end{aligned}$$

Then we proceed exactly as in the first case by using the fact that the last two relations imply that the operator $H_{\bar{z}\Phi}$ attains its norm. We will find unimodular $u \in f + c_0\bar{z} + H^\infty$ such that

$$u = \frac{\bar{g}}{z\Theta g}\,,$$

where Θ is inner and g is outer from H^2. Therefore, $v := zu$ will be in the coset $zf + H^\infty = F + H^\infty$ and will have the form $u = \frac{\bar{g}}{\Theta g}$ with the same Θ and g.

Again as in the first case $\|H_v\| = d(F, H^\infty) < 1$ (Nehari's theorem) ensures that T_v is left-invertible. And exactly as before we prove that $T_v^* = T_{\bar{v}}$ has trivial kernel. Hence T_v is invertible and we conclude once again by the Helson–Szegő theorem (see [**16**, Ch. VII]) that $v = \frac{\bar{h}}{h}$, where $|h|^2 \in A_2$. The AAK lemma is proved. □

It is easy to finish the second proof of our theorem. Let ϕ be strongly regular. This means that it is regular and so $f_0 := -\frac{\bar{\phi}\psi}{\bar{\psi}}$ is such that

$$f_0 + H^\infty = f_0 + \frac{\psi^2 e}{1 - \phi e},$$

where e runs over the unit ball of H^∞. But strong regularity means also that $\|H_{f_0}\| < 1$. Lemma 3.3 means that there exists an outer h such that $\frac{\bar{h}}{h} \in f_0 + H^\infty$ and $|h|^2 \in A_2$. We gather:

$$\frac{\bar{h}}{h} = -\frac{\bar{\phi}\psi}{\bar{\psi}} + \frac{\psi^2 e}{1 - \phi e}$$

for some e from the unit ball of H^∞. Then a.e. on the circle,

$$\left| -\frac{\bar{\phi}\psi}{\bar{\psi}} + \frac{\psi^2 e}{1 - \phi e} \right| = 1.$$

But the left hand side is

$$\left| \frac{e - \phi}{1 - \phi e} \right|,$$

and we conclude that e is an inner function. Then

$$\frac{\bar{h}}{h} = -\frac{\bar{\phi}\psi}{\bar{\psi}} + \frac{\psi^2 e}{1 - \phi e} = \frac{\psi \overline{1 - \phi e}}{\bar{\psi}(1 - \phi e)}.$$

Denote $g_e := \frac{1 - \phi e}{\psi}$. Then we have just got

$$\frac{\bar{h}}{h} = e \frac{\bar{g}_e}{g_e}.$$

But $|1/g_e|^2 = \frac{1 - |\phi e|^2}{|1 - \phi e|^2} \in L^1$ (see (1.11)), and so $1/g_e \in H^2$ (g_e is obviously an outer function). We then immediately conclude from the last displayed equality that $g = c \cdot h$ with some constant c. In fact, we have

$$e \frac{h}{g_e} = \frac{\bar{h}}{\bar{g}_e},$$

where the left hand side belongs to H^1 and the right hand side belongs to H^1_-. So both expressions are just constants. And $e = e^{ir}$, $g_e = c \cdot h$, as has been promised. Therefore, $\frac{|1 - e^{ir}\phi|^2}{1 - |\phi|^2} \in A_2$.

Now we use Theorem 3.5 proved in the Appendix to conclude that $\frac{|1-\phi|^2}{1-|\phi|^2} \in A_2$. We have finished the proof that regularity implies (3.1).

The following criterion was proposed in [**6**].

THEOREM 3.4. *ϕ is strongly regular if and only if the matrix weight*

$$W := \begin{bmatrix} 1 & \phi \\ \bar{\phi} & 1 \end{bmatrix} \tag{3.8}$$

satisfies the matrix A_2 condition.

The matrix A_2 condition was found in [**21**]. Let us notice that Theorem 3.1 has the following curious consequence.

THEOREM 3.5. *Let w be such that* (3.1) *holds. Consider a new positive harmonic function given by*

$$w_{e^{ic}} := \frac{1-|\phi|^2}{|1-e^{ic}\phi|^2}. \tag{3.9}$$

The positive harmonic function $w_{e^{ic}}$ does not have singular part in its Herglotz representation. Its absolutely continuous part on the circle (also called $w_{e^{ic}}$) is uniformly in A_2.

This is absolutely trivial from the point of view of Theorem 3.4: If a matrix function W is in matrix A_2, then also the following matrix is obviously in matrix A_2:

$$W_c := \begin{bmatrix} e^{ic/2} & 0 \\ 0 & e^{-ic/2} \end{bmatrix} \begin{bmatrix} 1 & \phi \\ \bar\phi & 1 \end{bmatrix} \begin{bmatrix} e^{-ic/2} & 0 \\ 0 & e^{ic/2} \end{bmatrix};$$

but $W_c = \left[\begin{smallmatrix} 1 & \phi_c \\ \bar\phi_c & 1 \end{smallmatrix}\right]$, where $\phi_c = e^{ic}\phi$.

The matrix A_2 condition for the specific weight (3.8) can be given as the following scalar condition (this is a slight modification of the condition given in [**6**]).

LEMMA 3.6. *A weight of the form* (3.8) *satisfies A_2 if and only if*

$$\sup_I \frac{1}{|I|}\int_I \frac{|\phi-\langle\phi\rangle_I|^2 + (1-|\langle\phi\rangle_I|^2)}{1-|\phi|^2}\, dm < \infty, \tag{3.10}$$

where for an arc $I \subset \mathbb{T}$ we put

$$\langle\phi\rangle_I := \frac{1}{|I|}\int_I \phi\, dm. \tag{3.11}$$

PROOF. If the matrix weight $\left[\begin{smallmatrix} 1 & \phi \\ \bar\phi & 1 \end{smallmatrix}\right]$ is in A_2, then

$$\begin{aligned} \frac{1}{|I|}\int_I \frac{\begin{bmatrix} 1 & -\phi \\ -\bar\phi & 1 \end{bmatrix}}{1-|\phi|^2}\, dm &\le C \begin{bmatrix} 1 & \langle\phi\rangle_I \\ \langle\bar\phi\rangle_I & 1 \end{bmatrix}^{-1} \\ &= C\left\{ \begin{bmatrix} 1 & 0 \\ \langle\bar\phi\rangle_I & \sqrt{1-|\langle\phi\rangle_I|^2} \end{bmatrix} \begin{bmatrix} 1 & \langle\phi\rangle_I \\ 0 & \sqrt{1-|\langle\phi\rangle_I|^2} \end{bmatrix} \right\}^{-1}, \end{aligned} \tag{3.12}$$

or

$$\frac{1}{|I|}\int_I \frac{\begin{bmatrix} 1-|\phi|^2+|\phi-\langle\phi\rangle_I|^2 & (\langle\phi\rangle_I-\phi)\sqrt{1-|\langle\phi\rangle_I|^2} \\ \overline{(\langle\phi\rangle_I-\phi)}\sqrt{1-|\langle\phi\rangle_I|^2} & 1-|\langle\phi\rangle_I|^2 \end{bmatrix}}{1-|\phi|^2}\, dm \le C, \tag{3.13}$$

which is equivalent to (3.10). □

REMARK 3.7. The latter form (3.10) of condition (3.1) indeed makes it evident that this condition is invariant under the rotation $\phi \mapsto \phi e^{ic}$. In fact, the condition is stable under any linear-fractional transform

$$\phi(\zeta) \mapsto e^{ic}\frac{\phi(\zeta)-a\zeta}{1-\bar a\phi(\zeta)/\zeta}, \quad |a|<1, \;\; c\in\mathbb{R}.$$

Since a proof of Theorem 3.4 is fairly simple and short, we give it here.

PROOF. First we note that in the case $I = \mathbb{T}$ inequality (3.10) has the form

$$\int_{\mathbb{T}} \frac{|\phi|^2 + 1}{1 - |\phi|^2}\, dm < \infty; \tag{3.14}$$

therefore, by Proposition 1.4, such a ϕ is regular.

Now, recall that a weight is in A_2 if and only if (there exists $Q > 0$ such that)

$$\langle W^{-1} P_+ X, P_+ X\rangle \le Q\langle W^{-1} X, X\rangle, \quad \forall X \in L^2_{W^{-1}}. \tag{3.15}$$

Note that

$$H^2 \to W \begin{bmatrix} 0 \\ H^2 \end{bmatrix}, \qquad H^2_- \to W \begin{bmatrix} H^2_- \\ 0 \end{bmatrix}$$

are unitary embeddings in $L^2_{W^{-1}}$.

The orthogonal complement to their sum (an alternative definition of K_ϕ) consists of vectors of the form

$$K_\phi = \left\{ X = \begin{bmatrix} x_+ \\ x_- \end{bmatrix}, \ x_\pm \in H^2_\pm \right\}.$$

The vectors

$$X = \begin{bmatrix} \psi x_+ \\ \bar{\psi} x_- \end{bmatrix}$$

evidently belong to K_ϕ. They are *dense* in K_ϕ if and only if $\hat{H}_\phi = \check{H}_\phi$, i.e., ϕ is regular, by Theorem 1.3.

Now we calculate the quadratic forms (3.15) for the test-vectors

$$X = \begin{bmatrix} \psi x_+ \\ \bar{\psi} x_- \end{bmatrix} + W \begin{bmatrix} y_- \\ y_+ \end{bmatrix}, \quad x_\pm \in H^2_\pm, \ y_\pm \in H^2_\pm.$$

We get

$$\begin{aligned}
\langle W^{-1} X, X\rangle &= \left\langle \frac{\begin{bmatrix} 1 & -\phi \\ -\bar{\phi} & 1 \end{bmatrix}}{1 - |\phi|^2} \begin{bmatrix} \psi x_+ \\ \bar{\psi} x_- \end{bmatrix}, \begin{bmatrix} \psi x_+ \\ \bar{\psi} x_- \end{bmatrix} \right\rangle + \left\langle \begin{bmatrix} 1 & \phi \\ \bar{\phi} & 1 \end{bmatrix} \begin{bmatrix} y_- \\ y_+ \end{bmatrix}, \begin{bmatrix} y_- \\ y_+ \end{bmatrix} \right\rangle \\
&= \left\langle \begin{bmatrix} 1 & \bar{f_0} \\ f_0 & 1 \end{bmatrix} \begin{bmatrix} x_+ \\ x_- \end{bmatrix}, \begin{bmatrix} x_+ \\ x_- \end{bmatrix} \right\rangle + \langle y_-, y_-\rangle + \langle y_+, y_+\rangle \\
&= \left\langle \begin{bmatrix} 1 & \Gamma^* \\ \Gamma & 1 \end{bmatrix} \begin{bmatrix} x_+ \\ x_- \end{bmatrix}, \begin{bmatrix} x_+ \\ x_- \end{bmatrix} \right\rangle + \langle y_-, y_-\rangle + \langle y_+, y_+\rangle.
\end{aligned}$$

Since

$$P_+ X = \begin{bmatrix} \psi x_+ \\ 0 \end{bmatrix} + \begin{bmatrix} \phi \\ 1 \end{bmatrix} y_+,$$

we get

$$\begin{aligned}
\langle W^{-1} P_+ X, P_+ X\rangle &= \left\langle \begin{bmatrix} 1 \\ -\bar{\phi} \end{bmatrix} \frac{x_+}{\bar{\psi}} + \begin{bmatrix} 0 \\ 1 \end{bmatrix} y_+, \begin{bmatrix} \psi x_+ \\ 0 \end{bmatrix} + \begin{bmatrix} \phi \\ 1 \end{bmatrix} y_+ \right\rangle \\
&= \langle x_+, x_+\rangle + \langle y_+, y_+\rangle.
\end{aligned}$$

Therefore, for such vectors, (3.15) is equivalent to

$$\|x_+\|^2 + \|y_+\|^2 \le Q\{\langle (I - \Gamma^*\Gamma) x_+, x_+\rangle + \|\Gamma x_+ + x_-\|^2 + \|y_-\|^2 + \|y_+\|^2\},$$

which is evidently equivalent to $(I - \Gamma^*\Gamma) > \epsilon I$ with $\epsilon > 0$.

Thus, A_2 implies the regularity and the bound $\|\Gamma\| < 1$. Conversely, if ϕ is regular and $\|\Gamma\| < 1$, then (3.15) holds on a dense set, and hence $W \in A_2$. $\square$

It is interesting that a similar condition on an arbitrary symbol of the Hankel operator guarantees the invertibility of $(I - \Gamma^*\Gamma)$.

PROPOSITION 3.8. *Let f satisfy the A_2-type condition*

$$\sup_I \frac{1}{|I|}\int_I \frac{|f-\langle f\rangle_I|^2 + (1-|\langle f\rangle_I|^2)}{1-|f|^2}\,dm < \infty, \tag{3.16}$$

where

$$\langle f\rangle_I := \frac{1}{|I|}\int_I f\,dm. \tag{3.17}$$

Then $\|\Gamma\| < 1$ *for* $\Gamma x := P_- f x$.

PROOF. We have

$$\left\langle \begin{bmatrix} 1 & \bar f \\ f & 1 \end{bmatrix} P_+X, P_+X \right\rangle \le Q\left\langle \begin{bmatrix} 1 & \bar f \\ f & 1 \end{bmatrix} X, X \right\rangle.$$

Put here

$$X = \begin{bmatrix} x \\ -\Gamma x \end{bmatrix}, \quad x \in H^2.$$

We get

$$\langle x, x\rangle \le Q\langle (I - \Gamma^*\Gamma)x, x\rangle. \qquad \square$$

Appendix

We wish to prove Theorem 3.5 without referring to strongly regular functions in the sense of Arov–Dym, namely, to give a direct proof of this curious result.

Let us recall that we start with $w \in A_2$. We extend it into the disc by harmonicity, get a positive harmonic function, and represent it, as usual, in the form

$$\frac{1-|\phi|^2}{|1-\phi|^2} = w\,, \tag{3.18}$$

where ϕ is holomorphic in the disc and has H^∞-norm at most 1. Such functions ϕ and positive harmonic functions are in one-to-one correspondence by (3.18).

Let $\phi_c := e^{ic}\phi$, $\psi_c := e^{ic}\psi$, $c \in \mathbb{R}$. Consider a new positive harmonic function given by

$$w_c := \frac{1-|\phi_c|^2}{|1-\phi_c|^2}\,. \tag{3.19}$$

THEOREM 3.9. *If $w \in A_2$, and $\phi \in H^\infty$, $\|\phi\|_\infty \le 1$ is given by* (3.18), *then a positive harmonic function w_c from* (3.19) *does not have singular part in its Herglotz representation. Its absolutely continuous part on the circle* (*also called w_c*) *is uniformly in A_2.*

PROOF. We have already seen that (3.18) implies

$$\frac{1}{1-|\phi|^2} \in L^1(\mathbb{T})\,.$$

Then, if we put $g_c := \frac{1-\phi_c}{\psi}$, we see that the outer function g_c is always in H^2.

Assume for a moment that we can prove the following:

$$\Gamma_{\frac{\bar g_c}{g_c}} = e^{-ic}\Gamma_{\frac{\bar g_0}{g_0}}\,. \tag{3.20}$$

The last operator has norm less than 1. In fact, $|g_0|^2$ is given to be in A_2 by (3.18). Then by Helson–Szegő's theorem we have this estimate of the norm of the Hankel operator; see [**16**] again. If the Hankel operator with unimodular symbol has norm strictly less than 1, then the Toeplitz operator with the same symbol is obviously bounded from below.

We conclude that for each real c the Toeplitz operator $T_{\frac{\bar{g}_c}{g_c}}$ is left invertible, and, combining this with the fact that $g_c \in H^2$, we conclude that its adjoint has trivial kernel. Then $T_{\frac{\bar{g}_c}{g_c}}$ is invertible. But unimodular symbols of invertible Toeplitz operators are $\frac{\bar{h}}{h}$ with $|h|^2 \in A_2$; see [**16**]. Now $1/g_c = \frac{\psi}{1-\phi_c}$ and $1/|g_c|^2 = \frac{1-|\phi_c|^2}{|1-\phi_c|^2} = \mathrm{Re}\,\frac{1+\phi_c}{1-\phi_c}$ in the unit disc. The latter function has positive real part, and so $1/|g_c|^2 \in L^1(\mathbb{T})$. As $1/g_c$ is outer, we conclude that $1/g_c \in H^2$.

Using that $1/g_c \in H^2$ and writing $\frac{\bar{g}_c}{g_c} = \frac{\bar{h}}{h}$ we conclude that $g_c = \mathrm{const} \cdot h$. Therefore, $\frac{|1-\phi_c|^2}{1-|\phi|^2} = |g_c|^2 \in A_2$.

To finish the proof of our Theorem 3.9, it remains to prove (3.20). To this end we prove that

$$\frac{\bar{g}_c}{g_c} \in e^{-ic}\frac{\bar{g}_0}{g_0} + H^\infty. \tag{3.21}$$

Let us write a chain of equalities:

$$\frac{\bar{g}_0}{g_0} = \frac{\overline{1-\phi}}{1-\phi}\cdot\frac{\psi}{\bar{\psi}} = \frac{(|\psi|^2+|\phi|^2)\psi - \bar{\phi}\psi}{(1-\phi)\bar{\psi}} = -\frac{\bar{\phi}\psi}{\bar{\psi}} + \frac{\psi^2}{1-\phi}.$$

Similarly,

$$\frac{\bar{g}_c}{g_c} = e^{-ic}\frac{\overline{1-\phi_c}}{1-\phi_c}\cdot\frac{\psi_{\frac{c}{2}}}{\bar{\psi}_{\frac{c}{2}}} = -e^{-ic}\left[\frac{\bar{\phi}_c\psi_{\frac{c}{2}}}{\bar{\psi}_{\frac{c}{2}}} + \frac{\psi_{\frac{c}{2}}^2}{1-\phi_c}\right] = -e^{-ic}\frac{\bar{\phi}\psi}{\bar{\psi}} + \frac{\psi^2}{1-\phi_c}.$$

Comparing these two equalities we obtain

$$\frac{\bar{g}_c}{g_c} = e^{-ic}\frac{\bar{g}_0}{g_0} + \left(\frac{\psi^2}{1-\phi_c} - e^{-ic}\frac{\psi^2}{1-\phi}\right).$$

Both functions in parentheses are in H^∞. In fact, they are obviously from the Smirnov class. Also their boundary values are at most 2 in absolute value: from the above chains of equalities one can see that they are simply sums of two unimodular functions each. □

References

[1] V. Adamjan, D. Arov and M. Kreĭn, *Infinite Hankel matrices and generalized problems of Carathéodory–Fejér and F. Riesz*, Funkcional. Anal. i Priložen. **2** (1968), no. 1, 1–19. (Russian)

[2] ———, *Infinite Hankel matrices and generalized Carathéodory–Fejér and I. Schur problems*, Funkcional. Anal. i Priložen. **2** (1968), no. 4, 1–17. (Russian)

[3] ———, *Analytic properties of the Schmidt pairs of a Hankel operator and the generalized Schur–Takagi problem*, Mat. Sb. (N.S.) **86 (128)** (1971), 34–75. (Russian)

[4] D. Arov, *Regular and singular J-inner matrix functions and corresponding extrapolation problems*, Funktsional. Anal. i Prilozhen. **22** (1988), no. 1, 57–59; English transl., Funct. Anal. Appl. **22** (1988), no. 1, 46–48.

[5] D. Arov and H. Dym, *On matricial Nehari problems, J-inner matrix functions and the Muckenhoupt condition*, J. Funct. Anal. **181** (2001), 227–299.

[6] ———, *Criteria for the strong regularity of J-inner functions and γ-generating matrices*, J. Math. Anal. Appl. **280** (2003), 387–399.

[7] A. Borichev and M. Sodin, *The Hamburger moment problem and weighted polynomial approximation on discrete subsets of the real line*, J. Anal. Math. **76** (1998), 219–264.

[8] ———, *Weighted polynomial approximation and the Hamburger moment problem*, Complex analysis and differential equations (Uppsala, 1997), pp. 110–122, Acta Univ. Upsaliensis Skr. Uppsala Univ. C Organ. Hist., 64, Uppsala Univ., Uppsala, 1999.

[9] M. Christ and A. Kiselev, *Scattering and wave operators for one-dimensional Schrödinger operators with slowly decaying nonsmooth potentials*, Geom. Funct. Anal. **12** (2002), 1174–1234.

[10] J. B. Garnett, *Bounded analytic functions*, Revised first edition, Graduate Texts in Mathematics, 236, Springer, New York, 2007.

[11] V. Katsnelson, *Left and right Blaschke–Potapov products and Arov-singular matrix-valued functions*, Integral Equations Operator Theory **13** (1990), no. 6, 836–848.

[12] A. Kheifets, *On regularization of γ-generating pairs*, J. Funct. Anal. **130** (1995), no. 2, 310–333.

[13] ———, *Nehari's interpolation problem and exposed points of the unit ball in the Hardy space H^1*, Proceedings of the Ashkelon Workshop on Complex Function Theory (1996), pp. 145–151, Israel Math. Conf. Proc., 11, Bar-Ilan Univ., Ramat Gan, 1997.

[14] A. Kheifets, F. Peherstorfer, and P. Yuditskii, *On scattering for CMV matrices*, arXiv: 0706.2970v1, 2007.

[15] N. Makarov and A. Poltoratski, *Beurling–Malliavin theory for Toeplitz kernels*, Preprint arXiv:math/0702497, 2007.

[16] N. Nikolskii, *Treatise on the Shift Operator*, Springer-Verlag, Berlin, 1986.

[17] Ch. Remling, *Schrödinger operators and de Branges spaces*, J. Funct. Anal. **196** (2002), no. 2, 323–394.

[18] B. Simon, *Orthogonal Polynomials on the Unit Circle, Part 1: Classical Theory*, AMS Colloquium Series, AMS, Providence, RI, 2005.

[19] ———, *Orthogonal Polynomials on the Unit Circle, Part 2: Spectral Theory*, AMS Colloquium Series, AMS, Providence, RI, 2005.

[20] B. Szőkefalvi-Nagy and C. Foiaş, *Harmonic analysis of operators on Hilbert space*. North-Holland, Amsterdam, 1970.

[21] S. Treil and A. Volberg, *The angle between past and future*, J. Funct. Anal. **143** (1997), no. 2, 269–308.

[22] T. Tao and Ch. Thiele, *Nonlinear Fourier Analysis*, IAS/Park City Mathematics Series (unpublished lectures).

[23] A. Volberg and P. Yuditskii, *On the inverse scattering problem for Jacobi matrices with the spectrum on an interval, a finite system of intervals or a Cantor set of positive length*, Comm. Math. Phys. **226** (2002), 567–605.

DEPARTMENT OF MATHEMATICS, MICHIGAN STATE UNIVERSITY, EAST LANSING, MI 48824, USA

E-mail address: volberg@math.msu.edu

INSTITUTE FOR ANALYSIS, JOHANNES KEPLER UNIVERSITY LINZ, A-4040 LINZ, AUSTRIA
E-mail address: Petro.Yudytskiy@jku.at

Titles in This Series

227 Selected Papers on Probability and Statistics

226 **Alexei Alexandrov, Anton Baranov, and Sergey Kislyakov, Editors,** Linear and Complex Analysis (Dedicated to V. P. Havin on the Occasion of His 75th Birthday)

225 **T. Suslina and D. Yafaev, Editors,** Spectral Theory of Differential Operators (M. Sh. Birman 80th Anniversary Collection)

224 **V. M. Buchstaber and I. M. Krichever, Editors,** Geometry, Topology, and Mathematical Physics (S. P. Novikov's Seminar: 2006–2007)

223 Selected Papers on Analysis and Related Topics

222 **N. N. Uraltseva, Editor,** Proceedings of the St. Petersburg Mathematical Society, XIII

221 **Yu. Neretin, M. Olshanetsky, and A. Rosly, Editors,** Seminar on Mathematical Physics, II

220 **M. S. Birman and N. N. Uraltseva, Editors,** Nonlinear Equations and Spectral Theory

219 **N. N. Uraltseva, Editor,** Proceedings of the St. Petersburg Mathematical Society, Volume XII

218 **N. N. Uraltseva, Editor,** Proceedings of the St. Petersburg Mathematical Society, Volume XI

217 **V. Kaimanovich and A. Lodkin, Editors,** Representation Theory, Dynamical Systems, and Asymptotic Combinatorics

216 **M. V. Karasev, Editor,** Quantum Algebras and Poisson Geometry in Mathematical Physics

214 **N. N. Uraltseva, Editor,** Proceedings of the St. Petersburg Mathematical Society, Volume X

213 **Ernest Vinberg, Editor,** Lie Groups and Invariant Theory

212 **V. M. Buchstaber and I. M. Krichever, Editors,** Geometry, Topology, and Mathematical Physics. S. P. Novikov's Seminar: 2002–2003

211 **K. Nomizu, Editor,** Selected Papers on Analysis and Differential Equations

210 **S. G. Gindikin, Editor,** Lie Groups and Symmetric Spaces. In Memory of F. I. Karpelevich

209 **N. N. Uraltseva, Editor,** Proceedings of the St. Petersburg Mathematical Society, Volume IX

208 **M. V. Karasev, Editor,** Asymptotic Methods for Wave and Quantum Problems

207 **Yu. M. Suhov, Editor,** Analytic Methods in Applied Probability

206 **M. A. Shubin and M. S. Agranovich, Editors,** Partial Differential Equations (Mark Viskik's Seminar)

205 **N. N. Uraltseva, Editor,** Proceedings of the St. Petersburg Mathematical Society, Volume VIII

204 **K. Nomizu, Editor,** Selected Papers on Classical Analysis

203 **N. N. Uraltseva, Editor,** Proceedings of the St. Petersburg Mathematical Society, Volume VII

202 **V. Turaev and A. Vershik, Editors,** Topology, Ergodic Theory, Real Algebraic Geometry. Rokhlin's Memorial

201 **Michael Semenov-Tian-Shansky, Editor,** L. D. Faddeev's Seminar on Mathematical Physics

200 **L. Lerman, G. Polotovskiĭ, and L. Shilnikov, Editors,** Methods of Qualitative Theory of Differential Equations and Related Topics

199 **N. N. Uraltseva, Editor,** Proceedings of the St. Petersburg Mathematical Society, Volume VI

TITLES IN THIS SERIES

198 **R. A. Minlos, Senya Shlosman, and Yu. M. Suhov, Editors,** On Dobrushin's Way. From Probability Theory to Statistical Physics

197 **Vladimir Arnold, Maxim Kontsevich, and Anton Zorich, Editors,** Pseudoperiodic Topology

196 **Ya. Eliashberg, D. Fuchs, T. Ratiu, and A. Weinstein, Editors,** Northern California Symplectic Geometry Seminar

195 **O. V. Belegradek et al.,** Model Theory and Applications

194 **A. Astashkevich and S. Tabachnikov, Editors,** Differential Topology, Infinite-Dimensional Lie Algebras, and Applications (D. B. Fuchs' 60th Anniversary Collection)

193 **N. N. Uraltseva, Editor,** Proceedings of the St. Petersburg Mathematical Society, Volume V

192 **Lev Beklemishev, Mati Pentus, and Nikolai Vereshchagin,** Provability, Complexity, Grammars

191 **A. Yu. Morozov and M. A. Olshanetsky, Editors,** Moscow Seminar in Mathematical Physics

190 **S. Tabachnikov, Editor,** Differential and Symplectic Topology of Knots and Curves

189 **V. Buslaev, M. Solomyak, and D. Yafaev, Editors,** Differential Operators and Spectral Theory (M. Sh. Birman's 70th Anniversary Collection)

188 **O. A. Ladyzhenskaya, Editor,** Proceedings of the St. Petersburg Mathematical Society, Volume IV

187 **M. V. Karasev, Editor,** Coherent Transform, Quantization, and Poisson Geometry

186 **A. Khovanskiĭ, A. Varchenko, and V. Vassiliev, Editors,** Geometry of Differential Equations

185 **B. Feigin and V. Vassiliev, Editors,** Topics in Quantum Groups and Finite-Type Invariants (Mathematics at the Independent University of Moscow)

184 **Peter Kuchment and Vladimir Lin, Editors,** Voronezh Winter Mathematical Schools (Dedicated to Selim Krein)

183 **K. Nomizu, Editor,** Selected Papers on Harmonic Analysis, Groups, and Invariants

182 **V. E. Zakharov, Editor,** Nonlinear Waves and Weak Turbulence

181 **G. I. Olshanski, Editor,** Kirillov's Seminar on Representation Theory

180 **A. Khovanskiĭ, A. Varchenko, and V. Vassiliev, Editors,** Topics in Singularity Theory

179 **V. M. Buchstaber and S. P. Novikov, Editors,** Solitons, Geometry, and Topology: On the Crossroad

178 **V. Kreinovich and G. Mints, Editors,** Problems of Reducing the Exhaustive Search

177 **R. L. Dobrushin, R. A. Minlos, M. A. Shubin, and A. M. Vershik, Editors,** Topics in Statistical and Theoretical Physics (F. A. Berezin Memorial Volume)

176 **E. V. Shikin, Editor,** Some Questions of Differential Geometry in the Large

175 **R. L. Dobrushin, R. A. Minlos, M. A. Shubin, and A. M. Vershik, Editors,** Contemporary Mathematical Physics (F. A. Berezin Memorial Volume)

174 **A. A. Bolibruch, A. S. Merkur′ev, and N. Yu. Netsvetaev, Editors,** Mathematics in St. Petersburg

173 **V. Kharlamov, A. Korchagin, G. Polotovskiĭ, and O. Viro, Editors,** Topology of Real Algebraic Varieties and Related Topics

172 **K. Nomizu, Editor,** Selected Papers on Number Theory and Algebraic Geometry